Student Study Guide/ Solutions Manual

for use with

Genetics
Analysis and Principles

Second Edition

Robert J. Brooker
University of Minnesota-Twin Cities

Prepared by

Robert J. Brooker
University of Minnesota-Twin Cities
and
Michael Windelspecht
Appalachian State University

Boston Burr Ridge, IL Dubuque, IA Madison, WI New York San Francisco St. Louis
Bangkok Bogotá Caracas Kuala Lumpur Lisbon London Madrid Mexico City
Milan Montreal New Delhi Santiago Seoul Singapore Sydney Taipei Toronto

The McGraw-Hill Companies

Student Study Guide/Solutions Manual for use with
GENETICS: ANALYSIS AND PRINCIPLES, SECOND EDITION
Robert J. Brooker

Published by McGraw-Hill Higher Education, an imprint of The McGraw-Hill Companies, Inc., 1221 Avenue of the Americas, New York, NY 10020. Copyright © 2005 by The McGraw-Hill Companies, Inc. All rights reserved.

No part of this publication may be reproduced or distributed in any form or by any means, or stored in a database or retrieval system, without the prior written consent of The McGraw-Hill Companies, Inc., including, but not limited to, network or other electronic storage or transmission, or broadcast for distance learning.

RECYCLED This book is printed on recycled, acid-free paper containing 10% postconsumer waste.

2 3 4 5 6 7 8 9 0 QPD QPD 0 9 8 7 6

ISBN-13: 978-0-07-284860-1
ISBN-10: 0-07-284860-X

www.mhhe.com

Contents

How to Use this Study Guide

This study guide is designed as a supplement to the textbook. While the problems in the textbook are designed to enhance your ability to solve problems relating to the study of genetics, the activities in the study guide are designed to assist you with learning the language of genetics. Students in an introductory genetics course learn more new terms than a first-year foreign language student. Most problems that students experience in the introductory genetics courses center on the terminology.

In order to use this study guide effectively, you must read the textbook. You should read each chapter prior to the associated lecture. While prereading the chapter, do not attempt to completely understand all of the terms and procedures. Instead, read the chapter as you would a work of fiction. Try to get a grasp of the plot and the major players.

Following the lecture on this material, you should reread the chapter for its content. This is the time to get the facts and terms straight. After each section, complete the exercises in the study guide to verify that you have a grasp of the important terms. You can then work the problems in the text and take the quiz in the study guide. If you find yourself missing questions, review those sections of the textbook and repeat the process. Don't forget that one of your greatest assets is your instructor!

Contents of the Study Guide

For each chapter the following information has been provided:

- **Student Learning Objectives**
  At the start of each chapter is a list of learning objectives. These objectives provide an overview of the major skills and information that you should master before proceeding to the next chapter of the text. Learning objectives are not intended as a comprehensive summary of the chapter content. Rather, they serve as a guide to indicate the key themes that are the focal points of the chapter. A knowledge of these objectives will greatly assist in your overall comprehension of the subject material.

- **Content Assistant**
  The chapters of this textbook are divided into major themes, which are indicated as numbered sections (1.1, 1.2, etc.). In this study guide, each of these sections further divided into a series of exercises and activities designed to test your comprehension of the chapter material.

  At the start of each major section there is a short narrative that explains, briefly, the importance of this material to the study of genetics. Helpful study tips and insight into potential problem areas for students are provided as needed. Following this is a list of key terms that appear in this section of the text. These are bold-faced or italicized terms that play an important role in the content of the section. One of the major stumbling blocks for genetics students is the learning of terminology. This section highlights those words and phrases for additional study.

Immediately after the key terms is a list of focal points for the chapter content. Typically these highlight important figures from the text. Often it is beneficial to visualize complex material as a diagram. The figures selected in these sections present an overview of the material. The other figures are important, but these allow you to focus on important concepts, and then build your knowledge of the remaining material from that point.

- **Chapter Quiz**
  The chapter review contains a short review quiz. Each quiz consists of a series of true/false, matching, and multiple choice questions that are designed to test your overall knowledge of the subject material in the chapter. If you incorrectly answer one of the questions, you should review the entire section in the text relating to that question. This ensures that you are not simply preparing for the test, but are developing an understanding of the science of genetics.

- **Answer Key for Study Guide Questions**
  The final section of each chapter is a list of answers and solutions for exercises and quizzes presented in this study guide.

- **Answers to Conceptual and Experimental Questions from the Text**
  At the end of each chapter in the text the author has provided sets of solved problems, conceptual questions, and experimental questions as a review of the chapter material. The solutions for the solved problems are provided following each of the problems. The textbook provides the solutions for the conceptual and experimental questions, the solutions for the even-numbered questions in these sections at the end of the text.

  This section of the study guide provides answers to all of the conceptual and experimental questions for the chapter. These answers/solutions are identical to those found in the text and at the Online Learning Center (www.mhhe.com/brooker).

About the Author

Michael Windelspecht is an assistant professor of biology at Appalachian State University. He is the author two books on the history of science and two books on human biology. He has served as the editor for several reference series. He teaches both transmission genetics and human genetics, and is actively involved in the genetics of behavior in *Drosophila*.

Chapter 1: An Overview of Genetics

Student Learning Objectives

Upon completion of this chapter you should be able to:

1. Understand the key biological molecules that are associated with the study of genetics.
2. Understand the relationships between genes and traits and the types of traits that are studied by geneticists.
3. Understand the three principle levels of genetic study.
4. Recognize the three major fields of genetics and the general characteristics of each field.

1.1 The Relationship Between Genes and Traits

Overview

The study of genetics involves a fundamental understanding of both organic molecules and cellular processes. For most students, these concepts were first introduced in their introductory biology classes. The text sections on biochemicals (pg. 4), proteins and enzymes (pg. 5), and DNA (pgs. 5-6) provide a quick refresher of this material. If a more comprehensive review is needed, you should consult an introductory text book.

One of the most important aspects of this section is the discussion of the relationship between genes and traits. While many students can provide a definition of these terms, the relationship is frequently not understood. The section on gene expression (pg. 6), and especially Figure 1.6, provides a visual link between the genetic information (DNA) and the functional protein that is at the core of the expression of traits. While educational materials frequently focus on morphological traits (pg. 6), primarily due to the fact that they are easier to describe, visualize, and illustrate, physiological traits (pg. 6) are also controlled by the expression of genes.

The science of genetics is directly linked to the study of biological evolution (pg. 10). Geneticists study variation at the molecular, cellular, organismal and population levels (pg. 6). This variation is the basis of the Theory of Natural Selection, the mechanism of evolutionary change first proposed by Charles Darwin in the 19th century.

Key Terms

Alleles
Amino acid
Biological evolution
Carbohydrates
Cellular level
Chromosomes
Deoxyribonucleic acid (DNA)
Diploid
Enzymes
- Anabolic Enzymes
- Catabolic Enzymes

Gene
Gene expression
Gene mutations
Genetic code
Genetic variation
Haploid
Homologues
Lipids
Macromolecules
Molecular level
Natural selection
Nucleic Acids

Nucleotides
Morphological traits
Morphs
Organismal level
Phenylketonuria (PKU)
Physiological traits
Proteins
Populational level
Ribonucleic acid (RNA)
Traits
Transcription
Translated (translation)

Focal Points

- Genes and traits (pg. 4)
- Gene expression (pg. 6)
- Genetic variation and evolution (pgs. 7-10)

Exercises

For questions 1 to 5, refer to the labels on the diagram below.

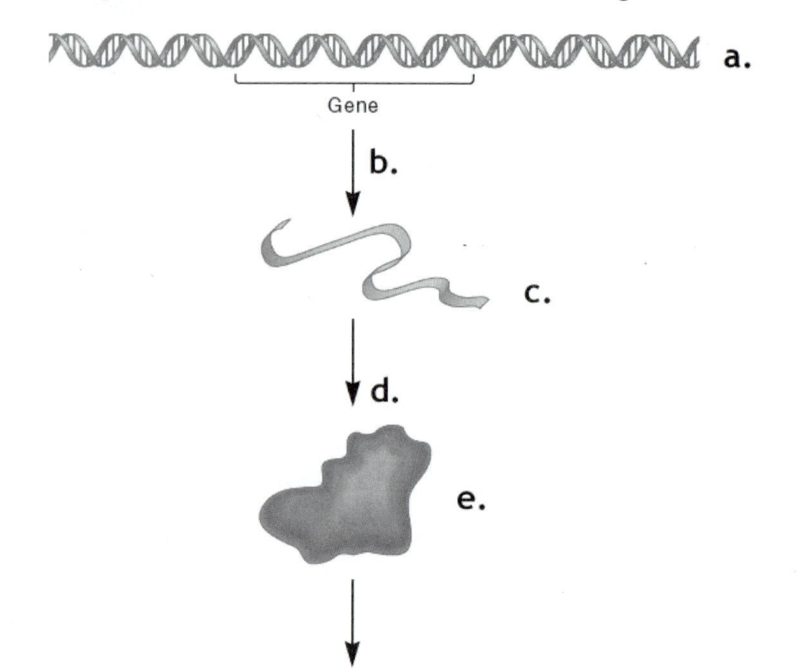

_____ 1. The process by which genetic information is converted into a functioning protein .

_____ 2. The functional protein responsible for the observed traits.

_____ 3. The temporary messenger that moves the genetic information from the transcription machinery to the site of translation.

_____ 4. The label that indicates the source of the information for producing a trait.

_____ 5. The process by which the genetic information is copied in RNA.

For questions 6 to 11, choose from one of the following levels of genetic study:

a. population level
b. molecular level
c. cellular level
d. organismal level

_____ 6. The action of proteins, such as enzymes, occurs at this level of organization.

_____ 7. Regulation of transcription and translation occurs at this level of organization.

_____ 8. An analysis of the prevalence of a trait in a given population.

_____ 9. It is at this level that the traits are most notably detectable for geneticists.

_____ 10. An allele is due to a change in the genetic material at this level.

_____ 11. Studies at this level examine how the occurrence of a trait influences habitat or survival.

For questions 12 to 16, choose from the following:
a. example of a physiological trait
b. example of a morphological trait

_____ 12. The flower colors and seed shapes of pea plants studied by Gregor Mendel in the 19th century.

_____ 13. Trunk length in elephants.

_____ 14. Metabolic efficiency of an organism under anaerobic conditions.

_____ 15. Enzyme deficiencies.

_____ 16. Wing size in insects.

1.2 Fields of Genetics

Overview

As is the case with any of the scientific disciplines, geneticists specialize in areas of study. Although these genetic specialists may differ in the *level* at which they study genetic principles, they frequently use similar methods. The purpose of this section is to first distinguish between the different areas of genetics, and then describe the importance of the scientific method to the study of genetics.

Key Terms

Genetic Approach
Genetic Cross
Loss-of-function allele
Loss-of-function mutation
Science
Scientific method

Exercises

For each of the following, indicate whether the research is at the (a) transmission, (b) molecular, or (c) population level of study. Some questions may have more than one answer.

1. An assessment of a loss-of-function allele on an organism's physiology.

2. Patterns of inheritance of a disease in a family.

3. The relationship between genetic variation and the environment.

4. Studies of the variations in DNA between species.

5. The movement of traits from parents to offspring.

6. An examination of how a group of animals is adapting to a new environment.

Chapter Quiz

1. The primary advantage of sexual reproduction is:
   a. it maintains the chromosome number for the species.
   b. it is necessary to protect against the effects of natural selection.
   c. it creates diploid gametes that are clones of the parents.
   d. it enhances genetic variation for the species.
   e. none of the above are correct.

2. Patterns of inheritance within a family would most likely be associated with which of the following:
   a. transmission genetics.
   b. population genetics.
   c. molecular genetics.
   d. none of the above.

3. Which of the following best describes the term genetic variation?
   a. differences in inherited traits within individuals of a population
   b. the changing of the genetic makeup of a population over time
   c. the basic unit of heredity
   d. the characteristics of an organism
   e. the effect of the environment on the expression of a trait

4. Which of the following best describes the type of molecule that would be involved in the generation of cellular energy?
   a. nucleotides
   b. nucleic acids
   c. anabolic enzymes
   d. catabolic enzymes
   e. lipids

5. In general, most enzymes belong to what class of macromolecules?
   a. lipids
   b. nucleic acids
   c. proteins
   d. carbohydrates
   e. none of the above

6. A morphological trait is one which affects the ability of an organism to function.
   a. true
   b. false

7. Which of the following describes the process by which scientists conduct scientific investigations?
   a. biological evolution
   b. genetic crosses
   c. natural selection
   d. scientific method
   e. none of the above

8. At which of the following levels of genetic analysis would a protein biochemist primarily examine?
   a. populational level
   b. molecular level
   c. organismal level
   d. cellular level

9. The basic unit of heredity is called the __________.
   a. protein
   b. trait
   c. gene
   d. enzyme
   e. morph

10. Phenylketonuria (PKU) is an example of __________.
   a. natural selection
   b. biological evolution
   c. genetic cross
   d. transcription
   e. environmental influence

Answer Key for Study Guide Questions

This answer key provides the answers to the exercises and chapter quiz for this chapter. Answers in parentheses ( ) represent possible alternate answers to a problem, while answers marked with an asterisk (*) indicate that the response to the question may vary.

1.1
1. d
2. e
3. c
4. a
5. b
6. c
7. b
8. a
9. d
10. b
11. a
12. b
13. b
14. a
15. a
16. b

1.2:
1. b
2. a
3. c
4. b
5. a
6. c

Quiz:
1. d
2. a
3. a
4. d
5. c
6. b
7. d
8. b
9. c
10. e

Answers to Conceptual and Experimental Questions

Conceptual Questions

C1. There are many possible answers. Some common areas to discuss might involve the impact of genetics in the production of new medicines, the diagnosis of diseases, the production of new kinds of food, and the use of DNA fingerprinting to solve crimes.

C2. A chromosome is a very long polymer of DNA. A gene is a specific sequence of DNA within that polymer; the sequence of bases creates a gene and distinguishes it from other genes. Genes are located in chromosomes, which are found within living cells.

C3. The structure and function of proteins govern the structure and function of living cells. The cells of the body determine an organism's traits.

C4. At the molecular level, a gene (which is a sequence of DNA) is first transcribed into RNA. The genetic code within the RNA is used to synthesize a protein with a particular amino acid sequence. This second process is called translation.

C5. A. Molecular level. This is a description of a how an allele affects protein function.

B. Cellular level. This is a description of how protein function affects cell structure.

C. Populational level. This is a description of how the two alleles affect members of a population.

D. Organismal level. This is a description of how the alleles affect the traits of an individual.

C6. Genetic variation involves the occurrence of genetic differences within members of the same species or different species. Within any population, there may be variation in the genetic material. There may be variation in particular genes so that some individuals carry one allele and other individuals carry a different allele. An example would be differences in coat color among mammals. There also may be variation in chromosome structure and number. In plants, differences in chromosome number can affect disease resistance.

C7. An extra chromosome (specifically an extra copy of chromosome 21) causes Down syndrome.

C8. You could pick almost any trait. For example, flower color in petunias would be an interesting choice. Some petunias are red and others are purple. There must be different alleles in a flower color gene that affect this trait in petunias. In addition, the amount of sunlight, fertilizer, and water also affects the intensity of flower color.

C9. The term *diploid* means that a cell has two copies of each type of chromosome. In humans, nearly all of the cells are diploid except for gametes (i.e., sperm and egg cells). Gametes have only one set of chromosomes.

C10. A DNA sequence is a sequence of nucleotides. Each nucleotide may have one of four different bases (i.e., A, T, G, or C). When we speak of a DNA sequence, we focus on the sequence of bases.

C11. The genetic code is the way in which the sequence of bases in RNA is read to produce a sequence of amino acids within a protein.

C12. A. A gene is a segment of DNA. For most genes, the expression of the gene results in the production of a functional protein. The functioning of proteins within living cells affects the traits of an organism.

B. A gene is a segment of DNA that usually encodes the information for the production of a specific protein. Genes are found within chromosomes. There are many genes within a single chromosome.

C. An allele is an alternative version of a particular gene. For example, suppose a plant has a flower color gene. One allele could produce a white flower, while a different allele could produce an orange flower. The white allele is an allele of the flower color gene.

D. A DNA sequence is a sequence of nucleotides. The information within a DNA sequence (which is transcribed into an RNA sequence) specifies the amino acid sequence within a protein.

C13. The statement in part A is not correct. Individuals do not evolve. Populations evolve because certain individuals are more likely to survive and reproduce and pass their genes to succeeding generations.

C14. A. How genes and traits are transmitted from parents to offspring.

B. How the genetic material functions at the molecular and cellular levels.

C. Why genetic variation exists in populations, and how it changes over the course of many generations.

Experimental Questions

E1. A genetic cross is a mating between two different individuals.

E2. This would be used primarily by molecular geneticists. The sequence of DNA is a molecular characteristic of DNA. In addition, as we will learn throughout this textbook, the sequence of DNA is interesting to transmission and population geneticists as well.

E3. We would see 47 chromosomes instead of 46. There would be three copies of chromosome 21 instead of two copies.

E4. A. Transmission geneticists. Dog breeders are interested in how genetic crosses affect the traits of dogs.

B. Molecular geneticists. This is a good model organism to study genetics at the molecular level.

C. Both transmission geneticists and molecular geneticists. Fruit flies are easy to cross and study the transmission of genes and traits from parents to offspring. Molecular geneticists have also studied many genes in fruit flies to see how they function at the molecular level.

D. Population geneticists. Most wild animals and plants would be the subject of population geneticists. In the wild, you cannot make controlled crosses. But you can study genetic variation within populations and try to understand its relationship to the environment.

E. Transmission geneticists. Agricultural breeders are interested in how genetic crosses affect the outcome of traits.

E5. You need to follow the scientific method. You can take a look at the experiment of Figure 25.14 to see how Kettlewell followed the scientific method to determine why the color of moth wings varies within a population of moths in England.

Chapter 2: Mendelian Inheritance

Student Learning Objectives

Upon completion of this chapter you should be able to:

1. Recognize the importance of Mendel's work to the study of inheritance.
2. Construct Punnett square diagrams of one and two factor crosses to predict phenotypic and genotypic ratios of offspring.
3. Analyze pedigree diagrams for patterns of inheritance.
4. Apply the laws of probability to the study of patterns of inheritance.
5. To use the chi square test to test the validity of a hypothesis.

Introduction

2.1 Mendel's Laws of Inheritance

Overview

The first section of this chapter provides an introduction to Mendel's contributions to the study of genetics. Mendel's success is based on his development of a mathematical, or empirical approach, to analyzing patterns of inheritance, coupled with an exceptional ability to design quality experiments to test his ideas.

Mendel's experimental approach centered on the use of the pea plant as a model organism. It is unlikely that Mendel was truly interested in the genetics of peas directly, but he recognized that this organism had the characteristics necessary to study inheritance. First, it was relatively easy to grow. Second, it had easily identifiable variants of traits, and third, he could easily self-fertilize the plants and produce true-breeding lines. These characteristics of a model organism are still recognized by geneticists in modern studies of inheritance.

Mendel's single-factor (or monohybrid) cross is recognized by many scientists as a classic model of the experimental method. Mendel hypothesized that inheritance patterns followed natural laws, and thus could be quantified using mathematical principles (pg. 21). His experimental system (pg. 22) provided clear evidence of several fundamental principles of genetics.

- The concept of dominance
- The particulate theory of inheritance
- The theory of segregation

It is important to note that earlier researchers had suggested many of these same principles, but it was Mendel's experimental system that provided the first direct evidence.

The two-factor crosses (or dihybrid crosses) are an extension of the monohybrid cross and were used by Mendel to further explore the results of the single-factor cross. It is important to note the relationship between Mendel's observations and the conceptual level of genes and alleles (see text Figure 2.8). The fact that during a two-factor cross the traits segregate from each other, producing non-parental offspring, is the basis of his second law of heredity, more often called the law of independent assortment.

Mendel's work was made possible by the fact that he could selectively breed the peas and then analyze large numbers of offspring to establish mathematical patterns. However, this is not the case with many organisms, including humans. Instead, human geneticists must observe patterns of inheritance in families to establish the genetics of a trait. This is called a pedigree analysis (pgs. 30-31). Pedigree studies differ from the crosses conducted by Mendel in that they do not involve large numbers of offspring and the analysis of ratios of phenotypes in the offspring. However, they do abide by the Mendelian concepts of dominance and the segregation of traits first described by Mendel.

While some of this material may be a review from previous courses, it is important that you completely understand the Mendelian laws of inheritance provided in the first section of this chapter, since much of the material in the next few chapters of the text will examine variations of these principles.

Key Terms

Allele
Characters
Dihybrid testcross
Dominant
Eggs
Empirical approach and laws
Fertilization
- Cross-fertilization
- Self-fertilization

Forked-line method
Gametes
Generations
- F_1 generation
- F_2 generation
- Parental (P) generation

Genotype
Heterozygous
Homozygous
Hybridization
Hybrids
Law of Independent Assortment
Law of Segregation
Loss-of-function alleles
Multiplication method
Nonparentals
Ovaries
Ovules
Pangenesis
Particulate theory of inheritance
Pedigree analysis
Phenotype
Pollen grains
Punnett square
Recessive
Segregate
Single-factor cross (monohybrid cross)
Sperm
Stamens
Stigma
Traits
True-breeding line
Two-factor cross (dihybrid cross)
Variants

Focal Points

- Single-factor crosses and the theory of segregation (pgs. 21-24)
- The use of Punnett squares to visualize genetic crosses (pgs. 24-25)
- Two-factor crosses and the theory of independent assortment (pgs. 25-29)

Exercises and Problems

For questions 1 to 10, match the definition on the left with the correct term on the right.

_____ 1. A genetic experiment involving a single trait

_____ 2. The offspring of two distinct individuals with different characteristics.

_____ 3. The gametes for a cross are derived from the same individual

_____ 4. The study of the mathematical relationships of an observation

_____ 5. A variation of a gene

_____ 6. The genetic composition of an individual

_____ 7. An individual who has two different alleles for a trait

_____ 8. The observable characteristics of an organism

_____ 9. An individual with two identical alleles for a trait

_____ 10. The basic unit of heredity

a. monohybrid cross
b. allele
c. genotype
d. hybrid
e. self-fertilization
f. empirical approach
g. phenotype
h. heterozygous
i. homozygous
j. gene

11. One of the pea characteristics studied by Mendel was seed shape. The two phenotypes that he studied were round and wrinkled seeds. In his experiments, Mendel obtained a pure-breeding line of round seeded plants and a pure-breeding line of wrinkled seeded plants. When these lines were crossed, the F_1 generation always yielded round seeds. Based on this information:

a. Which of the two traits is the dominant trait?
b. If two individuals from the F_1 generation were crossed, what would be the phenotypic and genotypic ratios of their offspring?

12. As a geneticist you decide to attempt to recreate some of Mendel's experiments. In your experiment you decide to use the traits of plant height (tall versus short) and pod color (yellow versus green). You obtain a pure-breeding line of plants that is tall (T) with yellow pods (g) and a second line that is short (t) with green pods (G).

a. Using the Punnett square below, outline the cross of these parental strains.

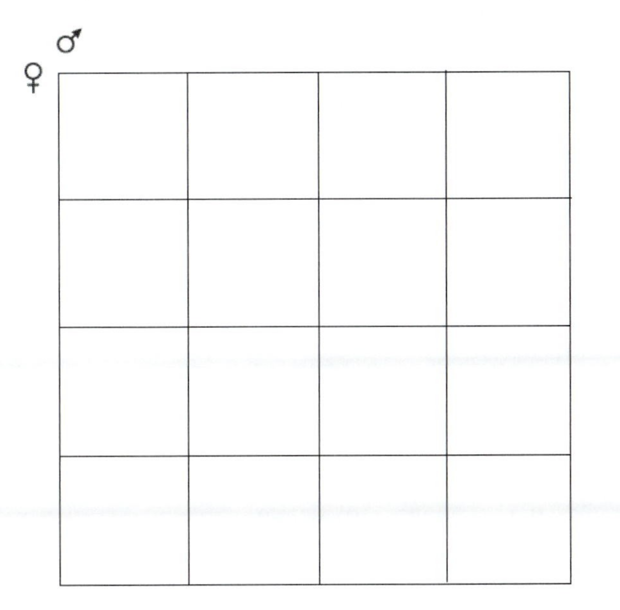

b. What is the genotypic and phenotypic ratios of the F_1 generation?

c. You now cross two members of the F_1 generation to obtain an F_2 generation. Diagram this cross using a Punnett square diagram.

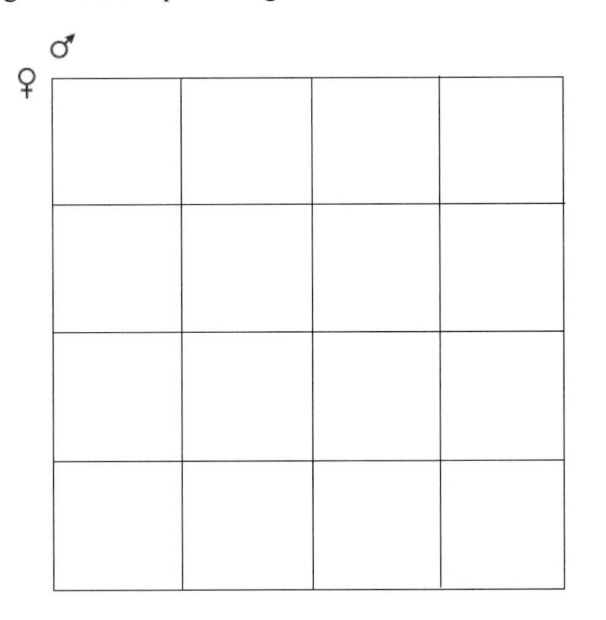

d. What is the phenotypic ratio of the offspring?

13. Using the information from question 12, if you obtained 1248 offspring in the F_2 generation, how many of each phenotypic class would you expect to have?

14. As a genetic counselor you have been asked by a family to study the inheritance of cystic fibrosis over the past three generations. The pedigree diagram for this family is below.

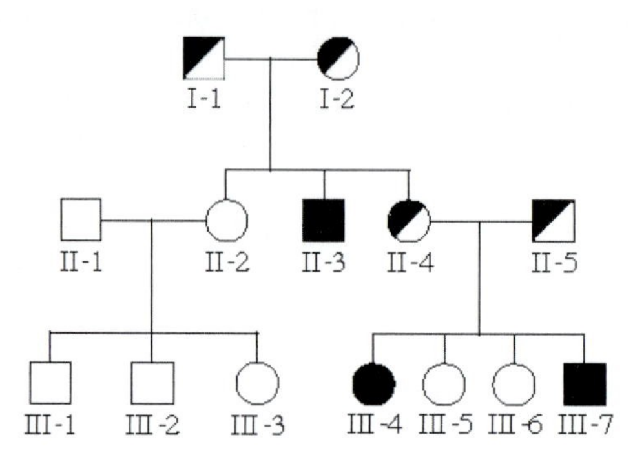

a. Based on this diagram, is cystic fibrosis a dominant or recessive trait?
b. Give the genotypes of the following individuals
   II-2
   II-3
   II-4
   III-5
c. Individual III-5 would like to start a family, what are the chances that she will pass a cystic fibrosis allele on to her offspring?

2.2 Probability and Statistics

Overview

One of Gregor Mendel's greatest contributions to the study of inheritance was the application of mathematical and statistical principles to observed patterns of inheritance. While the Punnett square analysis of the previous section is very useful to visualize a genetic cross, it has severe limitations in crosses involving more than three traits or when predicting the probability of certain genotypes or phenotypes in the offspring.

This section of the text introduces the concepts of probability and statistics to the study of inheritance. The challenge of this section is to identify which method to use to solve the problem at hand. The table below provides a quick summary of these methods.

Method	Application	Algebraic Expression
Sum rule	To determine the probability of one event OR another occurring.	$P(a) + P(b)$
Product rule	To determine the probability of two events occurring, in a specific order.	$P(a) \times P(b)$
Binomial Expansion	To determine the probability that a proportion of the events will have the selected outcome, regardless of order	$P = \frac{n!}{x!\,(n-x)!} p^x q^{n-x}$
Chi square test	To test the goodness of fit between the observation and the expected results.	$X^2 = \sum \frac{(O-E)^2}{E}$

The laws of probability may be applied to the study of inheritance. The Punnett square diagrams used in the previous sections are basically graphic representations of probable outcomes given a certain set of parents. However, at levels above a two-factor cross, Punnett squares become very cumbersome to use. The sum and product rules of probability analysis greatly aids the study of multifactorial crosses. To determine the probability of a three-factor cross, first determine the probability of each trait independently, then multiply the individual probability (product rule) to get the probability of all three events being present in the offspring.

If you are a visual learner, then the forked-line method may be applied. This is simply an extension of the product rule outlined above. It is presented on page 39 of the text.

Key Terms

Binomial expansion equation
Chi square test
Degrees of freedom
Goodness of fit
Hypothesis testing
Multinomial expansion equation
P values
Probability
Product rule
Random sampling error
Sum rule

Focal points

- The use of product and sum rules to determine probability (pgs. 32–33)
- Binomial expansions (pgs. 33–34)
- Chi square tests and hypothesis testing (pgs. 34–36)

Exercises and Problems

For questions 1 to 5, choose which of the following methods would best be applied to determine the required information.

a. Chi square test
b. Sum rule
c. Binomial expansion
d. product rule

_____ 1. The probability of a couple having three girls out of five offspring.

_____ 2. The probability that one or another event will occur in the next generation.

_____ 3. The probability of flipping a coin five times, and getting heads each time.

_____ 4. To determine if the observed phenotypic ratios of a two-factor cross fit the expected ratios by Mendelian inheritance.

_____ 5. The probability of an event occurring in a specific order each time.

For questions 6 to 10, match each of the following terms with its most appropriate answer.

_____ 6. Degrees of freedom
_____ 7. P values
_____ 8. Goodness of fit
_____ 9. Probability
_____ 10. Random sampling error

a. the chance that a future event will occur
b. the deviation between observed and expected outcomes.
c. the probability that the amount of variation is due to random chance.
d. a measure of how well the observation matches the expected values.
e. the measurement of the number of independent categories in a test.

11. Given the following parental genotypes, determine the probability of the genotypes of the offspring.

Parents AaBbCc x AaBbCc

a. AaBbCC
b. Aabbcc
c. AABbCC
d. aaBBCc

12. Assuming that the probability of having a boy and girl are equal (50% each), what would be the probability of each of the following:

a. children in the following order: boy – girl – boy – boy.
b. in six children, the first three are boys.
c. in six children, any three are boys.

13. In a two-factor cross using Mendelian traits in peas, two individuals who are heterozygous for flower color and height (see page 22) are crossed. What is the probability of getting each of the following phenotypic combinations in their offspring? Use either the forked-line method or the product rule to determine your answer.

a. a tall white plant followed by a purple - short plant
b. two purple – tall plants followed by a purple – short
c. two short-white plants in a row

14. In Mendel's two-factor cross (Figure 2.8) he obtained the following F_2 offspring:

315 round – yellow seeds
101 wrinkled – yellow seeds
108 round – green seeds
32 wrinkled – green seeds

Using a chi square test, determine if the observed offspring fit the expected ratios.

a. Propose a hypothesis to be tested.
b. Calculate the expected values for each of the phenotypic classes.
c. Calculate the chi square value.
d. Determine the number of degrees of freedom.
e. Determine the P value (see Table 2.1 on page 36).
f. Is your initial hypothesis accepted or rejected?

15. In *Drosophila melanogaster,* vestigial wings is a recessive trait to normal wings and ebony body color is recessive to honey body color. A fly who is heterozygous for both wings and body color is crossed to a vestigial wing fly who is heterozygous for body color. The following offspring are observed:

39 vestigial – ebony
125 vestigial – honey
43 normal wings – ebony
121 normal wings – honey

Using procedures a-f in question 14, determine if these traits fit the Mendelian model of inheritance.

Chapter Quiz

1. Which of the following methods of calculating probability is used if the order of a specific number of events occurring in the total is not important?
    a. chi square test
    b. binomial expansion
    c. product rule
    d. sum rule

2. Mendel's single factor crosses led to which of the following rules of inheritance?
    a. concept of dominance
    b. law of segregation
    c. support for the particulate theory of inheritance
    d. all of the above

3. In a Chi square test of a genetic cross there are *n* different phenotypic classes in the offspring. What is the value for the degrees of freedom for the test?
    a. n
    b. n – 1
    c. n + 1
    d. 2n
    e. none of the above

4. The pea plant, *Pisum sativum*, was used by Mendel as a model organism for which of the following reason(s):
    a. there were easily scorable traits such as flower color and seed shape.
    b. self-fertilization was possible.
    c. it was relatively easy to grow.
    d. all of the above are correct.

5. The observable characteristic of an organism is called its ____________.
    a. genotype
    b. phenotype
    c. pedigree
    d. allele
    e. none of the above

6. If an organism contains two identical alleles for the same trait, it is said to be which of the following?

- a. cross-fertilized
- b. heterozygous
- c. homozygous
- d. a hybrid
- e. none of the above

7. Two individuals who are heterozygous for a given trait are crossed. What will be the phenotypic ratio of their offspring?

- a. 75% will display the dominant trait, 25% will display the recessive trait
- b. 100 % will display the recessive trait
- c. 100% will be heterozygous individuals
- d. 50% will display the dominant trait, 50% will display the recessive trait
- e. all of the above are possible

8. In a two-factor cross, if two individuals who are heterozygous for both traits are crossed with one another, what will be the phenotypic ratio of their offspring based on Mendel's work?

- a. 3:1
- b. 1:2:1
- c. 9:3:3:1
- d. 15:1
- e. 9:6:1

9. The Punnett square is best used for which of the following?

- a. to predict the goodness of fit of observations to expectations
- b. to conduct crosses of four or more traits
- c. for pedigree analysis
- d. to visualize the possible outcomes of one and two-factor Mendelian crosses

10. Which of the following is correct concerning a pedigree analysis?

- a. it is useful when examining organisms that produce a small number of offspring
- b. it is useful when ethical problems prohibit controlled matings
- c. it is frequently used to study patterns of inheritance in humans
- d. it can be used to determine if a trait is dominant or recessive
- e. all of the above are correct

Answer Key for Study Guide Questions

This answer key provides the answers to the exercises and chapter quiz for this chapter. Answers in parentheses ( ) represent possible alternate answers to a problem, while answers marked with an asterisk (*) indicate that the response to the question may vary.

2.1
1. a
2. d
3. e
4. f
5. b
6. c
7. h
8. g
9. i
10. j

11. a. Round seeds

b. phenotypic ratio — 3 round : 1 wrinkled

genotypic ration — 1 homozygous round : 2 heterozygous : 1 homozygous wrinkled.

12. a. TTgg x ttGG

	Tg	Tg	Tg	Tg
tG	TtGg	TtGg	TtGg	TtGg
tG	TtGg	TtGg	TtGg	TtGg
tG	TtGg	TtGg	TtGg	TtGg
tG	TtGg	TtGg	TtGg	TtGg

b. phenotype: 100% tall with green pods
genotype: 100% TtGg

c. TtGg X TtGg

	TG	Tg	tG	Tg
TG	TTGG	TTGg	TtGG	TtGg
Tg	TTGg	TTgg	TtGg	Ttgg
tG	TtGG	TtGg	ttGG	ttGg
tg	TtGg	Ttgg	ttGg	ttgg

d. 9 tall with green pods: 3 tall with yellow pods : 3 short with green pods : 1 short with yellow pods

13. 702 tall with green pods
234 tall with yellow pods
234 short with green pods
78 short with yellow pods

14. a. recessive

b. II-2 is homozygous dominant, II-3 is homozygous recessive, II-4 is heterozygous, III-5 is homozygous dominant.

c. 0%

2.2

1. c
2. b
3. d
4. a
5. d
6. e
7. c
8. d
9. a
10. b

11. a. 1/16 (0.063)
b. 1/16 (0.063)
c. 1/32 (0.031)
d. 1/32 (0.031)

12. a. 1/16 (0.063)
b. 1/8 (0.125)
c. 0.313

13. a. 0.105
b. 0.059
c. 0.004

14. a. The traits of seed shape and seed color follow a Mendelian pattern of inheritance.
b. The total number of F_2 offspring is 556. Therefore, we can expect 312.75 round-yellow (9/16), 104.25 wrinkled yellow (3/16), 104.25 round-green (3/16), and 34.75 wrinkled green (1/16).
c. 0.470
d. df = 3
e. 0.95 > P > 0.80. The hypothesis is accepted.

15. Note: This is not a dihybrid cross of two heterozygous individuals. Therefore, you may not use the 9:3:3:1 ratio to calculate expected numbers.
a. The traits of wing shape and body color follow a Mendelian pattern of inheritance.
b. The total number of F2 offspring is 298. Therefore we can expect 37.25 vestigial-ebony flies (1/8), 111.75 vestigial-honey flies (3/8), 37.25 normal-ebony flies (1/8), and 111.75 normal-honey flies (3/8).
c. 3.306
d. df = 3
e. 0.50 > P > 4.642. The hypothesis is accepted.

Quiz

1. b
2. d
3. b
4. d
5. b
6. c
7. a
8. c
9. d
10. e

Answers to Conceptual and Experimental Questions

Conceptual Questions

C1. Mendel's work showed that genetic determinants are inherited in a dominant/recessive manner. This was readily apparent in many of his crosses. For example, when he crossed two true-breeding plants for a trait such as height (i.e., tall versus dwarf), all the F_1 plants were tall. This is inconsistent with blending. Perhaps more striking was the result obtained in the F_2 generation: 3/4 of the offspring were tall and 1/4 were short. In other words, the F_2 generation displayed phenotypes that were like the parental generation. There did not appear to be a blending to create an intermediate phenotype. Instead, the genetic determinants did not seem to change from one generation to the next.

C2. In the case of plants, cross-fertilization occurs when the pollen and eggs come from different plants while in self-fertilization they come from the same plant.

C3. The genotype is the type of genes that an individual inherits while the phenotype is the individual's observable traits. Tall pea plants, red hair in humans, and vestigial wings in fruit flies are phenotypes. Homozygous, *TT,* in pea plants; a heterozygous carrier of the cystic fibrosis allele; and homozygotes for the cystic fibrosis allele are descriptions of genotypes. It is possible to have different genotypes and the same phenotype. For example, a pea plant that is *TT* or *Tt* would both have a tall phenotype.

C4. A homozygote that has two copies of the same allele.

C5. Conduct a cross in which the unknown individual is mated to an individual that carries only recessive alleles for the genes in question.

C6. Diploid organisms contain two copies of each type of gene. When they make gametes, only one copy of each gene is found in a gamete. Two alleles cannot stay together within the same gamete.

C7. B. This statement is not correct because these are alleles of different genes.

C8. Genotypes: 1:1 *Tt* and *tt*

Phenotypes: 1:1 Tall and dwarf

C9. The recessive phenotype must be a homozygote. The dominant phenotype could be either homozygous or heterozygous.

C10. *c* is the recessive allele for constricted pods, *Y* is the dominant allele for yellow color. The cross is *ccYy* × *CcYy*. Follow the directions for setting up a Punnett square, as described in chapter 2. The genotypic ratio is 2 *CcYY* : 4 *CcYy* : 2 *Ccyy* : 2 *ccYY* : 4 *ccYy* : 2 *ccyy*. This 2:4:2:2:4:2 ratio could be reduced to a 1:2:1:1:2:1 ratio.

The phenotypic ratio is 6 inflated pods, yellow seeds : 2 inflated pods, green seeds: 6 constricted pods, yellow seeds : 2 constricted pods, green seeds. This 6:2:6:2 ratio could be reduced to a 3:1:3:1 ratio.

Note: smooth pods are the same as inflated pods.

C11. The genotypes are 1 *YY* : 2 *Yy* : 1 *yy*.

The phenotypes are 3 yellow : 1 green.

C12. Offspring with a nonparental phenotype are consistent with the idea of independent assortment. If two different traits were always transmitted together as unit, it would not be possible to get nonparental phenotypic combinations. For example, if a true-breeding parent had two dominant traits and was crossed to a true-breeding parent having the two recessive traits, the F_2 generation could not have offspring with one recessive and one dominant phenotype. However, because independent assortment can occur, it is possible for F_2 offspring to have one dominant and one recessive trait.

C13. A. It behaves like a recessive trait because unaffected parents sometimes produce affected offspring. In other words, we think that the unaffected parents are heterozygous carriers.

B. It behaves like a dominant trait. An affected offspring always has an affected parent. However, recessive inheritance cannot be ruled out.

C14. A. Barring a new mutation during gamete formation, the chance is 100% because they must be heterozygotes in order to produce a child with a recessive disorder.

B. Construct a Punnett square. There is a 50% chance of heterozygous children.

C. Use the product rule. The chance of being phenotypically normal is 0.75 (i.e., 75%), so the answer is $0.75 \times 0.75 \times 0.75 = 0.422$, which is 42.2%.

D. Use the binomial expansion equation where $n = 3$, $x = 2$, $p = 0.75$, $q = 0.25$. The answer is 0.422, or 42.2%.

C15. A. 100% because they are genetically identical.

B. Construct a Punnett square. We know the parents are heterozygotes because they produced a blue-eyed child. The fraternal twin is not genetically identical, but it has the same parents as its twin. The answer is 25%.

C. The probability that an offspring inherits the allele is 50% and the probability that this offspring will pass it on to his/her offspring is also 50%. We use the product rule: $(0.5)(0.5) = 0.25$, or 25%.

D. Barring a new mutation during gamete formation, the chance is 100% because they must be heterozygotes in order to produce a child with blue eyes.

C16. First construct a Punnett square. The chances are 75% of producing a solid pup and 25% of producing a spotted pup.

A. Use the binomial expansion equation where $n = 5$, $x = 4$, $p = 0.75$, $q = 0.25$. The answer is $0.396 = 39.6\%$ of the time.

B. You can use the binomial expansion equation for each litter. For the first litter, $n = 6$, $x = 4$, $p = 0.75$, $q = 0.25$; for the second litter, $n = 5$, $x = 5$, $p = 0.75$, $q = 0.25$. Since the litters are in a specified order, we use the product rule and multiply the probability of the first litter times the probability of the second litter. The answer is 0.070, or 7.0%.

C. To calculate the probability of the first litter, we use the product rule and multiply the probability of the first pup (0.75) times the probability of the remaining four. We use the binomial expansion equation to calculate the probability of the remaining four, where $n = 4$, $x = 3$, $p = 0.75$, $q = 0.25$. The probability of the first litter is 0.316. To calculate the probability of the second litter, we use the product rule and multiply the probability of the first pup (0.25) times the probability of the second pup (0.25) times the probability of the remaining five. To calculate the probability of the remaining five, we use the binomial expansion equation, where $n = 5$, $x = 4$, $p = 0.75$, $q = 0.25$. The probability of the second litter is 0.025. To get the probability of these two litters occurring in this order, we use the product rule and multiply the probability of the first litter (0.316) times the probability of the second litter (0.025). The answer is 0.008, or 0.8%.

D. Since this is a specified order, we use the product rule and multiply the probability of the firstborn (0.75) times the probability of the second born (0.25) times the probability of the remaining four. We use the binomial expansion equation to calculate the probability of the remaining four pups, where $n = 4$, $x = 2$, $p = 0.75$, $q = 0.25$. The answer is 0.040, or 4.0%.

C17. If *B* is the black allele, and *b* is the white allele, the male is *bb*, the first female is probably *BB*, and the second female is *Bb*. We are uncertain of the genotype of the first female. She could be *Bb*, although it is unlikely because she didn't produce any white pups out of a litter of eight.

C18. A. Use the product rule:

$$(1/4)(1/4)=1/16$$

B. Use the binomial expansion equation:

$$n = 4,\ p = 1/4,\ q = 3/4,\ x = 2$$

$$p = 0.21 = 21\%$$

C. Use the product rule:

$$(1/4)(3/4)(3/4) = 0.14, \text{ or } 14\%$$

C19. The parents must be heterozygotes so the probability is 1/4.

C20. A. 1/4

B. 1, or 100%

C. $(3/4)(3/4)(3/4) = 27/64 = 0.42$, or 42%

D. Use the binomial expansion equation where

$$n = 7,\ p = 3/4,\ q = 1/4,\ x = 3$$

$$P = 0.058, \text{ or } 5.8\%$$

E. The probability that the first plant is tall is 3/4. To calculate the probability that among the next four, any two will be tall, we use the binomial expansion equation, where $n = 4$, $p = 3/4$, $q = 1/4$, and $x = 2$.

The probability P equals 0.21.

To calculate the overall probability of these two events:

$$(3/4)(0.21) = 0.16, \text{ or } 16\%$$

C21. A. *T Y R, T y R, T Y r, T y r*

B. *T Y r, t Y r*

C. *T Y R, T Y r, T y R, T y r, t Y R, t Y r, t y R, t y r*

D. *t Y r, t y r*

C22. It violates the law of segregation because there are two copies of one gene in the gamete. The two alleles for the *A* gene did not segregate from each other.

C23. It is recessive inheritance. The pedigree is shown here. Affected individuals are shown with filled symbols.

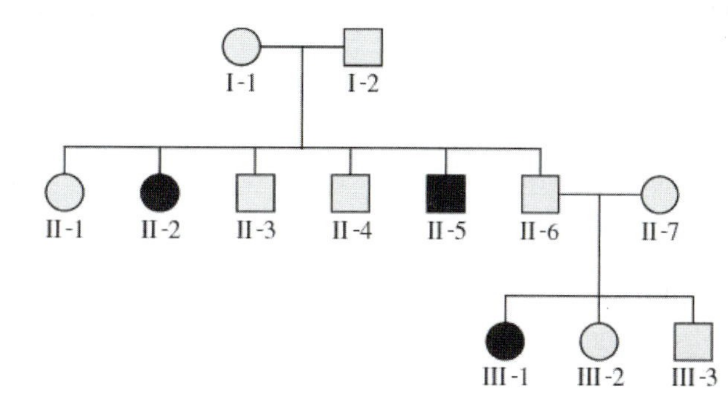

The mode of inheritance appears to be recessive. Unaffected parents (who must be heterozygous) produce affected children.

C24. Based on this pedigree, it is likely to be dominant inheritance because an affected child always has an affected parent. In fact, it is a dominant disorder.

C25. A. 3/16

B. (9/16)(9/16)(9/16) = 729/4096 = 0.18

C. (9/16)(9/16)(3/16)(1/16)(1/16) = 243/1,048,576 = 0.00023, or 0.023%

D. Another way of looking at this is that the probability that it will have round, yellow seeds is 9/16. Therefore, the probability that it will not is 1 – 9/16 = 7/16.

C26. It is impossible for the F_1 individuals to be true-breeding because they are all heterozygotes.

C27. This problem is a bit unwieldy, but we can solve it using the multiplication rule.

For height, the ratio is 3 tall : 1 dwarf.

For seed texture, the ratio is 1 round : 1 wrinkled.

For seed color, they are all yellow.

For flower location, the ratio is 3 axial : 1 terminal.

Thus, the product is

(3 tall + 1 dwarf)(1 round + 1 wrinkled)(1 yellow)(3 axial + 1 terminal)

Multiplying this out, the answer is

9 tall, round, yellow, axial
9 tall, wrinkled, yellow, axial
3 tall, round, yellow, terminal
3 tall, wrinkled, yellow, terminal
3 dwarf, round, yellow, axial
3 dwarf, wrinkled, yellow, axial
1 dwarf, round, yellow, terminal
1 dwarf, wrinkled, yellow, terminal

C28. 2 *TY*, *tY*, 2 *Ty*, *ty*, *TTY*, *TTy*, 2 *TtY*, 2 *Tty*

It may be tricky to think about, but you get 2 *TY* and 2 *Ty* because either of the two *T* alleles could combine with *Y* or *y*. Also, you get 2 *TtY* and 2 *Tty* because either of the two *T* alleles could combine with *t* and then combine with *Y* or *y*.

C29. The drone is *sB* and the queen is *SsBb*. According to the laws of segregation and independent assortment, the male can make only *sB* gametes, while the queen can make *SB*, *Sb*, *sB*, and *sb*, in equal proportions. Therefore, male offspring will be *SB*, *Sb*, *sB*, and *sb*, and female offspring will be *SsBB*, *SsBb*, *ssBB*, and *ssBb*. The phenotypic ratios, assuming an equal number of males and females, will be:

Males	*Females*
1 normal wings/black eyes	2 normal wings, black eyes

1 normal wings/white eyes	2 short wings, black eyes
1 short wings/black eyes	
1 short wings/white eyes	

C30. The genotype of the F_1 plants is *Tt Yy Rr*. According to the laws of segregation and independent assortment, the alleles of each gene will segregate from each other, and the alleles of different genes will randomly assort into gametes. A *Tt Yy Rr* individual could make eight types of gametes: *TYR, TyR, Tyr, TYr, tYR, tyR, tYr,* and *tyr,* in equal proportions (i.e., 1/8 of each type of gamete). To determine genotypes and phenotypes, you could make a large Punnett square that would contain 64 boxes. You need to line up the eight possible gametes across the top and along the side, and then fill in the 64 boxes. Alternatively, you could use one of the two approaches described in solved problem S3. The genotypes and phenotypes would be:

1 *TT YY RR*
2 *TT Yy RR*
2 *TT YY Rr*
2 *Tt YY RR*
4 *TT Yy Rr*
4 *Tt Yy RR*
4 *Tt YY Rr*
8 *Tt Yy Rr* = 27 tall, yellow, round

1 *TT yy RR*
2 *Tt yy RR*
2 *TT yy Rr*
4 *Tt yy Rr* = 9 tall, green, round

1 *TT YY rr*
2 *TT Yy rr*
2 *Tt YY rr*
4 *Tt Yy rr* = 9 tall, yellow, wrinkled

1 *tt YY RR*
2 *tt Yy RR*
2 *tt YY Rr*
4 *tt Yy Rr* = 9 dwarf, yellow, round

1 *TT yy rr*
2 *Tt yy rr* = 3 tall, green, wrinkled

1 *tt yy RR*
2 *tt yy Rr* = 3 dwarf, green, round

1 *tt YY rr*
2 *tt Yy rr* = 3 dwarf, yellow, wrinkled

1 *tt yy rr* = 1 dwarf, green, wrinkled

C31. Construct a Punnett square to determine the probability of these three phenotypes. The probabilities are 9/16 for round, yellow; 3/16 for round, green; and 1/16 for wrinkled, green. Use the multinomial expansion equation described in Solved problem S7, where $n = 5$, $a = 2$, $b = 1$, $c = 2$, $p = 9/16$, $q = 3/16$, $r = 1/16$. The answer is 0.007, or 0.7%, of the time.

C32. The wooly haired male is a heterozygote, because he has the trait and his mother did not. (He must have inherited the normal allele from his mother.) Therefore, he has a 50% chance of passing the wooly allele to his offspring; his offspring have a 50% of passing the allele to their offspring; and these grandchildren have a 50% chance of passing the allele to their offspring (the wooly haired man's great-grandchildren). Since this is an ordered sequence of independent events, we use the product rule: $0.5 \times 0.5 \times 0.5 = 0.125$, or 12.5%. Since there are no other Scandinavians on the island, there is an 87.5% chance of the offspring being normal (because they could not inherit the wooly hair allele from anyone else). We use the binomial expansion equation to determine the likelihood that one out of eight great-grandchildren will have wooly hair, where $n = 8$, $x = 1$, $p = 0.125$, $q = 0.875$. The answer is 0.393, or 39.3%, of the time.

C33. A. Construct a Punnett square. Since it is a rare disease, we would assume that the mother is a heterozygote and the father is normal. The chances are 50% that the man in his thirties will have the allele.

B. Use the product rule: 0.5 (chance that the man has the allele) times 0.5 (chance that he will pass it to his offspring), which equals 0.25, or 25%.

C. We use the binomial expansion equation. From part B, we calculated that the probability of an affected child is 0.25. Therefore the probability of an unaffected child is 0.75. For the binomial expansion equation, $n = 3$, $x = 1$, $p = 0.25$, $q = 0.75$. The answer is 0.422 or 42.2%.

C34. Use the product rule. If the woman is heterozygous, there is a 50% chance of having an affected offspring: $(0.5)^7 = 0.0078$, or 0.78%, of the time. This is a pretty small probability. If the woman has an eighth child who is unaffected, however, she has to be a heterozygote, since it is a dominant trait. She would have to pass a normal allele to an unaffected offspring. The answer is 100%.

Experimental Questions

E1. Pea plants are relatively small and hardy. They produce both pollen and eggs within the same flower. Since a keel covers the flower, self-fertilization is quite easy. In addition, cross-fertilization is possible by the simple manipulation of removing the anthers in an immature flower and later placing pollen from another plant. Finally, peas exist in several variants.

E2. The experimental difference depends on where the pollen comes from. In self-fertilization, the pollen and eggs come from the same plant. In cross-fertilization, they come from different plants.

E3. Two generations would take two growing seasons. About 1 and 1/2 years.

E4. According to Mendel's law of segregation, the genotypic ratio should be 1 homozygote dominant : 2 heterozygotes : 1 homozygote recessive. This data table considers only the plants with a dominant phenotype. The genotypic ratio should be 1 homozygote dominant : 2 heterozygotes. The homozygote dominants would be true-breeding while the heterozygotes would not be true-breeding. This 1:2 ratio is very close to what Mendel observed.

E5. In a monohybrid experiment, the experimenter is only concerned with the outcome of a single trait. In a dihybrid experiment, the experimenter follows the pattern of inheritance for two different traits.

E6. All three offspring had black fur. The ovaries from the albino female could only produce eggs with the dominant black allele (because they were obtained from a true-breeding black female). The actual phenotype of the albino mother does not matter. Therefore, all offspring would be heterozygotes (*Bb*) and have black fur.

E7. The data are consistent with two genes (let's call them gene 22 and gene 24) that exist in two alleles each, a susceptible allele and a resistant allele. The observed data approximate a 9:3:3:1 ratio. This is the expected ratio if two genes are involved, and if resistance is dominant to susceptibility.

E8. If we construct a Punnett square according to Mendel's laws, we expect a 9:3:3:1 ratio. Since a total of 556 offspring were observed, the expected number of offspring are

$556 \times 9/16 = 313$ round, yellow
$556 \times 3/16 = 104$ wrinkled, yellow
$556 \times 3/16 = 104$ round, green
$556 \times 1/16 = 35$ wrinkled, green

If we plug the observed and expected values into the chi square equation, we get a value of 0.51. There are four categories, so our degrees of freedom equal $n - 1$, or 3. If we look up our value in the chi square table (see Table 2.1), it is well within the range of expected error if the hypothesis is correct. Therefore, we accept the hypothesis. In other words, the results are consistent with the law of independent assortment.

E9. No, the law of independent assortment applies to transmission patterns of two or more genes. In a monohybrid experiment, you are monitoring only the transmission pattern of a single gene.

E10. A. If we let c^+ represent normal wings and c represent curved wings, and e^+ represents gray body and e represents ebony body:

Parental Cross: $cce^+e^+ \times c^+c^+ee$.

F_1 generation is heterozygous c^+ce^+e.

An F_1 offspring crossed to flies with curved wings and ebony bodies is

$$c^+ce^+e \times ccee.$$

The F_2 offspring would be a 1:1:1:1 ratio of flies.

$$c^+ce^+e : c^+cee : cce^+e : ccee$$

B. The phenotypic ratio of the F_2 flies would be a 1:1:1:1 ratio of flies.

normal wings, gray body : normal wings, ebony bodies : curved wings, gray bodies : curved wings, ebony bodies

C. From part B, we expect 1/4 of each category. There are a total of 444 offspring. The expected number of each category is
1/4 × 444, which equals 111.

$$\chi^2 = \frac{(114-111)^2}{111} + \frac{(105-111)^2}{111} + \frac{(111-111)^2}{111} + \frac{(114-111)^2}{111}$$

$$\chi^2 = 0.49$$

With 3 degrees of freedom, a value of 0.49 or greater is likely to occur between 95 and 80% of the time. So we accept our hypothesis.

E11. We would expect a ratio of 3 normal : 1 long neck. In other words, there should be 1/3 as many long-necked mice as normal mice. If we multiply 522 times 1/3 the expected value is 174. However, we observed only 62. Therefore, it appears that 174 – 62, or 112, mice died during early embryonic development; 112 divided by 174 gives us the percentage that died, which equals 0.644, or 64.4%.

E12. Follow through the same basic chi square strategy as before. We expect 3/4 of the dominant phenotype and 1/4 of the recessive phenotype.

The observed and expected values are as follows (rounded to the nearest whole number):

Observed*	Expected	$\frac{(O-E)^2}{E}$
5,474	5,493	0.066
1,850	1,831	0.197
6,022	6,017	0.004
2,001	2,006	0.012
705	697	0.092
224	232	0.276
882	886	0.018
299	295	0.054
428	435	0.113
152	145	0.338
651	644	0.076
207	215	0.298
787	798	0.152
277	266	0.455
		$\chi^2 = 2.15$

*Due to rounding, the observed and expected values may not add up to precisely the same number.

Because $n = 14$, there are 13 degrees of freedom. If we look up this value in the chi square table, we have to look between 10 and 15 degrees of freedom. In either case, we would expect such a low value or higher to occur greater than 99% of the time. Therefore, we accept the hypothesis.

E13. This means that a deviation value of 1.005 or greater (between the observed and expected data) would occur 80% of the time. In other words, it is fairly likely to obtain this value due to random sampling error. Therefore, we accept our hypothesis.

E14. The dwarf parent with terminal flowers must be homozygous for both genes since it is expressing these two recessive traits: *ttaa,* where *t* is the recessive dwarf allele, and *a* is the recessive allele for terminal flowers. The phenotype of the other parent is dominant for both traits. However, since this parent was able to produce dwarf offspring with axial flowers, it must have been heterozygous for both genes: *TtAa.*

E15. Our hypothesis is that disease sensitivity and herbicide resistance are dominant traits and they are governed by two genes that assort independently. According to this hypothesis, the F_2 generation should yield a ratio of 9 disease sensitive, herbicide resistant : 3 disease sensitive, herbicide sensitive : 3 disease resistant, herbicide resistant : 1 disease resistant, herbicide sensitive. Since there are a total of 288 offspring produced, the expected numbers would be

9/16 × 288 = 162 disease sensitive, herbicide resistant
3/16 × 288 = 54 disease sensitive, herbicide sensitive
3/16 × 288 = 54 disease resistant, herbicide resistant
1/16 × 288 = 18 disease resistant, herbicide sensitive

$$\chi^2 = \frac{(157-162)^2}{162} + \frac{(57-54)^2}{54} + \frac{(54-54)^2}{54} + \frac{(20-18)^2}{18}$$

$$\chi^2 = 0.54$$

If we look up this value in the chi square table under 3 degrees of freedom, the value lies between the 0.95 and 0.80 probability values. Therefore, we expect a value equal to or greater than 0.54, at least 80% of the time, due to random sampling error. Therefore, we accept the hypothesis.

E16. Our hypothesis is that blue flowers and purple seeds are dominant traits and they are governed by two genes that assort independently. According to this hypothesis, the F_2 generation should yield a ratio of 9 blue flowers, purple seeds : 3 blue flowers, green seeds : 3 white flowers, purple seeds : 1 white flower, green seeds. Since there are a total of 300 offspring produced, the expected numbers
would be

9/16 × 300 = 169 blue flowers, purple seeds
3/16 × 300 = 56 blue flowers, green seeds
3/16 × 300 = 56 white flowers, purple seeds
1/16 × 300 = 19 white flowers, green seeds

$$\chi^2 = \frac{(103-169)^2}{169} + \frac{(49-56)^2}{56} + \frac{(44-56)^2}{56} + \frac{(104-19)^2}{19}$$

$$\chi^2 = 409.5$$

If we look up this value in the chi square table under 3 degrees of freedom, the value is much higher than would be expected 1% of the time by chance alone. Therefore, we reject the hypothesis. The idea that the two genes are assorting independently seems to be incorrect. The F_1 generation supports the idea that blue flowers and purple seeds are dominant traits.

Chapter 3: Reproduction and Chromosome Transmission

Student Learning Objectives

Upon completion of this chapter you should be able to:

1. Know the stages of mitosis and recognize diagrams associated with this process.
2. Understand the process of cytokinesis and how it differs in plants and animals.
3. Know the end result of mitosis in terms of number of cells and their chromosome content.
4. Know the stages of meiosis and the cellular processes that are involved with each stage.
5. Understand the process of gamete formation in both plants and animals.
6. Recognize the importance of the chromosomal theory of inheritance to the study of genetics.
7. Understand the different methods of sex determination.
8. Understand the importance of Morgan's work as a proof of the chromosomal theory of inheritance.

3.1 General Features of Chromosomes

Overview

This chapter begins with a description of eukaryotic and prokaryotic cells. While the principles of genetics are basically the same for all organisms, there are some differences in the structure of the eukaryotic and prokaryotic genomes, as well as variations in the processes of gene structure and expression. Therefore, it is important that at this point in the course that you establish an understanding of the differences between these two classes of cells.

The concept of a karyotype is also introduced in this chapter as a mechanism of visualizing the chromosomal complement of a given cell. The visualization of chromosomes plays an important role in the identification of chromosomal abnormalities, which will be introduced again in Chapter 10. For now, you should concentrate on understanding of the purpose of the process.

This last portion of this first section introduces an important concept – the diploid nature of chromosomes in many eukaryotic organisms. It is crucial that you begin to distinguish between homologues (similar chromosomes) and the concept of sister chromatids (next section). Consider the chromosome to be a filing cabinet, where the individual files within the cabinet represent genes. As we will see in the following sections, we get one set of cabinets (homologues) from our mother, and a second set from our father. Since our parents are of the same species, it makes sense that the organization of the cabinets be the same, meaning that the physical location (locus) of each file (gene) is the same in each cabinet. But our parents are not clones of one another, therefore there may be minor variations in the content of the files. Figure 3.3 illustrates this concept.

Key Terms

Eukaryotes
Chromosomes
Cytogenetics
    Cytogeneticist
Diploid
Germ cells
Homologues
Karyotype
Locus (loci)
Nucleoid
Nucleus
Organelles
Prokaryotes
Somatic cells

Focal Points

- Differences between prokaryotic and eukaryotic cells (pgs. 46-47)
- The diploid nature of eukaryotic chromosomes (pgs. 47-49)

Exercises and Problems

For each of the following, match the information provided with its appropriate term. Answers may be used more than once, or not at all.

_____ 1. The site in the eukaryotic cell where the majority of the genetic information is located.

_____ 2. A photographic representation of the chromosomes in a cell.

_____ 3. A cell that does not have its genetic information contained in a nucleus.

_____ 4. The cells of humans belong to this general class of cells.

_____ 5. The structures of living cells that contain the genetic information.

_____ 6. Bacteria belong to this general class of cells.

_____ 7. Another term for gametes.

_____ 8. Membrane-bound structures that have specific functions in eukaryotic cells.

a. organelles
b. nucleus
c. somatic cell
d. germ cell
e. eukaryotic cells
f. prokaryotic cells
g. karyotype
h. nucleoid
i. chromosomes

For questions 9 to 14, match the following terms with their correct definition.

_____ 9. Loci

_____ 10. Homologues

_____ 11. Gene

_____ 12. Diploid

_____ 13. Homozygous

_____ 14. Heterozygous

a. The physical location of a gene on a chromosome.
b. An individual having two chromosomes with identical alleles for a trait.
c. A unit of genetic information, codes for a single trait.
d. Term used to identify organisms that have a pair of each chromosomes.
e. Chromosomes that are members of a pair.
f. An individual having two chromosomes with different alleles for a trait.

For questions 15 and 16, indicate whether the statement is true or false. If false, give a brief explanation as to why the statement is incorrect.

_____15. Homologous chromosomes are identical in the genetic information that they contain.

_____16. Alleles for a trait occupy the same loci on homologous chromosomes.

3.2 Cellular Division

Overview

This part of the chapter is divided into four major sections – prokaryotic division, the cell cycle, stages of mitosis, and cytokinesis. Since there are a number of important terms and concepts in each of these sections, you should address each of the sections individually and ensure that you have a firm grasp of the material before proceeding on to the next area.

After reviewing the section on prokaryotic cell division, or binary fission, it is important to spend some time reviewing the stages of the cell cycle (Figure 3.5). The majority of the chapter examines the cellular processes associated with cell division during the stages of mitosis and cytokinesis. However, before diving directly into mitosis, be sure that you understand the basic purpose of S phase. It is during S phase that the sister chromatids (Figure 3.6) are formed. Chapter 11 will examine this process in greater detail. Students frequently have a difficult time with the stages of mitosis, specifically with regards to chromosome number, if they do not comprehend the purpose of S phase.

Mitosis is nuclear division and involves five stages. The purpose of each of these stages is outlined in the following table.

Stage of Mitosis	General Characteristics
Prophase	Chromosomes condense. Nuclear membrane begins to disassociate. Mitotic spindle forms. Centrosomes (if present) begin to move to opposite poles of the cell.
Prometaphase	Nuclear envelope is completely disassociated. Spindle fibers interact with the kinetochore.
Metaphase	Chromsomes align along a central line (metaphase plate) of the cell.
Anaphase	Sister chromatids separate and begin to move towards opposite poles of the cell. Each chromatid is now considered to be a chromosome.
Telophase	Chromosomes arrive at the poles of the cell and decondense. Nuclear envelope reforms around the two sets of chromosomes.

It is sometimes easier to understand mitosis if you mark one of the chromosomes and follow it, and its homologue, through the stages. The animations provided with the text assist in the visualization of the process, and also help you to understand that mitosis is a fluid series of steps.

Cytokinesis is the last stage of the division process. Whereas mitosis was nuclear division, cytokinesis is cytoplasmic division. Note that attempts are not made to provide each cell with the same number of each organelle. Organelles will be replicated as needed as soon as the new cell enters into the G_1 phase of the cell cycle. You should also examine the differences between cytokinesis in plant and animal cells.

Key Terms

Anaphase
Asexual reproduction
Binary fission
Cell cycle
Cell plate
Centromere
Centrosomes
Chromatids
  Sister chromatids
Cleavage furrow
Cytokinesis
Interphase
Kinetochore
Metaphase
Metaphase plate
Mitosis
Mitotic spindle (apparatus)
Prometaphase
Prophase
Restriction point
Telophase

Focal Points

- The diagram of the cell cycle (Figure 3.5, pg. 50)
- The concept of sister chromatids (Figure 3.6, pg. 51)
- Stages of mitosis and cytokinesis (Figure 3.7, pgs. 51-53)

Exercises and Problems

In the diagram below, label each of the phases of the cell cycle.

__________ 1.

__________ 2.

__________ 3.

__________ 4.

__________ 5.

__________ 6.

__________ 7.

__________ 8.

__________ 9.

__________ 10.

__________ 11.

__________ 12.

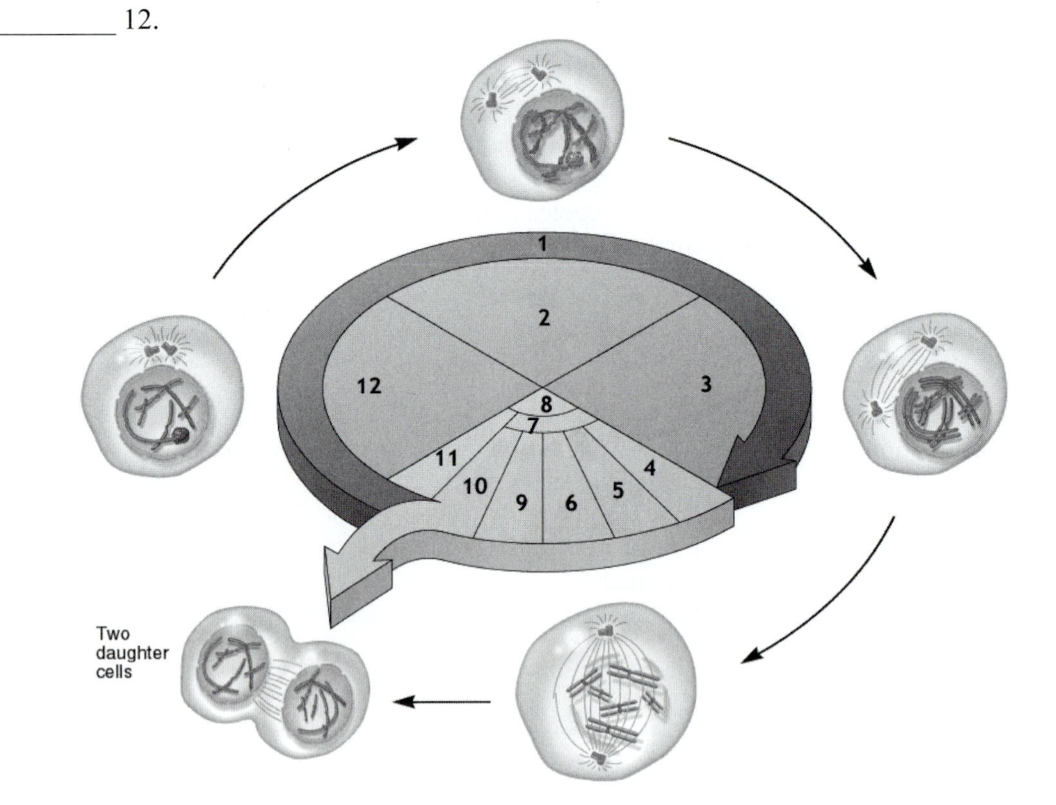

For questions 13 to 21, match each statement with the term that best describes the stage of the cell cycle or mitosis that it is associated with. Terms may be used more than once.

_____ 13. The chromosomes arrange themselves along a line in the center of the cell.

_____ 14. The sister chromatids are formed.

_____ 15. The sister chromatids separate, forming new chromosomes.

_____ 16. The nuclear envelope reforms around the chromosomes.

_____ 17. The spindle fibers attach to the kinetochore.

_____ 18. The centrosomes begin to move to opposite poles of the cell.

_____ 19. The chromosomes condense.

_____ 20. Gap phases of the cell cycle.

_____ 21. The nuclear envelope begins to disassociate.

a. prophase
b. S phase
c. prometephase
d. telophase
e. anaphase
f. metaphase
g. G_1 and G_2 phases

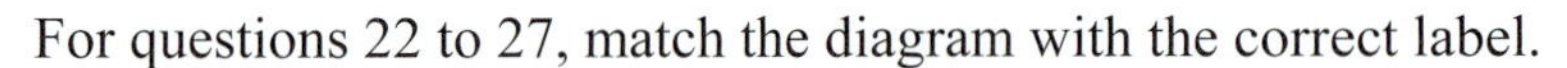

For questions 22 to 27, match the diagram with the correct label.

_____ 22.
_____ 23.
_____ 24.
_____ 25.
_____ 26.
_____ 27.

a. telophase
b. prometephase
c. interphase
d. anaphase
e. metaphase
f. prophase

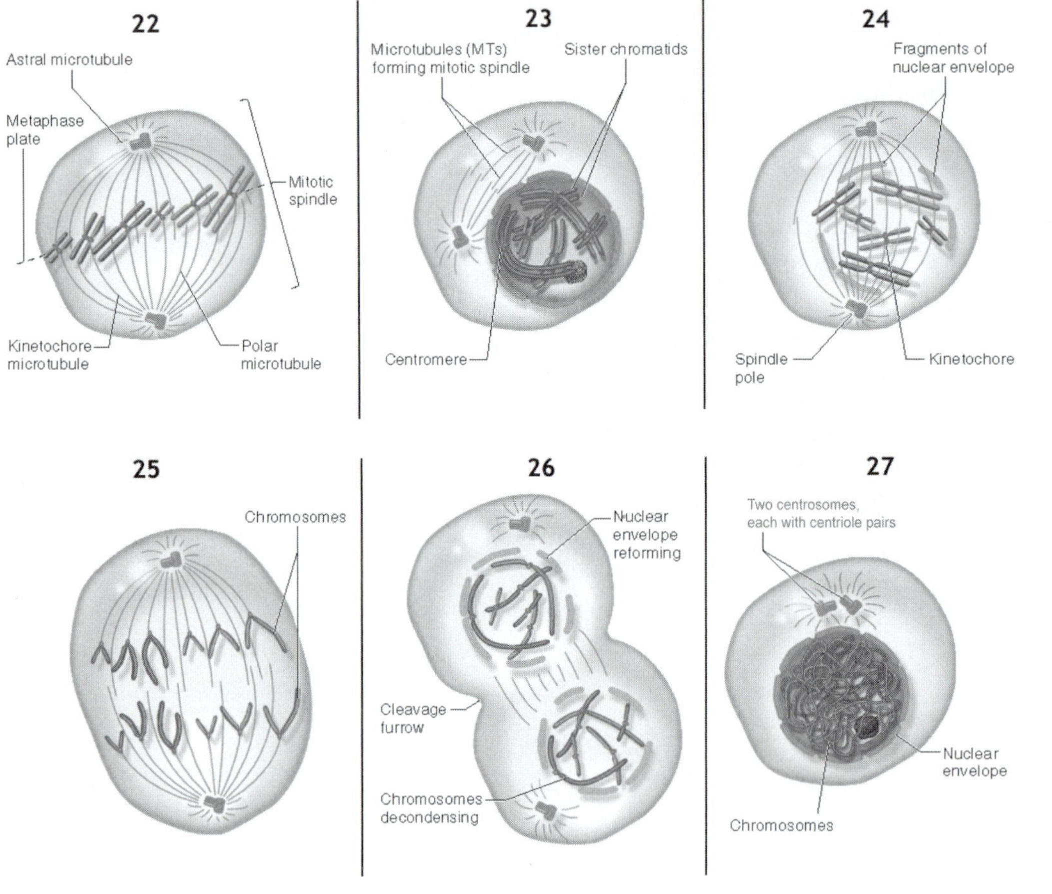

28. Using the number for each label, place these stages in their correct order.

29. Explain the difference between cytokinesis in a plant and animal cell.

3.3 Sexual Reproduction

Overview

For most organisms, the majority of the cell division in the organism following the formation of the zygote is by the process of mitosis. The daughter cells of mitotic division have the same chromosome number as the parent cell, and with some minor exceptions, contain identical genetic information. However, organisms also have a need to produce cells with a reduced number of chromosomes, as is the case with gamete formation in animals, require a modified form of cell division. That process is called meiosis.

Meiosis involves both a reduction in chromosome number, as well as providing variation. Reduction in chromosome number is achieved by conducting two consecutive cell divisions, *without an intervening interphase*. During meiosis I, the genetic composition of the chromatids is shuffled during prophase I. This is called crossing over, and is illustrated in Figure 3.11. This is followed by a chromosome shuffling during the first metaphase (Figure 3.12). These events introduce a significant amount of variation into the chromatids.

Typically students understand mitosis, but not meiosis. However, if you recognize that meiosis follows the same fundamental principles as mitosis, with the exception of a few changes to provide for variation and reduction of chromosome number, then the process becomes easier to understand.

The best means of studying meiosis is to follow the diagrams in the text, specifically Figure 3.12. As was the case with mitosis, it is often helpful to mark one of the chromatids and then follow it through the stages of meiosis.

It is also important to note that meiosis does not form gametes, but rather haploid cells. In animals, meiosis is followed by gametogenesis (Figure 3.14). However, the plants do not form gametes from meiosis. In plants, meiosis produces the gametophyte stage, which in turn produces gametes by mitosis (Figure 3.15).

An understanding of meiosis is crucial to the understanding of Mendelian inheritance and the chromosomal theory of inheritance (see section 3.4).

Key Words

Bivalents
Chiasma (chiasmata)
Crossing over
Diakinesis
Diplotene
Egg cells
Fertilization
Haploid
Heterogamous
Isogamous
Leptotene
Meiosis
Oogenesis
Pachytene
Spermatogenesis
Synapsis
Synaptonemal complex
Sperm cells
Tetrad
Zygotene

Focal Points

- The process of crossing over in prophase I (Figure 3.10, pg. 56)
- The stages of meiosis in animals (Figure 3.12, pg. 58)
- The process of gametogenesis in animals (Figure 3.14, pg. 59) and plants (Figure 3.15, pg. 60)

Exercises and Problems

For questions 1 to 5, match the events of prophase I to it correct function.

_____ 1. Diakinesis

_____ 2. Zygotena

_____ 3. Diplotena

_____ 4. Leptotena

_____ 5. Pachytena

a. Homologous chromosomes recognize each other by synapsis.
b. The chromosomes begin to condense, and are now visible using a microscope.
c. The stage in which crossing over occurs.
d. The synaptonemal complex disappears.
e. The final stage of prophase I.

6. Given the following chromosome number (n) is an organism, calculate the possible random chromosome alignments at metaphase I of meiosis.

a. n = 7
b. n = 2
c. n = 4

For questions 7 to 10, indicate whether the statement applies of oogenesis, spermatogenesis or both, in animals.

a. spermatogenesis
b. oogenesis
c. both

_____ 7. The process that produces haploid gametes.

_____ 8. Produces four functional gametes.

_____ 9. Produces three polar bodies and a single gamete.

_____ 10. Starts from a diploid cell.

For questions 11 to 15, match the following answers to their correct descriptions.

_____ 11. Embryo sac

_____ 12. Gametophyte

_____ 13. Endosperm

_____ 14. Sporophyte

_____ 15. Pollen grain

a. The female gametophyte.
b. The haploid generation of the plant.
c. A 3n structure that serves as the food-storage location for the embryo.
d. The represents the diploid generation of the plant.
e. The male gametophyte.

3.4 The Chromosome Theory of Inheritance and Sex Chromosomes

Overview

Chapter 2 introduced you to Mendel's principles of inheritance. In Chapter 3, we have examined the process of cell division. Although both meiosis and Mendel's principles were known at the start of the 20th century, it took the work of several researchers to understand that Mendel's laws were based on the stages of meiosis. The fact that genes are located on chromosomes, and that the segregation of these chromosomes during meiosis was finally worked out by Boveri and Sutton, although many other researchers contributed to this (Table 3.1).

The principles of the chromosomal theory of inheritance are provided on page 62 of the text. Although these seems obvious to us today, at the time they were developed they were revolutionary ideas. The important aspect of this first section of the text is how the chromosomal theory relates to the principles of inheritance first established by Mendel. Both the law of segregation (Figure 3.16) and law of independent assortment (Figure 3.17) are based upon the events of meiosis. You need to study and understand these figures before proceeding.

As humans, we are familiar with the fact that males have an X-Y combination of sex chromosomes, and females possess an X-X combination. What this means to geneticists is that human males are a heterogametic sex, which produce two different types of sperm. In addition, since males lack a second X chromosome, recessive traits on the X-chromosome appear to be dominant, since there is not a second gene to mask the effects of the recessive gene. The study of X-linked traits played an important role in the validation of the chromosomal theory (see below). Before leaving this section, notice that the X-Y system is not the only mechanism of sex determination. Other organisms use various methods of identifying the male and female of the species.

While Boveri and Sutton proposed the chromosomal theory, the first experimental proof was provided by Morgan in his study of X-linked traits in *Drosophila*. The experimental design of this proof is a classic example of how simple crosses were used to test the developing theories of genetics in the early 20th century. You should examine Figure 3.19 and understand how this experiment provided proof of the chromosomal theory.

The next section discusses the use of reciprocal crosses for studying sex-linked inheritance. In our previous studies of autosomal traits (Chapter 2), we did not identify differences between males and females when constructing a Punnett square. Notice on page 67 how this is changed with regards to sex-linked traits.

Finally, the chapter briefly introduces the fact that sex-linked inheritance in humans follows three basic models: X-linked, Y-linked, and pseudoautosomal. More on this will be provided later in the course.

Key Words

Autosomes
Chromosome theory of inheritance
Chromosomes
Hemizygous
Heterogametic sex
Holandric genes
Homogametic sex
Pseudoautosomal inheritance
Reciprocal cross
Sex chromosomes
Sex linkage
Testcross
X-linked alleles
X-linked genes
X-linked inheritance

Focal Points

- The relationship between independent assortment and meiosis (Figure 3.17)
- Morgan's experimental evidence supporting the chromosomal theory of inheritance (Experiment 3A and Figure 3.19)
- The concept of a reciprocal cross (pgs. 67-68)

Exercises and Problems

1. List the five basic principles of the chromosomal theory of inheritance.
    a.
    b.
    c.
    d.
    e.

2. Explain how the law of independent assortment relates to meiosis.

For questions 3 to 9, use the following answers.

a. X-Y system
b. X-O system
c. Z-W system
d. haplo-diplo system

_____ 3. In this system, the female is the heterogametic sex.

_____ 4. Females in this system are the homogametic sex.

_____ 5. Humans utilize this system of sex-determination.

_____ 6. In this system, the female possesses two sex chromosomes, while the male only has one sex chromosome.

_____ 7. This form of sex determination is common in birds.

_____ 8. In this system, sex is determined by the total chromosome number of the organism.

_____ 9. This sex determination system is common in insects.

10. What is the purpose of a testcross?

11. Explain how a reciprocal cross can be used to determine if a trait is sex-linked or autosomal.

For questions 12 to 16, choose the answer that is associated with each.

a. Y-linked inheritance
b. pseudoautosomal inheritance
c. X-linked inheritance

______ 12. *Mic2* in humans is an example of this.

______ 13. Holandric genes are associated with this pattern of inheritance.

______ 14. Refers to genes on the sex-chromosomes that display autosomal patterns of inheritance.

______ 15. The *Sry* gene in humans is an example of this.

______ 16. The term sex-linkage is usually associated with this type.

Chapter Quiz

1. A testcross is always conducted between an individual whose genotype is unknown, and which of the following?
   a. a heterozygous individual
   b. a homozygous dominant individual
   c. a homozygous recessive individual
   d. any of the above will work

2. A reciprocal cross is used to determine which of the following?
   a. if a trait is autosomal or sex-linked
   b. the number of autosomal chromosomes
   c. the dominance/recessive nature of a trait
   d. the sex of the organism
   e. none of the above

3. Thoman Hunt Morgan's experiment with white-eyed *Drosophila* provided proof of which of the following?
   a. law of independent assortment
   b. chromosomal theory of inheritance
   c. theory of natural selection
   d. law of segregation
   e. none of the above

4. An organism has 20 pairs of chromosomes at the start of mitosis, how many chromosome pairs with each cell have at the end of mitosis?
   a. 5
   b. 10
   c. 20
   d. 40

5. During which of the following stages of the cell cycle is the DNA of the organism replicated?
a. prophase
b. anaphase
c. S phase
d. G_2 phase

6. Bacteria reproduce by which of the following?
a. mitosis
b. meiosis
c. binary fission
d. all of the above

7. Before the start of meiosis, on which of the following would the genetic material be almost identical?
a. homologous chromosomes
b. sister chromatids
c. non-sister chromatids of homologous chromosomes
d. none of the above

8. During which of the following stages of meiosis does synapsis and crossing over occur?
a. metaphase
b. cytokinesis
c. prophase I
d. telophase II
e. prophase II

9. In which of the following is the male of the species the homogametic sex?
a. X-Y system
b. Z-W system
c. X-O system
d. haplo-diplo system
e. the male is never homogametic

10. The *Sry* gene of humans displays which of the following patterns of inheritance?
a. X- linked
b. pseudoautosomal
c. autosomal
d. Y-linked

Answer Key for Study Guide Questions

This answer key provides the answers to the exercises and chapter quiz for this chapter. Answers in parentheses ( ) represent possible alternate answers to a problem, while answers marked with an asterisk (*) indicate that the response to the question may vary.

3.1
1. b
2. g
3. f
4. e
5. i

6. f
7. d
8. a
9. a
10. e
11. c
12. d
13. b
14. f
15. False. Homologus chromosomes are similar, but not identical.
16. True

3.2

1. Interphase
2. S phase
3. G_2 phase
4. Prophase
5. Prometaphase
6. Metaphase
7. Mitosis
8. M
9. Anaphase
10. Telophase
11. Cytokinesis
12. G_1 phase
13. f
14. b
15. e
16. d
17. b
18. a
19. a
20. g
21. a
22. e
23. f
24. b
25. d
26. a
27. c
28. 27-23-24-22-25-26
29. In animal cells, cytokinesis is marked by the formation of a cleavage furrow, while plant cytokinesis involves the formation of a cell plate. (*)

3.3

1. e
2. a
3. d
4. b
5. c
6. a. 2^7 (128)
   b. 2^2 (4)
   c. 2^4 (16)

7. c
8. a
9. b
10. c
11. a
12. b
13. c
14. d
15. e

3.4 1. See text page 62.
2. The random alignment of honhomologous chromosomes at the metaphase plate of meiosis explains the law of independent assortment.
3. c
4. a
5. a
6. b
7. c
8. d
9. b (d)
10. The purpose of a testcross is the determine the genotype of an individual by crossing the individual with a homozygous recessive individual.
11. In a reciprocal cross, if the sexes in the F_1 generation differ for the trait being studied, then the trait is sex-linked.
12. b
13. a
14. b
15. a
16. c

Quiz

1. c
2. a
3. b
4. c
5. c
6. c
7. b
8. c
9. b
10. d

Answers to Conceptual and Experimental Questions

Conceptual Questions

C1. They are genetically identical, barring rare mutations, because they receive identical copies of the genetic material from the mother cell.

C2. The term *homologue* refers to the members of a chromosome pair. Homologues are usually the same size and carry the same types and order of genes. They may differ in that the genes they carry may be different alleles.

C3. Sister chromatids are identical copies derived from the replication of a parental chromosome. They remain attached to each other at the centromere. They are genetically identical, barring rare mutations and crossing over with homologous chromosomes.

C4. Metaphase is the organization phase and anaphase is the separation phase.

C5. In G_1, there should be six linear chromosomes. In G_2, there should be 12 chromatids that are attached to each other in pairs of sister chromatids.

C6. In metaphase I of meiosis, each pair of chromatids is attached to only one pole via the kinetochore microtubules. In metaphase of mitosis, there are two attachments (i.e., to both poles). If the attachment was lost, a chromosome would probably be lost and degraded because it would not migrate to a pole. Therefore, it would not become enclosed in a nuclear membrane after telophase. If left out in the cytoplasm, it would eventually be degraded.

C7. A. During mitosis and meiosis II

B. During meiosis I

C. During mitosis, meiosis I, and meiosis II

D. During mitosis and meiosis II

C8. The reduction occurs because there is a single DNA replication event but two cell divisions. Because of the nature of separation during anaphase I, each cell receives one copy of each type of chromosome.

C9.

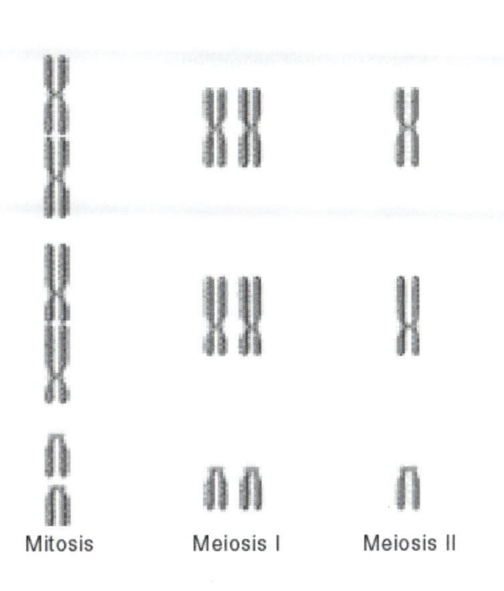

C10. It means that the arrangement of the maternally derived and paternally derived chromosomes is random during metaphase I. Refer to Figure 3.17.

C11. Mitosis—two diploid cells containing 10 chromosomes each (two complete sets)

Meiosis—four haploid cells containing 5 chromosomes each (one complete set)

C12. There are three pairs of chromosomes. So the possible number of arrangements equals 2^3, which is 8.

C13. $(1/2)^n = (1/2)^4 = 1/16$

C14. It would be much lower because pieces of maternal chromosomes would be mixed with the paternal chromosomes. Therefore, it is unlikely to inherit a chromosome that was completely paternally derived.

C15. Bacteria do not need to sort their chromosomes because they only have one type of chromosome. The attachment of the two copies of the chromosomes to the cell membrane prior to cell division ensures that each daughter cell receives one copy.

C16. During interphase, the chromosomes are greatly extended. In this conformation, they might get tangled up with each other and not sort properly during meiosis and mitosis. The condensation process probably occurs so that the chromosomes can easily align along the equatorial plate during metaphase without getting tangled up.

C17. To produce identical quadruplets, fertilization begins with one sperm and one egg cell. This fertilized egg then could divide twice by mitosis to produce four genetically identical cells. These four cells could then separate from each other to begin the lives of four distinct individuals. Another possibility is that mitosis could produce two cells that separate from each other. These two cells could then divide by mitosis to produce two pairs of cells, which also could separate to produce four individual cells.

C18. During prophase II, your drawing should show four replicated chromosomes (i.e., four structures that look like Xs). Each chromosome is one homologue. During prophase of mitosis, there should be eight replicated chromosomes (i.e., eight Xs). During prophase of mitosis, there are pairs of homologues. The main difference is that prophase II has a single copy of each of the four chromosomes whereas prophase of mitosis has four pairs of homologues. At the end of meiosis I, each daughter cell received only one copy of a homologous pair, not both. This is due to the alignment of homologues during metaphase I and their separation during anaphase I.

C19. The products of meiosis have only one copy of each type of chromosome. For example, one human gamete may contain the paternally derived copy of chromosome 11, whereas a different gamete may contain the maternally derived copy of chromosome 11. These two homologues may carry different alleles of the same genes and therefore are not identical. In contrast, mitosis produces genetically identical daughter cells that have both copies of all the pairs of homologous chromosomes.

C20. DNA replication does not take place during interphase II. The chromosomes at the end of telophase I have already replicated (i.e., they are found in pairs of sister chromatids). During meiosis II, the sister chromatids separate from each other yielding individual chromosomes.

C21.

Prophase	Telophase
Nuclear membrane vesiculates.	Nuclear membrane re-forms.
Spindle forms.	Spindle disassembles.
Chromosomes condense.	Chromosomes decondense.
Chromosomes attach to spindle.	Chromosomes detach from the spindle.

C22. A. 20

B. 10

C. 30

D. 20

C23. The hybrid offspring would have 44 chromosomes (i.e., 25 + 19). The reason for infertility is because each chromosome does not have a homologous partner. Therefore, the chromosomes cannot properly pair during metaphase I. The gametes do not receive one copy of each homologue. Gametes will be missing certain chromosomes, which makes them inviable.

C24. A. Dark males and light females; reciprocal: all dark offspring

B. All dark offspring; reciprocal: dark females and light males

C. All dark offspring; reciprocal: dark females and light males

D. All dark offspring; reciprocal: dark females and light males

C25. Male gametes are usually small and mobile. Animal and some plant male gametes often contain flagella, which make them motile. The goal of the male gamete is to locate the female gamete. Female gametes are usually much larger and contain nutrients to help the growth of the embryo after fertilization occurs.

C26. To produce sperm, a spermatogonial cell first goes through mitosis to produce two cells. One of these remains a spermatogonial cell and the other progresses through meiosis. In this way, the testis continues to maintain a population of spermatogonial cells.

C27. During oogenesis in humans, the cells are arrested in prophase I of meiosis for many years until selected primary oocytes progress through the rest of meiosis I and begin meiosis II. If fertilization occurs, meiosis II is completed.

C28. A. *A B C, A B c, A b C, A b c, a B C, a b C, a B c, a b c*

B. *A B C, A b C*

C. *A B C, A B c, a B C, a B c*

D. *A b c, a b c*

C29. A. X-linked recessive (unaffected mothers transmit the trait to sons)

B. Autosomal recessive (affected daughters and sons are produced from unaffected parents)

C30. The outcome is that all the daughters will be affected and all the sons will be unaffected. The ratio is 1:1.

C31. Duchenne muscular dystrophy is inherited as an X-linked recessive disorder. Affected individuals are males. Their mothers are related to males who have the disorder. The mothers of affected males are heterozygous carriers. Note: This pedigree is also consistent with autosomal recessive inheritance, although the pattern is more suggestive of X-linked inheritance.

C32. First set up the following Punnett Square:

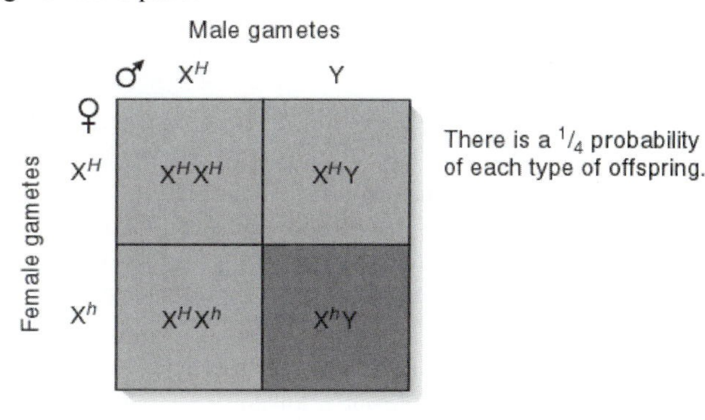

A. 1/4

B. (3/4)(3/4)(3/4)(3/4) = 81/256

C. 3/4

D. The probability of an affected offspring is 1/4 and the probability of an unaffected offspring is 3/4. For this problem, you use the binomial expansion where $x = 2$, $n = 5$, $p = 1/4$, and $q = 3/4$. The answer is 0.26, or 26%, of the time.

C33. The genotypes of the parents are that the mother is a heterozygous carrier and the father is hemizygous color-blind. By constructing a Punnett square, the probability of producing a color-blind offspring (male or female) is 1/2, or 50%. We use the product rule to compute ordered events. So, (1/2)(1/2) equals 1/4, or 25%.

C34. A. No. A male inherits the Y chromosome from his father, so he could not inherit an X chromosome from his paternal grandfather.

B. Yes. A male could inherit a maternal X chromosome from his maternal grandfather. It would first be transmitted from his grandfather to his mother, and then from his mother to him.

C. Yes. A female could inherit a maternal X chromosome from her maternal grandmother. It would first be transmitted from her grandmother to her mother, and then from her mother to her.

D. Yes. A female could inherit a maternal X chromosome from her maternal great-great-grandmother. It would first be transmitted from her great-great-grandmother to a maternal great-grandparent (male or female). It could then be transmitted to a grandparent, and then to the mother, and then from her mother to her.

Note: Parts B–D assume there is no crossing over between homologous X chromosomes.

C35. 1 affected daughter : 1 unaffected daughter : 1 affected son : 1 unaffected son

C36. A. The woman's mother must have been a heterozygote. So there is a 50% chance that the woman is a carrier. If she has children, 1/4 (i.e., 25%) will be affected sons if she is a carrier. However, there is only a 50% chance that she is a carrier. So we multiply 50% times 25%, which equals $0.5 \times 0.25 = 0.125$, or a 12.5% chance.

B. If she already had a color-blind son, then we know she must be a carrier, so the chance is 25%.

C. The woman is heterozygous and her husband is hemizygous for the color-blind allele. This couple will produce 1/4 offspring that are color-blind daughters. The rest are 1/4 carrier daughters, 1/4 normal sons, and 1/4 color-blind sons. Answer is 25%.

C37. There is a 1/2 chance that the mother will transmit her abnormal chromosome and a 1/2 chance that the father will. We use the product rule to calculate the chances of both events happening. So the answer is $1/2 \times 1/2 = 1/4$, or 25%. The probability that such a child will pass both chromosomes to an offspring is also 25% because that child had a 1/2 chance of passing either chromosome.

C38. A. The fly is a male because the ratio of X chromosomes to sets of autosomes is 1/2, or 0.5.

B. The fly is female because the ratio is 1.0.

C. The fly is male because the ratio is 0.5.

D. The fly is female because the ratio is 1.0.

C39. A. Female; there are no Y chromosomes.

B. Female; there are no Y chromosomes.

C. Male; the Y chromosome determines maleness.

D. Male; the Y chromosome determines maleness.

Experimental Questions

E1. A. G_2 phase (it could not complete prophase)

B. Metaphase (it could not enter anaphase)

C. Telophase (it could not divide into two daughter cells)

D. G_2 phase (it could not enter prophase)

E2. Perhaps the most convincing observation was that all of the white-eyed flies of the F_2 generation were males; not a single white-eyed female was observed. This suggests a link between sex determination and the inheritance of this trait. Since sex determination in fruit flies is determined by the number of X chromosomes, this suggests a relationship between the inheritance of the X chromosome and the inheritance of this trait.

E3. Actually, his data are consistent with this hypothesis. To rule out a Y-linked allele, he could have crossed an F_1 female with a wild-type male rather than an F_1 male. The same results would be obtained. Since the wild-type male would not have a white allele, this would rule out Y-linkage.

E4. The basic strategy is to set up a pair of reciprocal crosses. The phenotype of sons is usually the easiest way to discern the two patterns. If it is Y linked, the trait will only be passed from father to son. If it is X linked, the trait will be passed from mother to son.

E5. To be a white-eyed female, a fly must inherit two X chromosomes and both must carry the white-eye allele. This could occur only if both X chromosomes in the female stayed together and the male gamete contained a Y chromosome. The white-eyed females would be XXY. To produce a red-eyed male, a female gamete lacking any sex chromosomes could unite with a normal male gamete carrying the X^{W+}. This would produce an X0, red-eyed male.

E6. The 3:1 sex ratio occurs because the female produces 50% gametes that are XX (and must produce female offspring) and 50% that are X (and produce half male and half female offspring). The original female had one X chromosome carrying the red allele and two other X chromosomes carrying the eosin allele. Set up a Punnett square assuming that this female produces the following six types of gametes: $X^{W+}X^{W-e}$, $X^{W+}X^{W-e}$, $X^{W-e}X^{W-e}$, X^{W+}, X^{W-e}, X^{W-e}. The male of this cross is X^W Y.

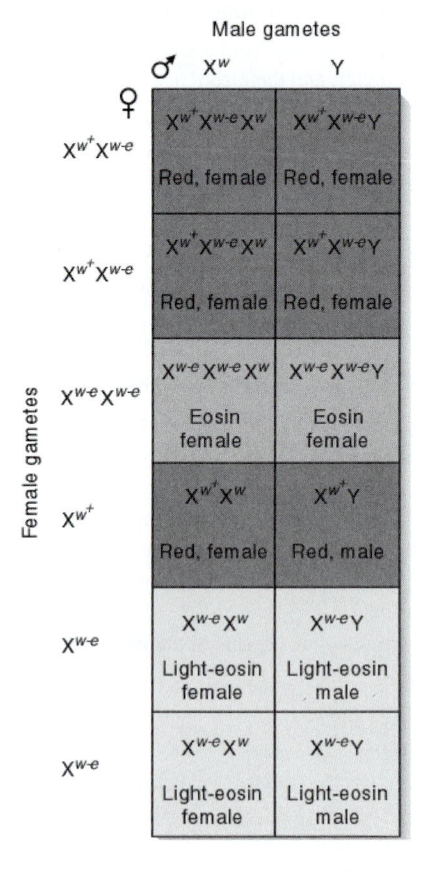

E7. During interphase, the chromosomes are longer, thinner, and much harder to see. In metaphase, they are highly condensed, which makes them thicker and shorter.

E8. If we use the data from the F_1 mating (i.e., F_2 results), there were 3,470 red-eyed flies. We would expect a 3:1 ratio between red- and white-eyed flies. Therefore, assuming that all red-eyed offspring survived, there should have been about 1,157 (i.e., 3,470/3) white-eyed flies. However, there were only 782. If we divide 782 by 1,157, we get a value of 0.676, or 67.6% have survived.

E9. In general, you cannot distinguish between autosomal and pseudoautosomal inheritance from a pedigree analysis. Mothers and fathers have an equal probability of passing the alleles to sons and daughters. However, if an offspring had a chromosomal abnormality, you might be able to tell. For example, in a family tree involving the *Mic2* allele, an offspring that was X0 would have less of the gene product and an offspring that was XXX or XYY or XXY would have extra amounts of the gene products. This may lead you to suspect that the gene is located on the sex chromosomes.

E10. You could karyotype other members of the family and see if affected members always carry the abnormal chromosome.

E11. You need to make crosses to understand the pattern of inheritance of traits (determined by genes) from parents to offspring. And you need to microscopically examine cells to understand the pattern of transmission of chromosomes. The correlation between the pattern of transmission of chromosomes during meiosis, and Mendel's laws of segregation and independent assortment, is what led to the chromosome theory of inheritance.

E12. These rare female flies would be XXY. Both X chromosomes would carry the white allele and the miniature allele. These female flies would have miniature wings because they would have inherited both X chromosomes from their mother.

E13. Originally, individuals who had abnormalities in their composition of sex chromosomes provided important information. In mammals, X0 individuals are females, while in flies, X0 individuals are males. In mammals, XXY individuals are males, while in flies, XXY individuals are females. These results indicate that the presence of the Y chromosome causes maleness in mammals, but it does not in flies. A further analysis of flies with abnormalities in the number of sets of autosomes revealed that it is the ratio between the number of X chromosomes and the number of sets of autosomes that determines sex.

E14. In X-linked recessive inheritance, it is much more common for males to be affected. In autosomal recessive inheritance, there is an equal chance of males and females being affected (unless there is a sex influence as described in Chapter 4). For X-linked dominant inheritance, affected males would produce 100% affected daughters and not transmit the trait to their sons. This would not be true for autosomal dominant traits, where there is an equal chance of males and females being affected.

Chapter 4: Extensions of Mendelian Inheritance

Student Learning Objectives

Upon completion of this chapter you should be able to:

1. Understand the relationship between gene expression and the recessive/dominant phenotype.
2. Understand the different types of Mendelian inheritance patterns and how to solve problems based upon those patterns.
3. Recognize how gene interactions can alter the predicted 9:3:3:1 ratio of a dihybrid cross.
4. Understand the process of epigenic inheritance.

4.1 Inheritance Patterns of Single Genes

Overview

The first part of this chapter is dedicated to variations of the Mendelian inheritance patterns first presented in Chapter 2. It is important to note that pure Mendelian inheritance is rare. Instead, patterns of inheritance are clouded by many factors, including the environment and protein function. One of the most important things to consider as you progress through this section is the relationship between the observed inheritance and the molecular basis of the pattern being described. By doing so, you will quickly come to realize that the behavior of the proteins produced by genes are the major factor in determining the phenotype of the individual.

An excellent summary of these deviations is presented in Table 4.1. By the end of the chapter you should be able to not only recognize these variations, but also be able to construct and analyze crosses that demonstrate these concepts.

Key Terms

Age of onset
Codominance
Conditional lethal alleles
Essential gene
Expressivity
Gene dosage effect
Heterozygote advantage
Incomplete dominance
Incomplete penetrance
Lethal allele
Mendelian inheritance
Multiple alleles
Mutant alleles
Nonessential genes
Overdominance
Semilethal alleles
Sex-influenced inheritance
Sex-limited traits
Simple Mendelian inheritance
Temperature-sensitive conditional allele
Temperature-sensitive (ts) lethals
Wild-type alleles

Focal Points

- Table 4.1 - Different Types of Mendelian Inheritance
- The experimental proof of gene dosage effect (Experiment 4A)

Exercises and Problems

For questions 1 to 7, match each of the following to its correct definition.

_____ 1. Wild-type allele

_____ 2. Conditional lethal alleles

_____ 3. Nonessential genes

_____ 4. Mutant allele

_____ 5. Essential gene

_____ 6. Semilethal alleles

_____ 7. Lethal allele

a. These are altered alleles that tend to be rare in natural populations.
b. An allele that kills an organism only under certain environmental conditions.
c. An allele that causes the death of an organism.
d. A lethal allele that only acts in some individuals.
e. The most prevalent form of an allele in a population.
f. A gene that encodes a protein that when absent causes the death of the organism.
g. Genes that are not absolutely required for survival.

8. Is incomplete dominance an example of blending? Explain your answer.

9. If a red and pink four o'clock plant are crossed, what will be the phenotypic and genotypic ratio of the F_1 generation?

For questions 10 to 18, match each of the following examples with its correct pattern of inheritance. Some answers may be used more than once. Some items may have more than one answer.

_____ 10. Heterozygous individuals for sickle-cell anemia

_____ 11. Human A, B and O blood groups

_____ 12. Flower color in the four-o'clock plant *Mirabilis jalapa*

_____ 13. Polydactyly in *Homo sapien*

_____ 14. Light eosin colored eyes in *Drosophila*

_____ 15. Pattern baldness in male *Homo sapiens*

_____ 16. PKU in humans

_____ 17. Coat color in rabbits

_____ 18. Hen feathering in chickens

a. incomplete dominance
b. incomplete penetrance
c. codominance
d. overdominance
e. gene dosage effect
f. multiple allele system
g. environmental influence
h. sex-influenced inheritance
i. sex-limited inheritance
j. X-linked inheritance

19. Refer to Figure 4.4 in the text (pg. 80). Given the following phenotypes of the parents, what coat colors are possible in the offspring?

a. full coat color x himalayan coat color
b. albino coat color x chincilla coat color
c. chincilla coat color x Himalayan coat color

For questions 20 to 25, select the molecular basis of each of the following patterns of inheritance.

_____ 20. Sex-influenced inheritance

_____ 21. Incomplete penetrance

_____ 22. Lethal alleles

_____ 23. Codomiance

_____ 24. Incomplete dominance

_____ 25. Overdominance

a. Heterozygotes produce proteins that function over a wider range of conditions.
b. The protein of a dominant allele does not exert its effects.
c. A loss of function mutation in an essential gene.
d. Hormones limit the molecular expression of a gene.
e. The proteins produced different alleles function in slightly different ways resulting in both being expressed in the heterozygote.
f. The amount of protein produced by a single dominant allele is not sufficient to produce the normal phenotype.

4.2 Gene Interactions

Overview

The previous section examined deviations from Mendelian inheritance based upon single genes. You should now also recognize that many of these patterns of inheritance were due to the proteins that were being produced by the genes. However, many traits are under the influence of more than one gene. In many cases, several genes produce enzymes that are involved in metabolic pathways. If a mutation in a gene early in the pathway produces a loss of function, then this may mask the phenotype of genes that contribute alleles later in the pathway (for example, see the bottom of pg. 90). This is called epistasis.

As you progress through this section, note how the gene interactions being described alter the 9:3:3:1 ratio of a Mendelian two-factor (dihybrid) cross. Before proceeding, carefully examine the table below of a two-factor cross involving two heterozygous individuals.

Gene A	Gene B	
A_	B_	9/16
aa	B_	3/16
A_	bb	3/16
aa	bb	1/16

Notice that the 9/16th aspect of ratio occurs if a dominant allele is inherited for both traits, while a 3/16th ratio occurs if a dominant allele is inherited for only one of the traits. 1/16th of the time the phenotype is recessive. By recognizing these numbers it is usually easy to identify epistatic interactions, since the masking of one of the traits will cause the ratios to be combined. For example, a 9:7 ratio indicates that the A_bb, aaB_, and aabb ratios have been combined together. This will be useful later in additional discussions of gene interactions.

As you have probably already discovered, the study of eye color in *Drosophila* was responsible for many of the early discoveries of non-Mendelian inheritance. The last experiment in this chapter examines the production of cream and eosin-colored eyes. This experiment was conducted by Calvin Bridges, who was trying to determine if the cream and eosin alleles were in the same gene. His experiments indicated that cream was an autosomal trait that modified the expression of eosin only, not the normal red eyes. The result was an 8:4:3:1 phenotypic ratio of eye color in the offspring (pg. 92).

Key Terms

Epistasis
Gene interaction

Focal Points

- Examples of gene interactions (Figures 4.17 and 4.18)
- Morgan and Bridges experimental work with cream-eye color in *Drosophila* (Experiment 4B)

Exercises and Problems

For questions 1 to 5, use the following answers.

a. flower color in the sweet pea
b. comb morphology in chickens
c. eosin and cream eye color in *Drosophila*
d. all of the above
e. neither a, b or c are correct

_____ 1. The F_2 generation produced four distinct phenotypic classes.
_____ 2. The offspring in the F_2 generation have a 9:7 phenotypic ratio.
_____ 3. An example of a gene interaction.
_____ 4. Demonstrated that an allele may act as a specific modifier of a second allele.
_____ 5. Produced an 8:4:3:1 phenotypic ration in the F_2 generation.
_____ 6. The only one that is not an example of an epistatic gene interaction.

7. Answer the following questions assuming that a pea-combed chicken and a single-combed chicken are crossed.

a. Are walnut-combed offspring possible from this cross?

b. If no single-combed chickens are produced, what was the genotype of the pea-combed parent?

8. In crosses involving the sweet pea, two white flowered plants are crossed. Indicate whether the following offspring are possible.

a. all white offspring

b. all purple offspring

c. a combination of purple and white offspring

9. Examine the cross from Bridge's work with eosin and cream eyes in Drosophila (pg. 92).

a. Based upon this information, what would the genotype of the parents have to be in order to produce a cream-eye colored female offspring?

b. Give an example of two parental phenotypes that could create this offspring.

c. What is the only possible genotype for this offspring?

Chapter Quiz

1. A trait that has a different level of expression in males than females is an example of which of the following?

a. sex-influenced inheritance

b. sex-limited inheritance

c. gene dosage

d. all of the above

e. none of the above

2. Expressivity is most often associated with which of the following?

a. multiple allele systems

b. heterozygote advantage

c. incomplete penetrance

d. incomplete dominance

e. none of the above

3. Coat color in rabbits and human blood groups are both examples of ____.

a. multiple allele systems

b. epistatic interactions

c. gene dosage

d. simple Mendelian inheritance

e. all of the above

4. The most common allele in the population is called the ________.

a. mutant allele

b. essential allele

c. dominant allele

d. wild-type allele

e. recessive allele

5. A trait that produces a 9:7 ratio in the F_2 generation of a two-factor cross is most likely exhibiting which of the following?

a. gene dosage
b. epistatic interactions
c. overdominance
d. multiple allele systems
e. sex-limited inheritance

6. Cream eye color in *Drosophila* is an example of _________.

a. incomplete inheritance
b. codominance
c. epistatic interactions
d. overdominance

7. Phenotypic blending is the result of ________.

a. incomplete dominance
b. incomplete penetrance
c. codominance
d. overdominance

8. A heterozygote that has a selective advantage over the homozygous dominant individual is an example of ______.

a. gene interaction
b. codominance
c. temperature sensitive lethals
d. overdominance

9. A trait that is only expressed in one sex of the species is an example of _____.

a. temperature-sensitive alleles
b. sex-influenced inheritance
c. sex-limited inheritance
d. X-linked inheritance

10. Phenylketonuria (PKU) in humans is an example of ___________.

a. overdominance
b. environmental influence on phenotype
c. gene dosage
d. epistatic interactions
e. none of the above

Answer Key for Study Guide Questions

This answer key provides the answers to the exercises and chapter quiz for this chapter. Answers in parentheses ( ) represent possible alternate answers to a problem, while answers marked with an asterisk (*) indicate that the response to the question may vary.

4.1
1. e
2. b
3. g
4. a
5. f

6. d
7. c
8. It is an example of phenotypic blending since the heterozygote produces an intermediate phenotype. However, the genotypes are not blended.
9. Phenotypic ratio is ½ red to ½ pink. The genotypic ratio is ½ C^RC^R t0 ½ C^RC^W.
10. d
11. c, f
12. a
13. b
14. e, j
15. h
16. g
17. f (g)
18. i
19. a. all phenotypes are possible
    b. all phenotypes except full coat color
    c. all phenotypes except full coat color
20. d
21. b
22. c
23. e
24. f
25. a

4.2
1. b (c)
2. a
3. d
4. c
5. c
6. b
7. a. no
   b. rrPP
8. a. yes
   b. yes
   c. yes
9. a. The parents must both possess at least one copy of the c^a allele. The father must be $X^{w-e}Y$ and the mother must carry at least one X^{w-e} allele.
   b. father C c^a X^{w-e} Y, mother C c^a X^{w-e} X (*)
   c. c^a c^a X^{w-e} X^{w-e}

Quiz

1. d
2. c
3. a
4. d
5. b
6. c
7. a
8. d
9. c
10. a

Answers to Conceptual and Experimental Questions

Conceptual Questions

C1. Dominance occurs when one allele completely exerts its phenotypic effects over another allele. Incomplete dominance is a situation in which two alleles in the heterozygote have an intermediate phenotype. Codominance is when both alleles exert their effects independently in the heterozygote. And overdominance is a case in which the heterozygote has a phenotype that is superior to either homozygote.

C2. Sex-influenced traits are influenced by the sex of the individual even though the gene that governs the trait may be autosomally inherited. Pattern baldness in people is an example. Sex-limited traits are an extreme example of sex influence. The expression of a sex-limited trait is limited to one sex. For example, colorful plumage in certain species of birds is limited to the male sex. Sex-linked traits involve traits whose genes are found on the sex chromosomes. Examples in humans include hemophilia and color blindness.

C3. The term *gene interaction* refers to the phenomenon that two or more different genes can have an impact on the same trait. This can occur, for example, if two genes encode enzymes in the same metabolic pathway. If the pathway is disrupted by a mutation in either of these two genes, the same net result may occur.

C4. If the normal allele is dominant, it tells you that one copy of the gene produces a saturating amount of the protein encoded by the gene. Having twice as much of this protein, as in the normal homozygote, does not alter the phenotype. If the allele is incompletely dominant, this means that one copy of the normal allele is not saturating.

C5. Recessive alleles are often loss-of-function alleles. It would generally be more likely for a recessive allele to eliminate a trait or function rather than create one. Therefore, the peach carries the dominant allele while the nectarine has a loss-of-function allele that prevents fuzz formation. The recessive allele is in a gene that is necessary for fuzz formation.

C6. There would be a ratio of 1 normal : 2 star-eyed individuals.

C7. The red and white seed packs should be from true-breeding (homozygous) strains. The pink pack should be seeds from a cross between white- and red-flowered plants.

C8. The amount of protein produced from a single gene is not a saturating amount. Therefore, additional copies produce more of the protein, whose functional consequences can be increased.

C9. If individual 1 is *ii,* individual 2 could be I^Ai, I^AI^A, I^Bi, I^BI^B, or I^AI^B.
If individual 1 is I^Ai or I^AI^A, individual 2 could be I^Bi, I^BI^B, or I^AI^B.
If individual 1 is I^Bi or I^BI^B, individual 2 could be I^Ai, I^AI^A, or I^AI^B.

Assuming individual 1 is the parent of individual 2:

If individual 1 is *ii*, individual 2 could be I^Ai or I^Bi.
If individual 1 is I^Ai, individual 2 could be I^Bi or I^AI^B.
If individual 1 is I^AI^A, individual 2 could be I^AI^B.
If individual 1 is I^Bi, individual 2 could be I^Ai or I^AI^B.
If individual 1 is I^BI^B, individual 2 could be I^AI^B.

C10. Types O and AB provide an unambiguous genotype. Type O can only be *ii,* and type AB can only be I^AI^B. It is possible for a couple to produce children with all four blood types. The couple would have to be I^Ai and I^Bi. If you construct a Punnett square, you will see that they can produce children with AB, A, B, and O blood types.

C11. The father could not be I^AI^B, I^BI^B, or I^AI^A. He is contributing the O allele to his offspring. Genotypically, he could be I^Ai, I^Bi, or *ii* and have type A, B, or O blood, respectively.

C12. A. 1/4

B. 0

C. (1/4)(1/4)(1/4) = 1/64

D. Use the binomial expansion:

$$P = \frac{n!}{x!(n-x)!} p^x q^{(n-x)}$$

$$n = 3,\ p = 1/4,\ q = 1/4,\ x = 2$$

$$P = 3/64 = 0.047, \text{ or } 4.7\%$$

C13. Perhaps it should be called codominant at the "hair level" because one or the other allele is dominant with regard to a single hair. However, this is not the same as codominance in blood types in which every cell can express both alleles.

C14. Let's begin with the assumption that the recessive alleles encode enzymes that are completely defective. If so, we could explain solved problem S1 in the following manner. Gene *A* might encode an enzyme that converts a colorless pigment precursor into a (black) pigmented molecule. If an animal is *aa,* no pigment is made and it becomes white. Gene *C* could encode an enzyme that converts some of this black pigment into a brown pigment to produce the agouti phenotype. If an animal is *cc,* no brown pigment is made, so the animal stays black.

C15. All the F_1 generation will be white because they have inherited the dominant white allele from their Leghorn parent. Construct a Punnett square. Let *W* and *w* represent one gene, where *W* is dominant and causes a white phenotype. Let *A* and *a* represent the second gene, where the recessive allele causes a white phenotype in the homozygous condition. The genotype of the F_1 birds is *WwAa.* The phenotypic ratio of the F_2 generation will be 13 white : 3 brown. The only brown birds will be 2 *wwAa,* 1 *wwAA.*

C16. We know that the parents must be heterozygotes for both genes.

The genotypic ratio of their children is 1 *BB* : 2 *Bb* : 1 *Bb*

The phenotypic ratio depends on sex.

1 *BB* bald male : 1 *BB* bald female : 2 *Bb* bald males : 2 *Bb* nonbald females : 1 *bb* nonbald male : 1 *bb* nonbald female

A. 50%

B. 1/8

C. (3/8)(3/8)(3/8)= 27/512 = 0.05, or 5%

C17. On a standard vegetarian diet, the results would be 50% white and 50% yellow.

On a xanthophyll-free diet, the results would be 100% white.

C18. A. The male offspring would be hemizygous for apricot and have 3% pigment. The female offspring would be heterozygous for apricot and white and also have 3% pigment.

B. Half of the male offspring would be hemizygous for coral and have 4% pigment, and the other half of the male offspring would be hemizygous for apricot and have 3% pigment. All of the female offspring would be heterozygous and have one copy of the dominant red allele. Therefore, all of the female offspring would have 100% pigment.

C. Half of the males would have red eyes with 100% pigment and the other half of the males would be hemizygous for the apricot allele and have 3% pigment. Half of the females would be heterozygous for the red allele and white allele and would have red eyes with 100% pigment (red is dominant). The other half of the females would be heterozygous for the white allele and apricot allele and have about 3% pigment.

D. All of the males would be hemizygous for the coral allele and have about 4% pigment. All of the females would be heterozygous for the coral and apricot alleles and have about 7% pigment.

C19. It probably occurred in the summer. Dark fur occurs in cooler regions of the body. If the fur grows during the summer, these regions are likely to be somewhat warmer, and therefore the fur will be lighter.

C20. Set up a Punnett square, but keep in mind that the eosin gene is X linked.

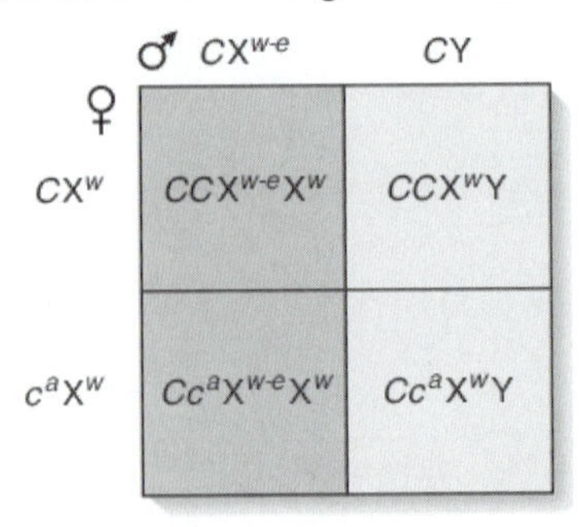

Because *C* is dominant, the phenotypic ratios are two light-eosin females and two white males.

C21. First, you would cross heterozygous birds to each other. This would yield an F_1 generation consisting of a ratio of 1 *HH* : 2 *Hh* : 1 *hh.* The male offspring that are *hh* would have cock-feathering. All the female offspring would have hen-feathering. You would then take cock-feathered males and cross them to F_1 females. If all the males within a brood were cock-feathered, it is likely that the mother was *hh.* If so, all the offspring would be *hh.* The offspring from such a brood could be crossed to each other. If they are truly *hh,* the males of the F_3 generation should all be cock-feathered.

C22. A. Could be.

B. No, because an unaffected father has an affected daughter.

C. No, because two unaffected parents have affected children.

D. No, because an unaffected father has an affected daughter.

E. No, because both sexes exhibit the trait.

F. Could be.

C23. A. Could be.

B. No, because an affected female has an unaffected son.

C. Could be.

D. No, because an affected male has an unaffected daughter.

E. No, because it affects both sexes.

C24. You would look at the pattern within families over the course of many generations. For a recessive trait, 25% of the offspring within a family are expected to be affected if both parents are unaffected carriers, and 50% of the offspring would be affected if one parent was affected. You could look at many families and see if these 25% and 50% values are approximately true. Incomplete penetrance would not necessarily predict such numbers. Also, for very rare alleles, incomplete penetrance would probably have a much higher frequency of affected parents producing affected offspring. For rare recessive disorders, it is most likely that both parents are heterozygous carriers. Finally, the most informative pedigrees would be situations in which two affected parents produce children. If they can produce an unaffected offspring, this would indicate incomplete penetrance. If all of their offspring were affected, this would be consistent with recessive inheritance.

C25. Molecular: The *ß*-globin gene for Hb^A homozygotes encodes a *ß*-globin polypeptide with a normal amino acid sequence compared to the Hb^S homozygotes whose *ß*-globin genes encode a polypeptide that has an abnormal structure. The abnormal structure affects the ability of hemoglobin to carry oxygen.

Cellular: Under conditions of low oxygen, Hb^SHb^S cells form a sickle shape compared to the normal biconcave disk shape of Hb^AHb^A cells.

Organismal: In Hb^SHb^S individuals, the sickle shape decreases the life span of the red blood cell, which causes anemia. Also, the clogging of red blood cells in the capillaries causes tissue damage and painful crises. This does not occur in Hb^AHb^A people.

C26. Since this is a rare trait, we assume the other parent is homozygous for the normal allele. The probability of a heterozygote passing the allele to his/her offspring is 50%. The probability of an affected offspring expressing the trait is 80%. We use the product rule to determine the likelihood of these two independent events.

$$(0.5)(0.8) = 0.4, \text{ or } 40\% \text{ of the time}$$

C27.

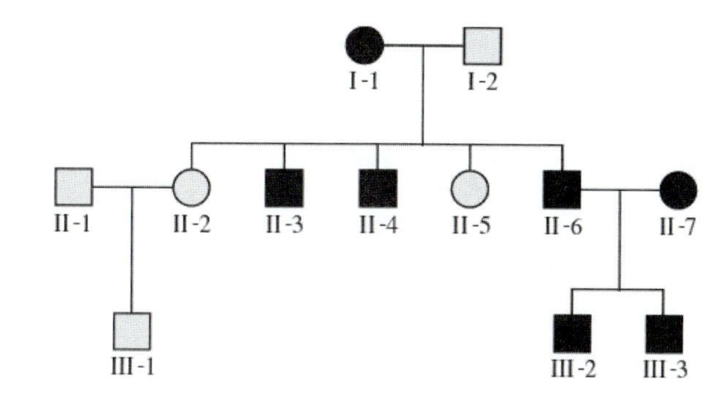

C28. This is an example of incomplete dominance. The heterozygous horses are palominos. For example, if *C* represents chestnut and *c* represents cremello, the chestnut horses are *CC,* the cremello horses are *cc,* and the palominos are *Cc.*

Experimental Questions

E1. Mexican hairless dogs are heterozygous for a dominant allele that is lethal when homozygous. In a cross between two Mexican hairless dogs, we expect 1/4 to be normal, 1/2 to be hairless, and 1/4 to die.

E2. Chinchilla 1 is heterozygous $c^{ch}c$.
Chinchilla 2 is heterozygous $c^{ch}c^{h}$.
Chinchilla 3 is heterozygous $c^{ch}c$.
Chinchilla 4 is probably $c^{ch}c^{ch}$ because it always produces chinchilla offspring when mated to chinchilla 1 or 2, which are heterozygous. However, it is possible that chinchilla 4 is also heterozygous $c^{ch}c^{h}$ or $c^{ch}c$.
Chinchilla 5 is $c^{ch}c^{h}$ or $c^{ch}c$.

As noted, we are not sure about the genotypes of chinchillas 4 and 5.

E3. There may be two redundant genes that are involved in feathering. The unfeathered Buff Rocks are homozygous recessive for the two genes. The Black Langhans are homozygous dominant for both genes. In the F_2 generation (which is a double heterozygote crossed to another double heterozygote), 1 out of 16 offspring will be doubly homozygous for both recessive genes. All the others will have at least one dominant allele for one of the two (redundant) genes.

E4. The first offspring must be homozygous for the horned allele. The father's genotype is still ambiguous; he could be heterozygous or homozygous for the horned allele. The mother's genotype must be heterozygous because her phenotype is polled (she cannot be homozygous for the horned allele) but she produced a horned daughter (who must have inherited a horned allele from its mother).

E5. The reason why all the puppies have black hair is because albino alleles are found in two different genes. If we let the letters *A* and *B* represent the two different pigmentation genes, then one of the dogs is *AAbb* and the other is *aaBB.* Their offspring are *AaBb* and therefore are not albinos because they have one dominant copy of each gene.

E6. It is a sex-limited trait where *W* (white) is dominant but expressed only in females. In this cross of two yellow butterflies, the male is *Ww* but is still yellow because the white phenotype is limited to females. The female is *ww* and yellow. The offspring would be 50% *Ww* and 50% *ww.* However, all the males would be yellow. Half of the females would be white (*Ww*) and half would be yellow (*ww*). Overall, this would yield 50% yellow males, 25% yellow females, and 25% white females.

E7. The sandy variation may be due to a homozygous recessive allele at one of two different genes in these two varieties of sandy pigs. Let's call them genes *A* and *B.* One variety of sandy pig could be *aaBB* and the other *AAbb.* The F_1 generation in this cross would be heterozygotes for both genes and are all red. This tells us that the *A* and *B* alleles are dominant. In the F_2 generation, 6 out of 16 will be homozygous for either the *aa* or *bb* alleles and become sandy. One out of 16 will be doubly homozygous and be white. The remaining 9 will contain at least one dominant allele for both genes.

E8. One parent must be *RRPp.* The other parent could be *RRPp* or *RrPp.* All the offspring would inherit (at least) one dominant *R* allele. With regard to the other gene, 3/4 would inherit at least one copy of the dominant *P* allele. These offspring would have a walnut comb. The other 1/4 would be homozygous *pp* and have a rose comb (because they would also have a dominant *R* allele).

E9. The yellow squash has to be *wwgg.* It has to be *ww* because it is colored, and it has to be *gg* because it is yellow. Since the cross produced 50% white and 50% green offspring, the other parent (i.e., the white squash) must be *WwGG.* The cross would produce 50% offspring that are *WwGg* (white) and 50% that are *wwGg* (green).

E10. Let's use the letters *A* and *B* for these two genes. Gene *A* exists in two alleles, which we will call *A* and *a.* Gene *B* exists in two alleles, *B* and *b.* The uppercase alleles are dominant to the lowercase alleles. The true-breeding long-shaped squash is *aabb* and the true-breeding disk-shaped is *AABB.* The F_1 offspring are *AaBb.* You can construct a Punnett square to determine the outcome of self-fertilization of the F_1 plants. The Punnett square will have 16 boxes.

To get the disk-shaped phenotype, an offspring must inherit at least one dominant allele from both genes.

1 *AABB* + 2 *AaBB* + 2 *AABb* + 4 *AaBb* = 9 disk-shaped offspring

To get the round phenotype, an offspring must inherit at least one dominant allele for one of the two genes but must be homozygous recessive for only one of the two genes.

$$1\ aaBB + 1\ AAbb + 2\ aaBb + 2\ Aabb = 6 \text{ round-shaped offspring}$$

To get the long phenotype, an offspring must inherit all recessive alleles: 1 *aabb*

E11. It is because they both have one copy of the $X^{w\text{-}e}$ allele, an allele that produces a little pigment. The X^{w} allele does not produce any pigment in the female.

E12. You would expect the alleles with intermediate pigmentation to exhibit a gene dosage effect because these are the ones that are not making a maximum amount of pigment. Alleles such as ivory, pearl, and apricot would be the best candidates. The alleles producing a darker reddish phenotype may not produce a gene dosage effect because a large amount of pigment is already made. To test this idea, you could set up a series of crosses, like the ones described in conceptual question C18, and then analyze the amount of pigment in the eyes of the offspring.

E13. For each of the three crosses we expect a 1:1 ratio, based on the hypothesis that the eosin and white alleles are X-linked alleles of the same gene, and that there is a gene dosage effect. The observed and expected values are as follows:

Observed	*Expected*
225 red eyes	216 red eyes
208 white eyes	216 white eyes
679 red eyes	713 red eyes
747 eosin eyes	713 eosin eyes
694 light-eosin eyes	636 light-eosin eyes
579 light-eosin eyes	636 light-eosin eyes

If we plug these values into the chi square equation (described in Chapter 2), we obtain a value of 14.3. With 5 degrees of freedom, this value lies between 0.05 and 0.01. At the 1% confidence limit, we would accept our hypothesis. At the 5% confidence limit, however, we would reject our hypothesis. Fortunately, many other experiments confirmed the idea that the white and eosin alleles are X-linked alleles in the same gene, and that they exhibit a gene dosage effect. Based on this experiment alone, however, the results would not be completely convincing.

E14. In this cross, we expect that there will be a 9:7 ratio between red and white. In other words, 9/16 will be red and 7/16 will be white. Since there are a total of 345 plants, the expected values are

$$9/16 \times 345 = 194 \text{ red}$$

$$7/16 \times 345 = 151 \text{ white}$$

$$\chi^2 = \sum \frac{(O-E)^2}{E}$$

$$\chi^2 = \frac{(201-194)^2}{194} + \frac{(144-151)^2}{151}$$

$$\chi^2 = 0.58$$

With 1 degree of freedom, our chi square value is too small to reject our hypothesis. Therefore, we accept that it may be correct.

E15. The results obtained when crossing two F_1 offspring appear to yield a 9:3:3:1 ratio, which would be expected if eye color is affected by two different genes that exist in dominant and recessive alleles. Neither gene is X linked. Let pr^+ represent the red allele of the first gene and *pr* the purple allele. Let sep^+ represent the red allele of the second gene and *sep* the sepia allele.

The first cross is: $prpr\ sep^+sep^+ \times pr^+pr^+sep\ sep$

All the F_1 offspring would be $pr^+pr\ sep^+sep$. They have red eyes because they have a dominant red allele for each gene. When the F_1 offspring are crossed to each other, the following results would be obtained:

♀ \ ♂	pr^+sep^+	pr^+sep	$pr\ sep^+$	$pr\ sep$
pr^+sep^+	pr^+pr^+ sep^+sep^+ Red	pr^+pr^+ sep^+sep Red	pr^+pr sep^+sep^+ Red	pr^+pr sep^+sep Red
pr^+sep	pr^+pr^+ sep^+sep Red	pr^+pr^+ $sep\ sep$ Sepia	pr^+pr sep^+sep Red	pr^+pr $sep\ sep$ Sepia
$pr\ sep^+$	pr^+pr sep^+sep^+ Red	pr^+pr sep^+sep Red	$pr\ pr$ sep^+sep^+ Purple	$pr\ pr$ sep^+sep Purple
$pr\ sep$	pr^+pr sep^+sep Red	pr^+pr $sep\ sep$ Sepia	$pr\ pr$ sep^+sep Purple	$pr\ pr$ $sep\ sep$ Pur/sepia

In this case, one gene exists as the red (dominant) or purple (recessive) allele and the second gene exists as the red (dominant) or sepia (recessive) allele. If an offspring is homozygous for the purple allele, it will have purple eyes. Similarly, if an offspring is homozygous for the sepia allele, it will have sepia eyes. An offspring that is homozygous for both recessive alleles has purplish sepia eyes. To have red eyes, it must have at least one copy of the dominant red allele for both genes. Based on an expected
9 red : 3 purple : 3 sepia : 1 purplish sepia, the observed and expected numbers of offspring are as follows:

Observed	*Expected*
146 purple eyes	148 purple eyes (791 × 3/16)
151 sepia eyes	148 sepia eyes (791 × 3/16)
50 purplish sepia eyes	49 purplish sepia eyes (791 × 1/16)
444 red eyes	445 red eyes (791 × 9/16)
791 total offspring	

If we plug the observed and expected values into our chi square formula, we obtain a chi square value of about 0.11. With 3 degrees of freedom, this is well within our expected range of values, so we cannot reject our hypothesis that purple and sepia alleles are in two different genes, and that these recessive alleles are epistatic to each other.

E16. Since the results of the first cross produce offspring with red eyes, it suggests that the vermilion allele and purple allele are not alleles of the same gene. The results of the second cross indicate that the vermilion allele is X linked, since all the male offspring had vermilion eyes, just like their mothers. Let pr^+ represent the red allele of the first gene and *pr* the purple allele. Let X^{V+} represent the red allele of the second gene and X^V the vermilion allele.

For the first cross: the male parents are $pr^+pr^+X^V Y$ and the female parents are $prpr\ X^{V+}X^{V+}$. The F_1 offspring are shown in this Punnett square.

♀ \ ♂	pr^+X^v	pr^+Y
prX^{v^+}	pr^+pr $X^{v^+}X^v$ Red	pr^+pr $X^{v^+}Y$ Red
prX^{v^+}	pr^+pr $X^{v^+}X^v$ Red	pr^+pr $X^{v^+}Y$ Red

For the second cross: the male parents are $prprX^{v+}Y$ and the female parents are $pr^+pr^+X^vX^v$. The F_1 offspring are shown in this Punnett square.

♀ \ ♂	prX^{v^+}	prY
pr^+X^v	pr^+pr $X^{v^+}X^v$ Red	pr^+pr X^vY Vermillion
pr^+X^v	pr^+pr $X^{v^+}X^v$ Red	pr^+pr X^vY Vermillion

Overall, the results of these two crosses indicate that the purple allele is autosomal, and the vermilion allele is X linked. To confirm this idea, the vermilion males could be crossed to homozygous wild-type females. The F_1 females should have red eyes. If these F_1 females are crossed to wild-type males, half of their sons should have vermilion eyes, if the gene is X linked. To verify that the purple allele is in an autosomal gene, a purple male could be crossed to a homozygous wild-type female. The F_1 offspring should all have red eyes. If these F_1 offspring are allowed to mate with each other, they should produce 1/4 purple offspring in the F_2 generation. In this case, the F_2 generation offspring with purple eyes would be both male and female.

E17. To see if the allele is X linked, the pink-eyed male could be crossed to a red-eyed female. All the offspring would have red eyes, assuming that the pink allele is recessive. When crossed to red-eyed males, the F_1 females will produce 1/2 red-eyed daughters, 1/4 red-eyed sons, and 1/4 pink-eyed sons if the pink allele is X linked.

If the pink allele is X linked, then one could determine if it is in the same X-linked gene as the white and eosin alleles by crossing pink-eyed males to white-eyed females. (Note: We already know that white and eosin are alleles of the same gene.) If the pink and white alleles are in the same gene, the F_1 female offspring should have pink eyes (assuming that the pink allele is dominant over white). However, if the pink and white alleles are in different genes, the F_1 females will have red eyes (assuming that pink is recessive to red). This is because the F_1 females will be heterozygous for two genes, $X^{w+p}X^{wp+}$, in which the X^{w+} and X^{p+} alleles are the dominant wild-type alleles that produce red eyes, and the X^w and X^p alleles are recessive alleles for these two different genes, which produce white eyes and pink eyes, respectively.

Questions for Student Discussion/Collaboration

1. At the cellular or molecular level, the issue of dominance relies on the level of gene expression. The level of expression is quantitatively related to the genotype. For example, if an individual has one normal copy of a gene and second copy that is completely defective, it is common that only 50% of the normal amount of protein will be made. At the cellular level, this 50% level is less than the amount of the protein that would be made in a normal homozygote. At the organismal level, however, 50% of the protein may be sufficient to give a normal phenotype.
2. The easiest way to solve this problem is to take one trait at a time. With regard to combs, all the F_1 generation would be *RrPp*, or walnut comb. With regard to shanks, they would all be feathered, because they would inherit one dominant copy of a feathered allele. With regard to hen- or cock-feathering: 1 male cock-feathered : 1 male hen-feathered : 2 females hen-feathered.
   Overall then, we would have a 1:1:2 ratio of

   walnut comb/feathered shanks/cock-feathered males
   walnut comb/feathered shanks/hen-feathered males
   walnut comb/feathered shanks/hen-feathered females
3. Let's refer to the alleles as *B* dominant, *b* recessive and *G* dominant, *g* recessive.

   The parental cross is *BBGG* × *bbgg*.

   All of the F_1 offspring are *BbGg*.

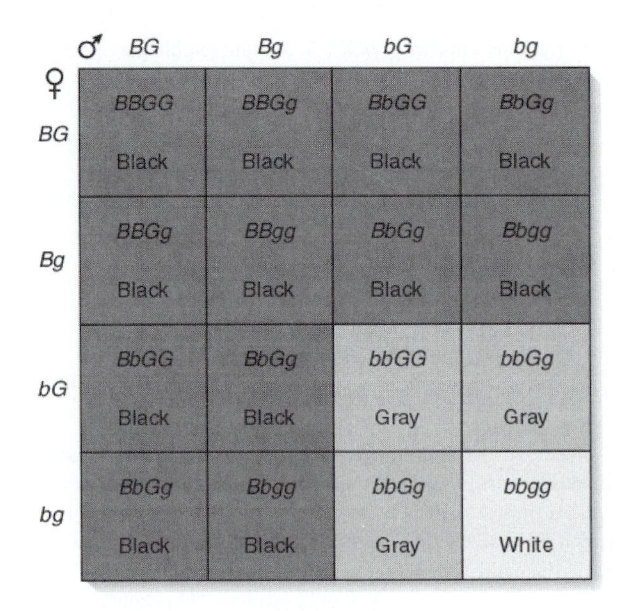

According to the Punnett square, the genotypes that are homozygous for the *b* allele and have at least one copy of the dominant *G* allele are gray. To explain this phenotype, we could hypothesize that the *B* allele encodes an enzyme that can make lots of pigment, whether or not the *G* allele is present. Therefore, you get a black phenotype when one *B* allele is inherited. The *G* allele encodes a somewhat redundant enzyme, but maybe it does not function quite as well or its pigment product may not be as dark. Therefore, in the absence of a *B* allele, the *G* allele will give a gray pigment.

Chapter 5: Linkage and Genetic Mapping in Eukaryotes

Student Learning Objectives

Upon completion of the chapter you should be able to:

1. Understand the concept of linkage and how it relates to patterns of inheritance.
2. Understand, and be able to solve problems of, genetic mapping in diploid plants and animals.
3. Understand the process of genetic linkage mapping in haploid eukaryotes.

5.1 Linkage and Crossing Over

Overview

This first section of the chapter presents some extremely important concepts regarding the processing of crossing over and genetic recombination. You should carefully examine Figure 5.1 and understand how the terms nonparental, recombinant, parental and nonrecombinant may be applied to the haploid cells at the bottom of the figure.

This chapter also presents two of the most important early experimental proofs of crossing over and genetic recombination. The first is the experiment by Thomas Hunt Morgan with X-linked traits in *Drosophila* (pgs. 103-106). In his trihybrid cross, genes that were assorting independently should produce equal numbers of offspring with each phenotype. Instead, his data (Figure 5.3) suggested that during meiosis that new combinations of traits were being generated. As you examine his data, you should be able to identify the parental and nonparental offspring, as well as those that are examples of double crossing over events.

The second experiment, by Creighton and McClintock, demonstrated what Morgan suspected, that homologous chromosomes exchange genetic information during the process of crossing over. They accomplished this by examining the assortment of chromosomes with known structural abnormalities, called translocations. Using traits that segregated with these abnormalities, these researchers were able to provide experimental proof that recombination involves the physical exchange of genetic information.

While in many cases it will be obvious that traits are not independently assorting in a genetic cross, frequently it is necessary to statistically test your dataset to determine linkage. This is most easily done using a chi-square analysis. Notice that the chi-square test uses as its hypothesis that the genes are assorting independently, or according to Mendelian principles. This makes it simple to calculate the expected values of the cross. If the chi-square test does not validate the hypothesis, then the result usually indicates linkage.

Key Terms

Bivalent
Crossing over
Dihybrid cross
Genetic recombination
Linkage
Linkage groups
Mitotic recombination
Nonparental cells
Nonrecombinant
Parental
Recombinant cells
Translocation
Trihybrid cross

Focal Points

- Experimental procedure of Morgan (pgs. 103-106, especially Figure 5.4)
- Chi square test of linkage (pgs. 106-107)
- Experimental procedure of Creighton and McClintock (Experiment 5A)

Exercises and Problems

1. Bateson and Punnett's experiment with sweet peas suggested linkage of traits. Using a chi square analysis, test this concept.

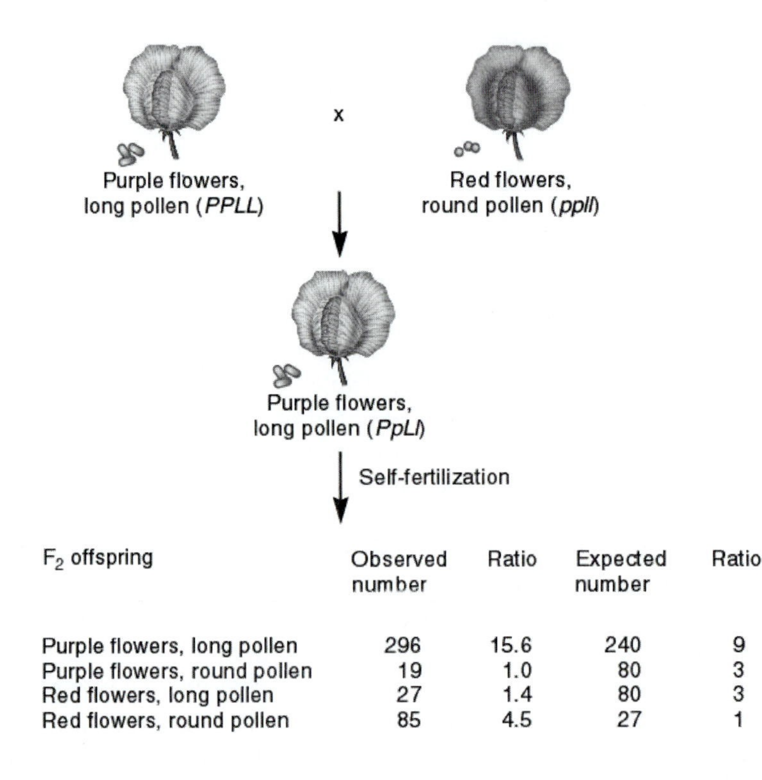

F_2 offspring	Observed number	Ratio	Expected number	Ratio
Purple flowers, long pollen	296	15.6	240	9
Purple flowers, round pollen	19	1.0	80	3
Red flowers, long pollen	27	1.4	80	3
Red flowers, round pollen	85	4.5	27	1

a. Develop a hypothesis for the test.
b. Determine the chi-square value.
c. Determine the degrees of freedom.
d. Determine the P value (text pg. 36).
e. Accept or reject the hypothesis.

2. Distinguish between recombinant and nonrecombinant chromosomes at the end of meiosis.

For questions 3 to 10, match each of the following to its correct description.

_____ 3. Morgan
_____ 4. Mitotic recombination
_____ 5. Nonrecombinant offspring
_____ 6. Creighton and McClintock
_____ 7. Linkage
_____ 8. Linkage groups
_____ 9. Bateson and Punnett
_____ 10. Recombinant offspring

a. Individuals that do not contain the original combination of alleles that were present on the parental chromosomes.
b. The inheritance of genes as a unit.
c. The first to suggest that the closeness of genes on a chromosome influences the rate of crossing over between the genes.
d. The process that may produce somatic cells that differ in their allele combinations.
e. Used translocations on chromosomes to prove the physical exchange of genetic material during meiosis.
f. The first to describe that linkage upsets the expected Mendelian ratios.
g. Individuals that contain the combination of alleles that were present on the parental chromosomes.
h. All of the genes on a given chromosome.

5.2 Genetic Mapping in Plants and Animals

Overview

In the previous section, Morgan suggested that the observed recombination between genes was a factor of the distance between the genes. This chapter expands on this observation by examining the relationship between recombination rates and map distance. The purpose of calculating map distances is to obtain the relative position of genes on a chromosome, it does not establish exact physical distances.

The concept of a testcross is first introduced, as well as the definitions of map distance and map units (or centimorgans). A testcross involves the use of an individual who is homozygous recessive for all alleles being mapped. In essence what this means is that we are mapping the traits of one of the parents, since we will not observe recombination in a homozygous individual. At this stage it is also important that you recognize that the term map distance is equivalent to recombination frequency.

The next major section outlines the work of Sturtevant in the construction of the first genetic map. Using sex-linked recessive traits, he was able to map genes on the X chromosome in *Drosophila.* The procedure that he used remains valid today, and is worth examining. You should especially note why map distances greater than 50 are considered unreliable by geneticists.

The final section of the chapter applies the earlier concepts in the mapping of traits using a trihybrid cross. It is important to mention that prior to constructing the map you should use a chi-square analysis to ensure that your traits are linked (see previous section). Familiarize yourself with the steps and calculations, especially with the identification of recombinant and nonrecombinant offspring. Finally, review the material on interference, since this helps to explain the usual decrease in the number of observed double-crossovers.

Key Terms

Centimorgans
Genetic linkage map
Genetic mapping
Locus
Map distance
Map units
Positive interference
Testcross

Focal Points

- Mapping using a trihybrid cross (pgs. 117-120)

Exercises and Problems

For questions 1 to 5, match the term to its correct definition.

_____ 1. Interference

_____ 2. Testcross

_____ 3. Linkage map

_____ 4. Map unit

_____ 5. Locus

a. Changes in the observed incidences of double crossovers in a trihybrid cross.
b. The physical location of a gene on a chromosome.
c. A unit of measurement based on a 1% recombination frequency.
d. Uses recombination rates to determine relative positions of genes on a chromosome.
e. Utilizes a homozygous recessive individual to map genes of interest.

6. Explain why recombination rates greater than 50% are unreliable in the construction of a genetic linkage map.

For questions 7 to 12, place the steps in the use of a trihybrid cross to construct a genetic map in their correct order by placing a number 1 next to the first step, a number 2 by the second step, etc.

_____ 7. Calculate the map distance between the pairs of genes.

_____ 8. Conduct a testcross using a strain that is homozygous recessive for all traits being mapped.

_____ 9. Calculate interference values.

_____ 10. Collect data for the F_2 generation.

_____ 11. Construct the map.

_____ 12. Perform a chi-square analysis to test for independent assortment.

13. After conducting a trihybrid cross to construct a genetic linkage map, you determine that the interference value for the cross is 0.45. Explain what this indicates.

5.3 Genetic Mapping in Haploid Eukaryotes

Overview

To this point in the text we have been primarily associating ourselves with diploid organisms. However, it is possible to construct a genetic map of a haploid organism. Most of these studies have involved research into fungal genetics, therefore, it is necessary to review the fungal sexual life cycle (Figure 5.12) before proceeding. Be sure to note the differences in the use of the word tetrad and octad with regards to haploid organisms.

Genetic mapping using haploid fungi can be divided into two major classes. Ordered tetrad analysis can be used to determine the distance between a gene of interest and a centromere, which serves as a form of fixed reference point. Note that instead of scoring spores as recombinants, the pattern of the spores within the asci is described. The second type of cross, an unordered tetrad analysis, can be used to determine if genes are linked, and the map distance between the linked genes. This is demonstrated in Figure 5.17.

The challenge in this section is keeping the terminology correct, since the terms and calculations differ significantly from those used in the diploid trihybrid cross.

Key Terms

Ascus	Ordered tetrad (octad)	Tetrad
Octad	Spores	Unordered
tetrad (octad)		

Focal Points

- Figures 5.13, 5.15, 5.17
- Calculations of map distance on pgs. 125-126

Exercises and Problems

For questions 1 to 5, identify the following asci patterns.

_____ 1. T

_____ 2. NPD

_____ 3. FDS

_____ 4. SDS

_____ 5. PD

a. An ascus that contains a 2:2:2:2 or 2:4:2 pattern of spores.
b. An ascus that contains two spores with parental allele combinations and two spores with nonparental allele combinations.
c. A tetrad that contains four spores with parental allele combinations.
d. An ascus that contains four spores with nonparental allele combinations.
e. An ascus that contains a 4:4 pattern of spores.

For questions 6 to 10, match the definition to its correct term.

_______ 6. May be used to map genes.

______ 7. A group of four spores.

______ 8. Can be used to determine the distance between a trait and the centromere.

______ 9. Contains the products of a single mitotic division.

______ 10. The haploid cells of a fungi.

a. tetrad
b. ordered tetrad
c. spores
d. ascus
e. unordered tetrad

Chapter Quiz

1. Which of the following best defines interference?
   a. the suppression of recombination during mitosis
   b. the inhibition of one crossover by another crossover in the region
   c. the statistical determination of linkage
   d. the rate of recombination between two genes

2. The researcher(s) that provided the first physical proof that crossing over involves an exchange of genetic information.
   a. Morgan
   b. Sturtevant
   c. Creighton and McClintock
   d. Bateson and Punnett
   e. Stern

3. When testing for linkage using a chi-square analysis, what should be the hypothesis?
   a. that the traits are linked
   b. that the traits are assorting independently
   c. that each trait is located near the centromere
   d. that all mutations produce the same phenotype

4. When mapping traits in a haploid fungi, which of the following would allow you to establish the distance between the trait and the centromere of the chromosome?
   a. trihybrid cross
   b. testcross
   c. ordered tetrad analysis
   d. unordered tetrad analysis
   e. none of the above

5. A testcross typically uses an individual who is heterozygous for the traits of interest and which of the following?

a. a homozygous dominant individual
b. a heterozygous individual
c. a homozygous recessive individual
d. all of the above may be used

6. The first genetic linkage map was constructed by _____.

a. Sturtevant
b. Bridges
c. Bateson and Punnett
d. Creighton and McClintock
e. Stern

7. Recombination only occurs between homologous chromosomes in prophase I of meiosis.

a. True
b. False

8. Recombination rates above what percent are considered unreliable due to the presence of undetected double crossovers?

a. 5%
b. 10%
c. 20 %
d. 50%

9. One map unit is equal to which of the following?

a. 100 nucleotides
b. 1% of the chromosome
c. 1% recombination rate
d. 1% mutation rate
e. none of the above

10. A tetrad that contains two cells with parental alleles, and two cells with nonparental alleles, is called a(n) _________.

a. parental ditype
b. nonparental ditype
c. ascus
d. tetratype
e. octad

Answer Key for Study Guide Questions

This answer key provides the answers to the exercises and chapter quiz for this chapter. Answers in parentheses ( ) represent possible alternate answers to a problem, while answers marked with an asterisk (*) indicate that the response to the question may vary.

5.1

1. a. The traits of flower color and pollen shape are assorting independently.
   b. $X^2 =$ 219.286
   c. d.f = 3
   d. P > 0.01
   e. Reject the hypothesis that the traits are assorting independently.
2. Recombinant chromosomes contain new combinations of alleles, which were not present in the parental chromosomes that began meiosis. Nonrecombinant chromosomes contain the same allelic combinations as the parental chromosomes. (*)
3. c
4. d
5. g
6. e
7. b
8. a
9. f
10. a

5.2

1. a
2. e
3. d
4. c
5. b
6. As recombination rates approach 50%, there is a greater chance that undetected double-crossovers may be influencing the values.
7. 4
8. 2
9. 6
10. 3
11. 5
12. 1
13. An interference value of 0.45 indicates that only 45% of the double-crossovers were detected in the phenotype of the offspring. This is an example of positive interference.

5.3

1. b
2. d
3. e
4. a
5. c
6. e
7. a
8. b
9. d
10. c

Quiz

1. b
2. c
3. b
4. c
5. c
6. a
7. b
8. d
9. c
10. d

Answers to Conceptual and Experimental Questions

Conceptual Questions

C1. Genetic recombination is a term that refers to a new combination of alleles in an offspring. Crossing over is a physical event that involves the exchange of pieces of genetic material between two chromosomes.

C2. An independent assortment hypothesis is used because it enables us to calculate the expected values based on Mendel's ratios. Using the observed and expected values, we can calculate whether or not the deviations between the observed and expected values are too large to occur as a matter of chance. If the deviations are very large, we reject the hypothesis of independent assortment.

C3. Mitotic recombination is crossing over between homologous chromosomes during mitosis in somatic cells. Mitotic recombination is one explanation for a blue patch. Following mitotic recombination, the two chromosomes carrying the *b* allele could segregate into the same cell and produce the blue color. Another reason could be chromosome loss; the chromosome carrying the *B* allele could be lost during mitosis.

C4.

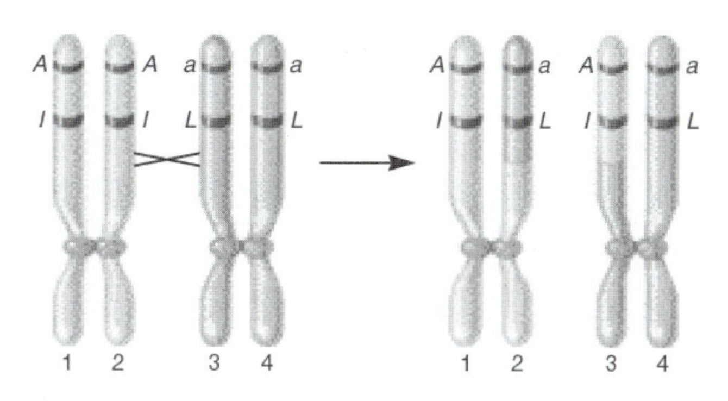

If the chromosomes labeled 2 and 4 move into one daughter cell, that will lead to a patch that is albino and has long hair. The other cell will receive chromosomes 1 and 3, which will produce a patch that has dark, short hair.

C5. A. 1 and 4

B. 2 and 3

C. 2 and 4, or 3 and 1

C6. A single crossover produces *A B C, A b c, a B C,* and *a b c.*

A. Between 2 and 3, between genes *B* and *C*

B. Between 1 and 4, between genes *A* and *B*

C. Between 1 and 4, between genes *B* and *C*

D. Between 2 and 3, between genes *A* and *B*

C7. There are 7 chromosomes per haploid genome. If we divide 20,000 by 7, there are about 2,857 genes per chromosome, on average.

C8. The likelihood of scoring a basket would be greater if the basket was larger. Similarly, the chances of a crossover initiating in a region between two genes is proportional to the size of the region between the two genes. There are a finite number (usually a few) that occur between homologous chromosomes during meiosis, and the likelihood that a crossover will occur in a region between two genes depends on how big that region is.

C9. If there are seven linkage groups, this means there are seven chromosomes per set. The sweet pea is diploid, meaning it has two sets of chromosomes. Therefore, the sweet pea has 14 chromosomes in leaf cells.

C10. The pedigree suggests a linkage between the dominant allele causing nail-patella syndrome and the I^B allele of the ABO blood type gene. In every case, the individual who inherits the I^B allele also inherits this disorder.

C11. There are four phenotypic categories for the F_2 offspring: brown fur, short tails; brown fur, long tails; white fur, short tails; and white fur, long tails. The recombinants are brown fur, long tails and white fur, short tails. The F_2 offspring will occur in a 1:1:1:1 ratio if the two genes are not linked. In other words, there will be 25% of each of the four phenotypic categories. If the genes are linked, there will be a lower percentage of the recombinant offspring.

C12. *Ass-1* 43 *Sdh-1* 5 *Hdc* 9 *Hao-1* 6 *Odc-2* 8 *Ada-1*

C13. We use the product rule. The likelihood of a double crossover is 0.1 × 0.1, which equals 0.01, or 1%. The likelihood of a triple crossover is 0.1 × 0.1 × 0.1 = 0.001, or 0.1%. Positive interference would make these values lower.

C14. The inability to detect double crossovers causes the map distance to be underestimated. In other words, there are more crossovers occurring in the region than we realize. When we have a double crossover, we do not get a recombinant offspring (in a dihybrid cross). Therefore, the second crossover cancels out the effects of the first crossover.

C15. The map distance between genes *A* and *B* would appear to be greater than between *C* and *D* because of the higher rate of crossing over. We would obtain more recombinant offspring with regard to genes *A* and *B,* so the computed map distance would be higher, even though the physical distance between *A* and *B,* and between *C* and *D,* is identical.

C16. The key feature is that all the products of a single meiosis are contained within a single sac. The spores in this sac can be dissected out, and then their genetic traits can be analyzed individually. In fungi, the traits most commonly analyzed are nutritional growth requirements (as described in Chapter 5), as well as pigmentation and antibiotic resistance.

C17. A tetrad contains four spores; an octad contains eight. In a tetrad, meiosis produces four spores. In an octad, meiosis produces four cells, and then they all go through mitosis to double the number to eight cells.

C18. In an unordered ascus, the products of meiosis are free to move around. In an ordered octad (or tetrad), they are lined up according to their relationship to each other during meiosis and mitosis. An ordered octad can be used to map the distance between a single gene and its centromere.

C19. It could be a 2:2:2:2 or a 2:4:2 arrangement.

C20. They would be higher with respect to gene *A*. First-division segregation patterns occur when there is not a crossover between the centromere and the gene of interest. Since gene *A* is closer to the centromere compared to gene *B,* it would be less likely to have a crossover between gene *A* and the centromere. This would make it more likely to observe first-division segregation.

Experimental Questions

E1. If we hypothesize two genes independently assorting, the predicted ratio is 9:3:3:1. There is a total of 427 offspring. The expected numbers of offspring are

9/16 × 427 = 240 purple flowers, long pollen
3/16 × 427 = 80 purple flowers, round pollen
3/16 × 427 = 80 red flowers, long pollen
1/16 × 427 = 27 red flowers, round pollen

Plugging these values into our chi square formula,

$$\chi^2 = \frac{(296-240)^2}{240} + \frac{(19-80)^2}{80} + \frac{(27-80)^2}{80} + \frac{(85-27)^2}{27}$$

$$\chi^2 = 13.1 + 46.5 + 35.1 + 124.6$$

$$\chi^2 = 219.3$$

Looking up this value in the chi square table under 3 degrees of freedom, we find that such a large value is expected by chance less than 1% of the time. Therefore, we reject the hypothesis that the genes assort independently.

E2. They could have used a strain with two abnormal chromosomes. In this case, the recombinant chromosomes would either look normal or have abnormalities at both ends.

E3. The top of the Conceptual Level column in Figure 5.6 shows the chromosomes of McClintock's cross. This experiment could be modified to a standard testcross in the following way. In the heterozygous parent, the *C* (colored) and *Wx* (starchy) alleles could be on the knobbed, translocation chromosome and the *c* (colorless) and *wx* (waxy) alleles on a normal chromosome. The other parent would have two cytologically normal copies of chromosome 9 and be homozygous for the recessive alleles (i.e., *cc wxwx*). If the cross were done in this way, nonrecombinant offspring would be colored and starchy, or colorless and waxy; recombinant offspring would be colored and waxy, or colorless and starchy. The recombinant offspring should inherit a chromosome with a knob but no translocation, or a translocation but no knob.

E4. A gene on the Y chromosome in mammals would only be transmitted from father to son. It would be difficult to genetically map Y-linked genes because a normal male has only one copy of the Y chromosome, so you do not get any crossing over between two Y chromosomes. Occasionally, abnormal males (XYY) are born with two Y chromosomes. If such males were heterozygous for alleles of Y-linked genes, one could examine the normal male offspring of XYY fathers and determine if crossing over has occurred.

E5. The rationale behind a testcross is to determine if recombination has occurred during meiosis in the heterozygous parent. The other parent is usually homozygous recessive, so we cannot tell if crossing over has occurred in the recessive parent. It is easier to interpret the data if a testcross does use a completely homozygous recessive parent. However, in the other parent, it is not necessary for all of the dominant alleles to be on one chromosome and all of the recessive alleles on the other. The parental generation provides us with information concerning the original linkage pattern between the dominant and recessive alleles.

E6. The answer is explained in solved problem S5. We cannot get more than 50% recombinant offspring because the pattern of multiple crossovers can yield an average maximum value of only 50%. When a testcross does yield a value of 50% recombinant offspring, it can mean two different things. Either the two genes are on different chromosomes or the two genes are on the same chromosome but at least 50 mu apart.

E7. The reason why the percentage of recombinant offspring is more accurate when the genes are close together is because there are fewer double crossovers. The inability to detect double crossover causes the map distance to be underestimated. If two genes are very close together, there are very few double crossovers so that the underestimation due to double crossovers is minimized.

E8. If two genes are at least 50 mu apart, you would need to map genes in between them to show that the two genes were actually in the same linkage group. For example, if gene *A* was 55 mu from gene *B,* there might be a third gene (e.g., gene *C*) that was 20 mu from *A* and 35 mu from *B.* These results would indicate that *A* and *B* are 55 mu apart, assuming dihybrid testcrosses between genes *A* and *B* yielded 50% recombinant offspring.

E9. He determined this by analyzing the data in gene pairs. This analysis revealed that there were fewer recombinants between certain gene pairs (e.g., body color and eye color) than between other gene pairs (e.g., eye color and wing shape). From this comparison, he hypothesized that genes that are close together on the same chromosome will produce fewer recombinants than genes that are farther apart.

E10. Sturtevant used the data involving the following pairs: *y* and *w,* *w* and *v,* *v* and *r,* and *v* and *m.*

E11. Map distance:

$$= \frac{64 + 58}{333 + 64 + 58 + 380} \times 100$$

$$= 15.1 \text{ mu}$$

E12. A. Since they are 12 mu apart, we expect 12% (or 120) recombinant offspring. This would be approximately 60 *Aabb* and 60 *aaBb* plus 440 *AaBb* and 440 *aabb.*

B. We would expect 60 *AaBb,* 60 *aabb,* 440 *Aabb,* and 440 *aaBb.*

E13. We consider the genes in pairs: there should be 10% offspring due to crossing over between genes *A* and *B* and 5% due to crossing over between *A* and *C*.

A. This is due to a crossover between *B* and *A*. The parentals are *Aa bb Cc* and *aa Bb cc*. The 10% recombinants are *Aa Bb Cc* and *aa bb cc*. If we assume there is an equal number of both types of recombinants, there are 5% *Aa Bb Cc*.

B. This is due to a crossover between *A* and *C*. The parentals are *Aa bb Cc* and *aa Bb cc*. The 5% recombinants are *aa Bb Cc* and *Aa bb cc*. If we assume there are an equal number of both types of recombinants, there are 2.5% *aa Bb Cc*.

C. This is also due to a crossover between *A* and *C*. The parentals are *Aa bb Cc* and *aa Bb cc*. The 5% recombinants are *aa Bb Cc* and *Aa bb cc*. If we assume there are an equal number of both types of recombinants, there are 2.5% *Aa bb cc*.

E14. Due to the large distance between the two genes, they will assort independently even though they are actually on the same chromosome. According to independent assortment, we expect 50% parental and 50% recombinant offspring. Therefore, this cross will produce 150 offspring in each of the four phenotypic categories.

E15. A. One basic strategy to solve this problem is to divide the data up into gene pairs and determine the map distance between two genes.

184 tall, smooth
13 tall, peach
184 dwarf, peach
12 dwarf, smooth

$$\text{Map distance} = \frac{13+12}{184+13+184+12} = 6.4 \text{ mu}$$

153 tall, normal
44 tall, oblate
155 dwarf, oblate
41 dwarf, normal

$$\text{Map distance} = \frac{44+41}{153+44+155+41} = 21.6 \text{ mu}$$

163 smooth, normal
33 smooth, oblate
31 peach, normal
166 peach, oblate

$$\text{Map distance} = \frac{33+31}{163+33+31+166} = 16.3 \text{ mu}$$

Use the two shortest distances to compute the map:

Tall, dwarf   6.4   Smooth, peach   16.3   Normal, oblate

E16. A. If we hypothesize two genes independently assorting, then the predicted ratio is 1:1:1:1. There are a total of 390 offspring. The expected number of offspring in each category is about 98. Plugging the figures into our chi square formula,

$$\chi^2 = \frac{(117-98)^2}{98} + \frac{(115-98)^2}{98} + \frac{(78-98)^2}{98} + \frac{(80-98)^2}{98}$$

$$\chi^2 = 3.68 + 2.95 + 4.08 + 3.31$$

$$\chi^2 = 14.02$$

Looking up this value in the chi square table under 3 degrees of freedom, we reject our hypothesis, since the chi square value is above 7.815.

B. Map distance:

$$\text{Map distance} = \frac{78 + 80}{117 + 115 + 78 + 80}$$
$$= 40.5 \text{ mu}$$

Because the value is relatively close to 50 mu, it is probably a significant underestimate of the true distance between these two genes.

E17. In the backcross, the two parental types would be the homozygotes that cannot make either enzyme and the heterozygotes that can make both enzymes. The recombinants would make one enzyme but not both. Because the two genes are 12 mu apart, 12% would be recombinants and 88% would be parental types. Because there are two parental types produced in equal numbers, we would expect 44% of the mice to be unable to make either enzyme.

E18. The percentage of recombinants for the green, yellow and wide, narrow is 7%, or 0.07; there will be 3.5% of the green, narrow and 3.5% of the yellow, wide. The remaining 93% parentals will be 46.5% green, wide and 46.5% yellow, narrow. The third gene assorts independently. There will be 50% long and 50% short with respect to each of the other two genes. To calculate the number of offspring out of a total of 800, we multiply 800 by the percentages in each category.

(0.465 green, wide)(0.5 long)(800) = 186 green, wide, long
(0.465 yellow, narrow)(0.5 long)(800) = 186 yellow, narrow, long
(0.465 green, wide)(0.5 short)(800) = 186 green, wide, short
(0.465 yellow, narrow)(0.5 short)(800) = 186 yellow, narrow, short
(0.035 green, narrow)(0.5 long)(800) = 14 green, narrow, long
(0.035 yellow, wide)(0.5 long)(800) = 14 yellow, wide, long
(0.035 green, narrow)(0.5 short)(800) = 14 green, narrow, short
(0.035 yellow, wide)(0.5 short)(800) = 14 yellow, wide, short

E19. A. If we represent *B* (bushy tail) and *b* (normal tail) for one gene, and *Y* (yellow) and *y* (white) for the second gene:

Parent generation: *BBYY* × *bbyy*

F_1 generation: All *BbYy* (NOTE: if the two genes are linked, *B* would be linked to *Y* and *b* would be linked to *y*.)

Testcross: F_1 *BbYy* × *bbyy*

Nonrecombinant offspring from testcross: *BbYy* and *bbyy*

BbYy males—bushy tails, yellow
bbyy males—normal tails, white
BbYy females—normal tails, yellow
bbyy females—normal tails, white

Recombinant offspring from testcross: *Bbyy* and *bbYy*

Bbyy males—bushy tails, white
bbYy males—normal tails, yellow
Bbyy females—normal tails, white
bbYy females—normal tails, yellow

We cannot use the data regarding female offspring, because we cannot tell if females are recombinant or nonrecombinant, because all females have normal tails. However, we can tell if male offspring are recombinant.

If we use the data on males to conduct a chi-square analysis, we expect a 1:1:1:1 ratio among the male offspring. Since there are 197 male offspring total, we expect 1/4, or 49 (rounded to the nearest whole number), of the four possible phenotypes. To compute the chi square:

$$\chi^2 = \frac{(28-49)^2}{49} + \frac{(72-49)^2}{49} + \frac{(68-49)^2}{49} + \frac{(29-49)^2}{49}$$
$$\chi^2 = 9.0 + 10.8 + 7.4 + 8.2$$
$$\chi^2 = 35.4$$

If we look up the value of 35.4 in our chi square table, with 3 degrees of freedom, the value lies far beyond the 0.01 probability level. Therefore, it is very unlikely to get such a large deviation if our hypothesis of independent assortment is correct. Therefore, we reject our hypothesis and conclude that the genes are linked.

B. To compute map distance:

$$\frac{28 + 29}{28 + 72 + 68 + 29} \times 100 = 28.9 \text{ mu}$$

E20. Let's use the following symbols: *G* for green pods, *g* for yellow pods, *S* for green seedlings, *s* for bluish green seedlings, *C* for normal plants, *c* for creepers.

The parental cross is *GG SS CC* crossed to *gg ss cc*.

The F_1 plants would all be *Gg Ss Cc*. If the genes are linked, the alleles *G, S,* and *C* would be linked on one chromosome and the alleles *g, s,* and *c* would be linked on the homologous chromosome.

The testcross is F_1 plants, which are *Gg Ss Cc,* crossed to *ggsscc.*

To measure the distances between the genes, we can separate the data into gene pairs.

Pod color, seedling color

2,210 green pods, green seedlings—nonrecombinant

296 green pods, bluish green seedlings—recombinant

2,198 yellow pods, bluish green seedlings—nonrecombinant

293 yellow pods, green seedlings—recombinant

$$\text{Map distance} = \frac{296 + 293}{2{,}210 + 296 + 2{,}198 + 293} \times 100 = 11.8 \text{ mu}$$

Pod color, plant stature

2,340 green pods, normal—nonrecombinant

166 green pods, creeper—recombinant

2,323 yellow pods, creeper—nonrecombinant

168 yellow pods, normal—recombinant

$$\text{Map distance} = \frac{166 + 168}{2{,}340 + 166 + 2{,}323 + 168} \times 100 = 6.7 \text{ mu}$$

Seedling color, plant stature

2,070 green seedlings, normal—nonrecombinant

433 green seedlings, creeper—recombinant

2,056 bluish green seedlings, creeper—nonrecombinant

438 bluish green seedlings, normal—recombinant

$$\text{Map distance} = \frac{433 + 438}{2{,}070 + 433 + 2{,}056 + 438} \times 100 = 17.4 \text{ mu}$$

The order of the genes is seedling color, pod color, plant stature or you could say the opposite order. Pod color is in the middle. If we use the two shortest distances to construct our map:

S 11.8 *G*
6.7 *C*

E21. Let's use the following symbols: *S* for normal nose, *s* for snubnose, *p* for normal tail, *P* for pintail, *J* for normal gait, *j* for jerker.

The parental cross is *ss PP jj* crossed to *SS pp JJ.*

The F_1 offspring would all be *Ss Pp Jj.* If the genes are linked, the alleles *s, P,* and *j* would be linked on one chromosome and the alleles *S, p,* and *J* would be linked on the homologous chromosome.

The testcross is F_1 mice, which are *Ss Pp Jj,* crossed to *ss pp jj* mice.

To measure the distances between the genes, we can separate the data into gene pairs.

Nose shape, tail length

631 snubnose, pintail—nonrecombinant
111 snubnose, normal tail—recombinant
625 normal nose, normal tail—nonrecombinant
115 normal nose, pintail—recombinant

$$\text{Map distance} = \frac{111+115}{631+111+625+115} \times 100 = 15.2 \text{ mu}$$

Nose shape, walking gait

662 snubnose, jerker—nonrecombinant
80 snubnose, normal—recombinant
652 normal nose, normal—nonrecombinant
88 normal nose, jerker—recombinant

$$\text{Map distance} = \frac{80+88}{662+80+652+88} \times 100 = 11.3 \text{ mu}$$

Tail length, walking gait

571 pintail, jerker—nonrecombinant
175 pintail, normal—recombinant
557 normal tail, normal gait—nonrecombinant
179 normal tail, jerker—recombinant

$$\text{Map distance} = \frac{175+179}{571+175+557+179} \times 100 = 23.9 \text{ mu}$$

The order of the genes is tail length, nose shape, walking gait or you could say the opposite order. Nose shape is in the middle.

If we use the two shortest distances to construct our map:

P 15.2 *S* 11.3 *J*

E22. According to a hypothesis of independent assortment, we would expect an equal proportion of all four phenotypes. (You should construct a Punnett square if this is not apparent.) There are a total number of 1,074 offspring. Therefore, the expected number of each of the four categories is 1/4 × 1,074, which equals 268 (rounded to the nearest whole number). We use the value of 268 as our expected value in the chi square calculation.

$$\chi^2 = \frac{(368-268)^2}{268} + \frac{(160-268)^2}{268} + \frac{(194-268)^2}{268} + \frac{(352-268)^2}{268}$$

$$\chi^2 = 37.3 + 43.5 + 20.4 + 26.3$$

$$\chi^2 = 127.5$$

If we look up the value of 127.5 in our chi square table, with 3 degrees of freedom, the value lies far beyond the 0.01 probability level. Therefore, it is very unlikely to get such a large deviation if our hypothesis of independent assortment is correct. Therefore, we reject our hypothesis and conclude that the genes are linked. In the parental (true-breeding) generation, black is linked to chinchilla and brown is linked to Himalayan. Therefore, the recombinant offspring are black Himalayan and brown chinchilla.

To compute map distance:

$$\frac{160+194}{368+160+194+352} \times 100 = 33.0 \text{ mu}$$

E23. To answer this question, we can consider genes in pairs. Let's consider the two gene pairs that are closest together. The distance between the wing length and eye color genes is 12.5 mu. From this cross, we expect 87.5% to have long wings and red eyes or short wings and purple eyes, and 12.5% to have long wings and purple eyes or short wings and red eyes. Therefore, we expect 43.75% to have long wings and red eyes, 43.75% to have short wings and purple eyes, 6.25% to have long wings and purple eyes, and 6.25% to have short wings and red eyes. If we have 1,000 flies, we expect 438 to have long wings and red eyes, 438 to have short wings and purple eyes, 62 to have long wings and purple eyes, and 62 to have short wings and red eyes (rounding to the nearest whole number).

The distance between the eye color and body color genes is 6 mu. From this cross, we expect 94% to have a parental combination (red eyes and gray body or purple eyes and black body) and 6% to have a nonparental combination (red eyes and black body or purple eyes and gray body). Therefore, of our 438 flies with long wings and red eyes, we expect 94% of them (or about 412) to have long wings, red eyes, and gray body, and 6% of them (or about 26) to have long wings, red eyes, and black bodies. Of our 438 flies with short wings and purple eyes, we expect about 412 to have short wings, purple eyes, and black bodies, and 26 to have short wings, purple eyes, and gray bodies.

Of our 62 flies with long wings and purple eyes, we expect 94% of them (or about 58) to have long wings, purple eyes, and black bodies, and 6% of them (or about 4) to have long wings, purple eyes, and gray bodies. Of our 62 flies with short wings and red eyes, we expect 94% (or about 58) to have short wings, red eyes, and gray bodies, and 6% (or about 4) to have short wings, red eyes, and black bodies.

In summary:

Long wings, red eyes, gray body	412
Long wings, purple eyes, gray body	4
Long wings, red eyes, black body	26
Long wings, purple eyes, black body	58
Short wings, red eyes, gray body	58
Short wings, purple eyes, gray body	26
Short wings, red eyes, black body	4
Short wings, purple eyes, black body	412

The flies with long wings, purple eyes, and gray bodies, or short wings, red eyes, and black bodies, are produced by a double crossover event.

E24. A.

Parent		Parent
$\underline{b\ 7\ A\ 4\ C}$	×	$\underline{B\ 7\ a\ 4\ c}$
$\underline{b\ 7\ A\ 4\ C}$		$\underline{B\ 7\ a\ 4\ c}$

$\underline{b\ 7\ A\ 4\ C}$ Offspring
$\underline{B\ 7\ a\ 4\ c}$

B. A heterozygous F_2 offspring would have to inherit a chromosome carrying all of the dominant alleles. In the F_1 parent (of the F_2 offspring), a crossover in the 7 mu region between genes *b* and *A* (and between *B* and *a*) would yield a chromosome that was *B A C* and *b a c*. If an F_2 offspring inherited the *BAC* chromosome from its F_1 parent and the *b a c* chromosome from the homozygous parent, it would be heterozygous for all three genes.

C. If you look at the answer to part B, a crossover between genes *b* and *A* (and between *B* and *a*) would yield *B A C* and *b a c* chromosomes. If an offspring inherited the *b a c* chromosome from its F_1 parent and the *b a c* chromosome from its homozygous parent, it would be homozygous for all three genes. The chances of a crossover in this region are 7%. However, half of this 7% crossover event yields chromosomes that are *B A C* and the other half yields chromosomes that are *bac*. Therefore, the chances are 3.5% of getting homozygous F_2 offspring.

E25. Yes. Begin with females that have one X chromosome that is X^{Nl} and the other X chromosome that is X^{nL}. These females have to be mated to $X^{NL}Y$ males because a living male cannot carry the *n* or *l* allele. In the absence of crossing over, a mating between $X^{Nl}X^{nL}$ females to $X^{NL}Y$ males should not produce any surviving male offspring. However, during oogenesis in these heterozygous female mice there could be a crossover in the region between the two genes and this will produce an X^{NL} chromosome and an X^{nl} chromosome. Therefore, male offspring inheriting these recombinant chromosomes will be either $X^{NL}Y$ or $X^{nl}Y$ (whereas nonrecombinant males will be $X^{nL}Y$ or $X^{Nl}Y$). Only the male mice that inherit $X^{NL}Y$ will live. The living males represent only half of the recombinant offspring. (The other half are $X^{nl}Y$, which are born dead.)
To compute map distance:

$$\text{Map distance} = \frac{2\ (\text{Number of male living offspring})}{\text{Number of males born dead} + \text{Number of males born alive}}$$

E26. A. The first thing to do is to determine which asci are parental ditypes (PD), nonparental ditypes (NPD), and tetratypes (T). A parental ditype will contain a 2:2 combination of spores with the same genotypes as the original haploid parents. The combination of the 502 asci are the parental ditypes. The nonparental ditypes are those containing a 2:2 combination of genotypes that are unlike the parentals. The combination of four asci fits this description. Finally, the tetratypes contain a 1:1:1:1 arrangement of genotypes, half of which have a parental genotype and half of which do not. There are 312 tetratypes in this case. Computing the map distance:

$$\begin{aligned}\text{Map Distance} &= \frac{\text{NPD} + (1/2)(\text{T})}{\text{Total number of asci}} \times 100\\ &= \frac{4 + (1/2)(312)}{818}\\ &= 19.6 \text{ mu}\end{aligned}$$

If we use the more accurate equation:

$$\begin{aligned}\text{Map distance} &= \frac{\text{T} + 6\text{NPD}}{\text{Total number of asci}} \times 0.5 \times 100\\ &= \frac{312 + (6)(4)}{818}\\ &= 20.5 \text{ mu}\end{aligned}$$

B. The frequency of single crossovers is 0.205 if we use the more accurate equation.

C. Nonparental ditypes are produced from a double crossover. To compute the expected number, we multiply $0.205 \times 0.205 = 0.042$, or 4.2%. Since we had a total of 818 asci, we would expect 34.3 asci to be the product of a double crossover. However, as described in Figure 5.17, only 1/4 of them would be a nonparental ditype. Therefore, we multiply 34.3 by 1/4, obtaining a value of 8.6 nonparental ditypes due to a double crossover. Since we observed only four, this calculation tells us that positive interference is occurring.

E27.

$$\begin{aligned}\text{Map distance} &= \frac{(1, 2)(\text{SDS})}{\text{Total}} \times 100\\ &= \frac{(1, 2)(22 + 21 + 21 + 23)}{22 + 21 + 21 + 451 + 23 + 455} \times 100\\ &= 4.4 \text{ mu}\end{aligned}$$

E28.

A. Types	B. Number
pro-1 pro-1 pro-1 pro-1 pro^+ pro^+ pro^+ pro^+	402
pro^+ pro^+ pro^+ pro^+ *pro-1 pro-1 pro-1 pro-1*	402
pro^+ pro^+ *pro-1 pro-1* pro^+ pro^+ *pro-1 pro-1*	49
pro-1 pro-1 pro^+ pro^+ *pro-1 pro-1* pro^+ pro^+	49
pro^+ pro^+ *pro-1 pro-1 pro-1 pro-1* pro^+ pro^+	49
pro-1 pro-1 pro^+ pro^+ pro^+ pro^+ *pro-1 pro-1*	49

Questions for Student Discussion/Collaboration

1. The basic strategy is first to determine the percentage of recombination between the gene pairs that are linked. The percentage of recombinants between normal, droopy ears and normal, flaky tail is 6%. There will be 3% normal ears, flaky tail and 3% droopy ears, nonflaky tail. The remaining 94% will be 47% normal ears, nonflaky tail and 47% droopy ears, flaky tail. The other two genes will assort independently of these two linked genes. For each phenotypic category, there will be 50% long tails versus 50% short tails, and 50% normal gait versus 50% jerker. We multiply these percentages together and with the total value of 400 to determine the relative numbers of each type of offspring.

   (0.47 normal ears, nonflaky)(0.5 long tail)(0.5 normal gait)(400) = 47 normal ears, nonflaky tail, normal gait, long tail

   (0.47 droopy ears, flaky)(0.5 long tail)(0.5 normal gait)(400) = 47 droopy ears, flaky tail, normal gait, long tail

   (0.47 normal ears, nonflaky)(0.5 long tail)(0.5 jerker)(400) = 47 normal ears, nonflaky tail, jerker, long tail

   (0.47 droopy ears, flaky)(0.5 long tail)(0.5 jerker)(400) = 47 droopy ears, flaky tail, jerker, long tail

   (0.47 normal ears, nonflaky)(0.5 short tail)(0.5 normal gait)(400) = 47 normal ears, nonflaky tail, normal gait, short tail

   (0.47 droopy ears, flaky)(0.5 short tail)(0.5 normal gait)(400) = 47 droopy ears, flaky tail, normal gait, short tail

   (0.47 normal ears, nonflaky)(0.5 short tail)(0.5 jerker)(400) = 47 normal ears, nonflaky tail, jerker, short tail

   (0.47 droopy ears, flaky)(0.5 short tail)(0.5 jerker)(400) = 47 droopy ears, flaky tail, jerker, short tail

   (0.03 normal ears, flaky)(0.5 long tail)(0.5 normal gait)(400) = 3 normal ears, flaky tail, normal gait, long tail

   (0.03 droopy ears, nonflaky)(0.5 long tail)(0.5 normal gait)(400) = 3 droopy ears, nonflaky tail, normal gait, long tail

   (0.03 normal ears, flaky)(0.5 long tail)(0.5 jerker)(400) = 3 normal ears, flaky tail, jerker, long tail

   (0.03 droopy ears, nonflaky)(0.5 long tail)(0.5 jerker)(400) = 3 droopy ears, nonflaky tail, jerker, long tail

   (0.03 normal ears, flaky)(0.5 short tail)(0.5 normal gait)(400) = 3 normal ears, flaky tail, normal gait, short tail

   (0.03 droopy ears, nonflaky)(0.5 short tail)(0.5 normal gait)(400) = 3 droopy ears, nonflaky tail, normal gait, short tail

   (0.03 normal ears, flaky)(0.5 short tail)(0.5 jerker)(400) = 3 normal ears, flaky tail, jerker, short tail

   (0.03 droopy ears, nonflaky)(0.5 short tail)(0.5 jerker)(400) = 3 droopy ears, nonflaky tail, jerker, short tail

2. The X and Y chromosomes are not completely distinct linkage groups. One might describe them as overlapping linkage groups having some genes in common, but most genes of which are not common to both.

3. The discussion could lead into many areas, such as bioethics, the probability of this happening by chance, and so forth.

Chapter 6: Genetic Transfer and Mapping in Bacteria and Bacteriophages

Student Learning Objectives

Upon completion of this chapter you should be able to:

1. Understand the mechanisms by which gene transfer can occur in bacteria.
2. Understand how an understanding of gene transfer in bacteria can be used to map prokaryotic genes.
3. Understand the key experiments that initiated gene mapping in bacteria and bacteriophages.
4. Understand the process of intragenic mapping in bacteriophages.

6.1 Genetic Transfer and Mapping in Bacteria

Overview

The previous chapters in this course have primarily examined the patterns of inheritance between parents and offspring, or what is called vertical gene transfer (see below). This chapter explores gene transfer in bacteria. Bacteria may transfer genetic information primarily by one of three methods: conjugation, transformation, and transduction. In addition, genetic mapping in bacteria differs significantly from the earlier discussions of mapping in eukaryotic organisms. Rather than construct genetic crosses, bacterial geneticists rely on procedures using the three mechanisms of gene transfer to map genes in bacteria.

Conjugation is the transfer of genetic information between two bacteria that are in contact with one another. This transfer of genetic information usually involves the movement of small circular pieces of DNA, called plasmids. While bacteria do not have sexes, they are classified as donor and recipient strains, which differ slightly in their genetic composition. In addition, some strains of bacteria, called *Hfr* strains, actually transfer portions of the bacterial chromosome during conjugation. You should familiarize yourself with the mechanism of transfer of the genetic material for both conjugation and *Hfr*-mediation conjugation (pgs. 139-140) before leaving this section. The process by which conjugation can be used to map bacterial genes is outlined in Experiment 6A.

The next mechanism of gene transfer to be discussed is transduction. This involves the transfer of genetic material from one bacteria to another using a virus (or phage) as an intermediate. In order to understand how this is possible, it is important to review the life cycles of a virus (Figure 6.9). A rare form of transduction, called cotransduction, can be used to map bacterial genes that are closely linked.

The third major form of gene transfer in bacteria is transformation. During transformation, bacteria acquire DNA from their environment. Not all bacteria are capable of transformation. Those that possess the ability are called competent cells. Transformation is a natural event, but has been exploited by geneticists to introduce genes of interest into bacteria. This process is called artificial transformation. Transformation may be used to map two genes that are closely linked by a process called cotransformation.

Your challenge for this section will be to understand the terms and basic processes by which the bacteria uses these three methods to introduce variation into its genome.

Key Terms

Acquired antibiotic resistance
Artificial transformation
Bacteriophages (phages)
Competence factors
Competent cells
Conjugation
Conjugative plasmids
Cotransduction
Cotransformation
Episome
F factor
F' factors
Generalized transduction
Genetic transfer
Heteroduplex
Hfr strain
Homologous recombination
Horizontal gene transfer
Illegitimate recombination
Interrupted mating
Lysogenic cycle
Lytic cycle
Minutes
Natural transformation
Nonhomologous recombination
Nucleoprotein
Origin of transfer
Plasmid
Prophage
Relaxosome
Sex pili
T DNA
Temperate phage
Transduction
Transformation
Vertical gene transfer
Virulent phages

Focal Points

- Process of conjugation (Figure 6.4)
- Bacteriophage life cycles (Figure 6.9)
- Process of transduction (Figure 6.10)
- Process of transformation (Figure 6.12)

Exercises and Problems

For questions 1 to 7, match the definition provided with its label on the diagram below. Some answers may be used more than once.

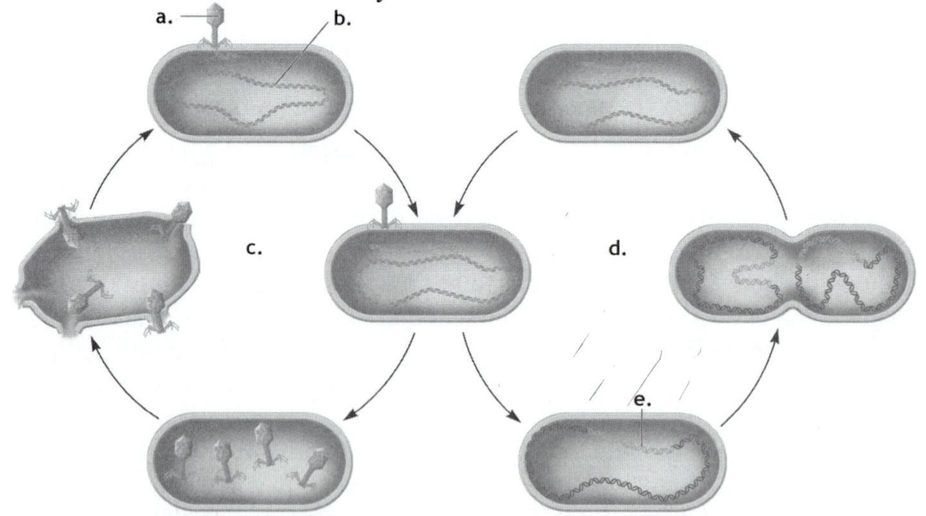

_____ 1. This life cycle immediately begins to reproduce new phage.
_____ 2. The label for the circular bacterial chromosome.
_____ 3. A bacteriophage
_____ 4. The lysogenic life cycle
_____ 5. Virulent phages are active only in this cycle.
_____ 6. The prophage of the lysogenic cycle
_____ 7. The life cycle that is used exclusively by temperate phages.

For questions 8 to 13, state whether the question is true (T) or false (F). If false, correct the statement so that is it correct.

_____ 8. Genetic distance obtained from conjugation studies is measured in total nucleotides.

_____ 9. Horizontal gene transfer is the movement of genetic information from parent to offspring.

_____ 10. Bacteria that can uptake genetic material from the environment are called competent cells.

_____ 11. Artificial transduction may be used to introduce DNA into a bacteria from its environment using laboratory procedures.

_____ 12. The movement of genetic material by conjugation begins at a site on the plasmid called the start site.

_____ 13. Interupted mating is a form of conjugation study that can be used to map genes in bacteria.

For questions 14 to 21, state which of the following forms of gene transfer the statement represents. Some statements may have more than one answer.

a. transduction
b. conjugation
c. cotransduction
d. transformation
e. cotransformation

_____ 14. Involves the use of a structure called a sex pilli.

_____ 15. Utilizes a phage to transfer material from one bacteria to another.

_____ 16. Used to map closely linked genes in bacteria.

_____ 17. *Hfr* strains use this process to transfer portions of bacterial chromosomes.

_____ 18. Used to determine the relative order of genes on a bacterial chromosome.

_____ 19. Competent cells are required for this to occur.

_____ 20. The donor strain is indicated by an F factor.

_____ 21. The size of the DNA being mapped is limited by the physical characteristics of the phage.

For questions 22 to 30, match each of the following with its correct definition.

_____ 22. A complex formed between the relaxase enzyme and the T DNA during conjugation.

_____ 23. A protein encoded by the F factor that recognizes the origin of transfer in conjugation.

_____ 24. The site that is cut by the relaxosome, it will be the first transferred into the recipient cell.

_____ 25. A phage that has been integrated into the bacterial chromosome.

_____ 26. A cut strand of DNA that will be transferred to the recipient cell.

_____ 27. A complex between the host DNA and the foreign DNA.

_____ 28. Proteins that bind DNA to the surface of the bacteria.

_____ 29. Circular pieces of DNA, located outside the bacterial chromosome.

_____ 30. A segment of DNA that can either be found in a plasmid or the bacterial chromosome.

a. heteroduplex
b. relaxosome
c. T DNA
d. competence factors
e. nucleoprotein
f. prophage
g. episome
h. plasmid
i. origin of transfer

6.2 Intragenic Mapping in Bacteriophages

Overview

The genomes of bacteriophages, also known as phages and viruses, is much smaller than that of the bacteria. In many cases, the genome consists of less than 100 genes. While not living, viruses are important to biologists due to their ability to infect cells and cause disease. Thus, there is an interest in mapping viral genomes. This section examines several of the historical processes that have been used in viral gene mapping.

It is important to recognize that even though viruses are $1/100^{th}$ the size of bacteria, they have phenotypes that can be scored for mutations. This has to do with the way that viral colonies, called plaques, look under a microscope. The differences in plaque phenotype are due to variations in certain genes within the viral genome.

Complementation tests can determine if similar plaque phenotypes are associated with the same gene. Figure 6.16 demonstrates the process by which a complementation test is performed. If a noncomplementation occurs, meaning that the mutations are in the same gene, then it is possible to map the location of the mutation using either intragenic mapping (Figure 6.17) or deletion mapping (Figure 6.19). It should be noted that the technological advances in DNA sequencing have reduced the use of these procedures, since it is often faster simply to sequence the entire genome. However, an understanding of these procedures is important in recognizing the structure of viral genomes.

Key Terms

Complementation
Deletion mapping
Fine structure mapping
Homoallelic
Hot spots
Intragenic mapping
Noncomplementation
Plaque

Focal Points

- Complementation test (Figure 6.16)
- Intragenic recombination (Figure 6.17)
- Deletion mapping (Figure 6.19)

Exercises and Problems

For questions 1 to 5, explain which experimental procedure the statement applies to.

a. intragenic mapping
b. complementation test
c. deletion mapping
d. all of the above

_____ 1. Can be used to determine if two mutations are homoallelic.

_____ 2. Can be used to determine the location of mutation hot spots in a gene.

_____ 3. The procedure that establishes whether two similar phenotypes are the result of mutations in the same, or different, genes.

_____ 4. Examines the rate of rare crossovers within a gene to determine the location of mutations.

_____ 5. Used to map genes within viruses.

For questions 6 to 10, indicate whether the statement is true of false. If false, correct the statement so that is it true.

_____ 6. The determination of the distance between two different genes is called intergenic mapping.

_____ 7. Hot spots are areas of a gene that have a higher rate of mutation than is found in neighboring areas of the gene.

_____ 8. Two mutations that are located at the same location in the same gene are called homoallelic.

_____ 9. Intragenic mapping involves the removal of segments of a gene to isolate the location of a specific mutation.

_____ 10. Two strains of mutated bacteriophages that coinfect bacteria and cause lysis are said to be in complementation.

Chapter Quiz

1. Which of the following mechanisms of gene transfer in bacteria involves the use of a viral intermediate?
    a. transduction
    b. transformation
    c. conjugation
    d. intragenic transfer
    e. none of the above

2. Which of the following viral life cycles is unique to a temperate phage?
    a. lysogenic
    b. lytic
    c. homoallelic
    d. cotransduction

3. In which of the following do the bacteria have to be in physical contact for gene transfer to occur?
    a. transduction
    b. transformation
    c. lysogenic cycle
    d. complementation
    e. conjugation

4. Bacterial cells that are able to take up DNA from the environment are called ___.
    a. episomes
    b. prophages
    c. F factors
    d. competent cells
    e. none of the above

5. Which of the following procedures is not utilized for genetic mapping in viruses?
    a. complementation
    b. intragenic mapping
    c. cotransformation
    d. deletion mapping
    e. all of the above are used

6. The transfer of genetic material between two different species is called ____.
    a. vertical gene transfer
    b. horizontal gene transfer
    c. conjugation
    d. sexual reproduction

7. The name of the small circular piece of DNA that is transferred during conjugation is the____.

a. relaxosome
b. nucleoprotein
c. plasmid
d. prophage
e. none of the above

8. ______ bacteria are capable of transferring portions of the bacterial chromosome during conjugation.

a. All
b. No
c. *Hfr* strain
d. Competent

9. Interrupted mating is used to study the process of bacterial _______.

a. transformation
b. conjugation
c. transduction
d. complementation
e. all of the above

10. A _____ is a virus that has integrated into the bacterial chromosome.

a. episome
b. plasmid
c. F factor
d. prophage

Answer Key for Study Guide Questions

This answer key provides the answers to the exercises and chapter quiz for this chapter. Answers in parentheses ( ) represent possible alternate answers to a problem, while answers marked with an asterisk (*) indicate that the response to the question may vary.

6.1

1. c
2. b
3. a
4. d
5. c
6. e
7. d
8. F, measured in minutes
9. F, vertical gene transfer
10. T
11. F, artificial transformation
12. F, called the origin of transfer
13. T
14. b
15. a , c
16. c, e

17. b
18. b
19. d, e
20. b
21. c
22. e
23. b
24. i
25. f
26. c
27. a
28. d
29. h
30. g

6.2

1. a
2. c
3. b
4. a
5. d
6. F, intragenic
7. T
8. T
9. F, deletion mapping
10. T

Quiz

1. a
2. a
3. e
4. d
5. c
6. b
7. d
8. c
9. b
10. d

Answers to Conceptual and Experimental Questions

Conceptual Questions

C1. All of these processes are similar in that a segment of genetic material has been transferred from one bacterial cell to another. The main difference is the underlying mechanism whereby this transfer occurs. In conjugation, two living cells make direct contact with each other, and genetic material is replicated and transferred from one cell to another. In transduction, a virus infects a cell, makes new virus particles that contain some genetic material from the bacterium, and then these virus particles are transferred to a new bacterial cell. Finally, in transformation, segments of DNA are released from a dead bacterium and later taken up by a living cell.

C2. It is not a form of sexual reproduction whereby two distinct parents produce gametes that unite to form a new individual. However, conjugation is similar to sexual reproduction in the sense that the genetic material from two cells are somewhat mixed. In conjugation, there is not the mixing of two genomes, one from each gamete. Instead, there is a transfer of genetic material from one cell to another. This transfer can alter the combination of genetic traits in the recipient cell.

C3. If neither cell has a selective growth advantage, we would expect that the F^+ cells would eventually overrun the population. This is because a mating starts with an F^- and F^+ cell and ends with two F^+ cells. Therefore, F^+ cells can convert F^- cells into F^+ cells but the opposite cannot occur.

C4. An F^+ strain contains a separate, circular piece of DNA that has its own origin of transfer. An *Hfr* strain has its origin of transfer integrated into the bacterial chromosome. An F^+ strain can transfer only the DNA contained on the F factor. If given enough time, an *Hfr* strain can actually transfer the entire bacterial chromosome to the recipient cell.

C5. The role of the origin of transfer is to provide a starting site where two important events occur. The DNA is nicked and one strand begins its transfer into a recipient cell. The direction of transfer in Hfr-mediated transfer will determine the order of transfer of the genes. For example, if the origin is between gene *A* and *B,* it could be oriented so that gene *A* will be transferred first. Alternatively, it could be oriented in the opposite direction so that gene *B* will be transferred first.

C6. Sex pili promote the binding of donor and recipient cells and provide a passageway for the transfer of genetic material from the donor to the recipient cell.

C7. Originally, there may have been a segment of the normal bacterial chromosome that was accidentally excised to form a circle. F factors have their own origin of replication and carry bacterial genes.

C8. Though exceptions are common, interspecies genetic transfer via conjugation is not as likely because the cell surfaces do not interact correctly. Interspecies genetic transfer via transduction is also not very likely because each species of bacteria is sensitive to particular bacteriophages. The correct answer is transformation. A consequence of interspecies genetic transfer is that new genes can be introduced into a bacterial species from another species. For example, interspecies genetic transfer could provide the recipient bacterium with a new trait such as resistance to an antibiotic. Evolutionary biologists call this horizontal gene transfer, while the passage of genes from parents to offspring is termed vertical gene transfer.

C9. Briefly, the lytic cycle involves the production of new viruses. The phage takes over the synthetic machinery of the cells to make many more additional virus particles. Eventually, the cell lyses to release these new particles. The lysogenic cycle involves the integration of the viral DNA into the host cell's chromosome. This integrated DNA is called a prophage. It will replicate whenever the host chromosome is replicated. At a later time, a prophage can be induced to progress through the lytic cycle.

C10. Cotransduction is the transduction of two or more genes. The distance between the genes determines the frequency of cotransduction. When two genes are close together, the cotransduction frequency would be higher compared to two genes that are relatively farther apart.

C11. Most of the P1 phages would contain P1 genetic material. However, an occasional phage contains a piece of the bacterial chromosome.

C12. If a site that frequently incurred a breakpoint was between two genes, the cotransduction frequency of these two genes would be much less than expected. This is because the site where the breakage occurred would separate the two genes from each other.

C13. The steps that occur during transformation are described in Figure 6.12. A competent cell is able to take up DNA from the environment. Several factors affect competency including temperature, ionic conditions, and nutritional growth conditions.

C14. The transfer of conjugative plasmids such as F factor DNA.

C15. One advantage of the uptake of DNA is that it could be used as a nutritional source. The bacterial cells could use the nucleotides for their own growth and metabolism. The integration of DNA into the bacterial chromosome (i.e., transformation) has the advantage of possibly giving the bacterium a new trait that might be beneficial to its survival. For example, if a bacterium that cannot metabolize lactose is transformed with lac^+ genes, the transformed bacterium can survive in a medium containing lactose as the sole carbon source for growth.

C16. A. If it occurred in a single step, transformation is the most likely mechanism because conjugation does not usually occur between different species, particularly distantly related species, and different species are not usually infected by the same bacteriophages.

B. It could occur in a single step, but it may be more likely to have involved multiple steps.

C. The use of antibiotics selects for the survival of bacteria that have resistance genes. If a population of bacteria is exposed to an antibiotic, those carrying resistant genes will survive and their relative numbers will increase in subsequent generations.

C17. The term refers to the outcome when two mutations are combined in a single individual. If two defective mutations combined in a single individual result in a nondefective phenotype, complementation has occurred. This is because the two mutations are in different genes, and the corresponding wild-type versions are dominant. If complementation does not occur, this means that the defective mutations are in the same gene. When combined in the same individual, complementation cannot occur because the individual does not have a normal (functional) copy of the gene.

C18. The term *allele* means alternative forms of the same gene. Therefore, mutations in the same gene among different phages are alleles of each other; the mutations may be at different positions within the same gene. When we map the distance between mutations in the same gene, we are mapping the distance between the mutations that create different alleles of the same gene. An intragenic map describes the locations of mutations within the same gene.

C19. Perhaps the mutations are at exactly the same spot (i.e., the same nucleotide) in the gene. For example, if a gene is composed of 500 nucleotides, perhaps two different strains both have a mutation at nucleotide number 377 (from one end of the gene).

Experimental Questions

E1. Because there were no colonies when 10^9 cells of the $bio^- met^- phe^+ thr^+$ or $bio^+ met^+ phe^- thr^-$ strains were plated alone.

E2. Mix the two strains together and then put some of them on plates containing streptomycin and some of them on plates without streptomycin. If mated colonies are present on both types of plates, then the phe^+ and thr^+ genes were transferred to the $bio^+ met^+ phe^- thr^-$ strain. If colonies are found only on the plates that lack streptomycin, then the bio^+ and met^+ genes are being transferred to the $bio^- met^- phe^+ thr^+$ strain. This answer assumes a one-way transfer of genes from a donor to a recipient strain.

E3. The U-tube can distinguish between conjugation and transduction because of the pore size. Since the pores are too small for the passage of bacteria, this prevents direct contact between the two bacterial strains. However, viruses can pass through the pores and infect cells on either side of the U-tube.

It might be possible to use a U-tube to distinguish between transduction and transformation if a filter was used that had a pore size that was too small for viruses but large enough to allow the passage of DNA fragments. However, this might be difficult since the difference in sizes between DNA fragments and phages are relatively small compared to the differences in sizes between bacteria and phages.

E4. An interrupted mating experiment is a procedure in which two bacterial strains are allowed to mate, and then the mating is interrupted at various time points. The interruption occurs by agitation of the solution in which the bacteria are found. This type of study is used to map the locations of genes. It is necessary to interrupt mating so that you can vary the time and obtain information about the order of transfer; which gene transferred first, second, etc.

E5. The time of entry is the time it takes for a gene to be initially transferred from one bacterium to another. To determine this time, we make many measurements at various lengths of time and then extrapolate these data back to the x-axis.

E6. Mate unknown strains *A* and *B* to the F^- strain in your lab that is resistant to streptomycin and cannot use lactose. This is done in two separate tubes (i.e., strain *A* plus your F^- strain in one tube, and strain *B* plus your F^- strain in the other tube). Plate the mated cells on growth media containing lactose plates plus streptomycin. If you get growth of colonies, the unknown strain had to be the F^+ strain that had lactose utilization genes on its F factor.

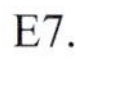

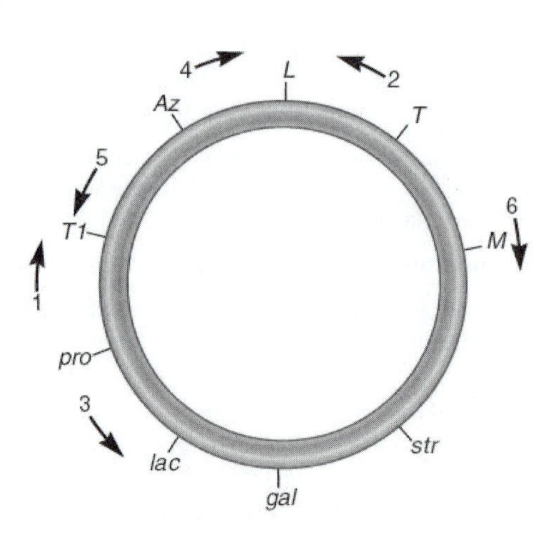

E8. A. If we extrapolate these lines back to the x-axis, the *hisE* intersects at about 3 minutes and the *pheA* intersects at about 24 minutes. These are the values for the times of entry. Therefore, the distance between these two genes is 21 minutes (i.e., 24 minus 3).

B. ____

↑ 4 ↑ 17 ↑

hisE ***pabB*** ***pheA***

E9. First mate the streptomycin-resistant strain to the strain that has the genes that allow the cell to metabolize lactose. You can select for mated cells on growth media containing lactose and streptomycin. These cells will initially have an F factor with the streptomycin-resistant gene. Expose these cells to acridine orange in a media that also contains streptomycin. This will select for the survival of rare cells in which the F factor has become integrated into the chromosome to become *Hfr* cells.

E10. One possibility is that you could treat the P1 lysate with Dnase I, an enzyme that digests DNA. (Note: If DNA were digested with Dnase I, the function of any genes within the DNA would be destroyed.) If the DNA were within a P1 phage, it would be protected from Dnase I digestion. This would allow you to distinguish between transformation (which would be inhibited by Dnase I) versus transduction (which would not be inhibited by Dnase I). Another possibility is that you could try to fractionate the P1 lysate. Naked DNA would be smaller than a P1 phage carrying DNA. You could try to filter the lysate to remove naked DNA, or you could subject the lysate to centrifugation and remove the lighter fractions that contain naked DNA.

E11. You could infect *E. coli* with lambda and then grow under conditions that promote the lysogenic cycle (i.e., on minimal media). You could then infect the lysogenic *E. coli* strain with P1. On occasion, the P1 phage packages a piece of the bacterial chromosome and therefore it could package the lambda DNA that has been integrated into the bacterial chromosome. (Note: The size of the lambda genome is small enough to fit inside of P1.) The P1 phages isolated from the lysogenic *E. coli* strain could then be used to infect a nonlysogenic strain.

E12. Cotransduction frequency = $(1 - d/L)^3$

For the normal strain:

Cotransduction frequency = $(1 - 0.7/2)^3 = 0.275$, or 27.5%

For the new strain:

Cotransduction frequency = $(1 - 0.7/5)^3 = 0.64$, or 64%

The experimental advantage is that you could map genes that are farther than 2 minutes apart. You could map genes that are up to 5 minutes apart.

E13. Cotransduction frequency = $(1 - d/L)^3$

Cotransduction frequency = $(1 - 0.6 \text{ minutes}/2 \text{ minutes})^3$

Cotransduction frequency = 0.34, or 34%

E14. Cotransduction frequency = $(1 - d/L)^3$

$$0.53 = (1 - d/2 \text{ minutes})^3$$

$$(1 - d/2 \text{ minutes}) = \sqrt[3]{0.53}$$

$$(1 - d/2 \text{ minutes}) = 0.81$$

$$d = 0.38 \text{ minutes}$$

E15. You would conclude that the two genes are farther apart than the length of 2% of the bacterial chromosome.

E16. A. We first need to calculate the cotransformation frequency, which equals 2/70, or 0.029.

Cotransformation frequency = $(1\text{-}d/L)^3$

$$0.029 = (1 - d/2 \text{ minutes})^3$$

$$d = 1.4 \text{ minutes}$$

B.

$$\begin{aligned}\text{Cotransformation frequency} &= (1 - d/L)^3 \\ &= (1 - 1.4/4)^3 \\ &= 0.27\end{aligned}$$

As you may have expected, the cotransformation frequency is much higher when the transformation involves larger pieces of DNA.

E17. A P1 plaque mostly contains P1 bacteriophages that have a phage coat and P1 DNA. On occasion, however, a phage coat contains a segment of the bacterial chromosome. It would also contain material from the *E. coli* cells that had been lysed.

E18. Benzer could use this observation as a way to evaluate if intragenic recombination had occurred. If two *rII* mutations recombined to make a wild-type gene, the phage would produce plaques in this *E. coli K12(λ)* strain.

E19. In *E. coli B,* both *rII* mutants and wild-type strains could produce plaques. In *E. coli K12(λ)* only the rare recombinants that produced the wild-type gene would be able to form plaques. If the phage preparation was not diluted enough, the entire bacterial lawn would be lysed so it would be impossible to count the number of plaques.

E20. *rIIA:* L47, L92

rIIB: L33, L40, L51, L62, L65, and L91

E21. You would basically follow the strategy of Benzer (see Figure 6.18) except that you would not need to use different *E. coli* strains. Instead, you would grow the cells at different temperatures. You would coinfect an *E. coli* strain with two of the temperature-sensitive phages and then grow at 32°C. You would then reisolate phages as shown in step 4 of Figure 6.18. You would take some of the phage preparation, dilute it greatly (10^{-8}), infect *E. coli,* and grow at 32°C. This would tell you how many total phages you had. You would also take some of the phage preparation, dilute it somewhat (10^{-6}), infect *E. coli,* and then grow at 37°C. At 37°C, only intragenic recombinants that produced wild-type phages would form plaques.

E22. Benzer first determined the individual nature of each gene by showing that mutations within the same gene did not complement each other. He then could map the distance between two mutations within the same gene. The map distances defined each gene as a linear, divisible unit. In this regard, the gene is divisible due to crossing over.

E23. You can first narrow down a mutation to one of several different regions by doing pairwise crosses with just a few deletion strains. After that, you then make pairwise crosses only with strains that have also been narrowed down to the same region. You do not have to make pairwise crosses between all the mutant strains.

Questions for Student Discussion/Collaboration

1. There are many advantages to microorganisms. They are typically haploid and easy to grow. In addition, they grow fast and have a short generation time. Fungi, which include yeasts and molds, have provided model organisms to investigate eukaryotic genetic phenomena, while bacteria and phages are wonderful to study basic molecular genetic phenomena. The most common phenotypic traits that researchers study in microorganisms are those that affect nutritional growth requirements, colony or plaque morphologies, and resistance to toxic molecules, such as antibiotics.

2. Epistasis is another example. When two different genes affect flower color, the white allele of one gene may be epistatic to the purple allele of another gene. A heterozygote (e.g., *CcPp*) for both genes would have purple flowers. In other words, the two purple alleles (*C* and *P*) are complementing the two white alleles (*c* and *p*). Other similar examples are discussed in Chapter 4 in section 4.2 on Gene Interactions.

Chapter 7: Non-Mendelian Inheritance

Student Learning Objectives

Upon completion of this chapter you should be able to:

1. Understand how maternal effect influences the phenotype of the offspring and the molecular basis of this pattern of inheritance.
2. Understand the process of epigenetic inheritance, dosage compensation, genomic imprinting, and X inactivation in mammals.
3. Know the origins of extranuclear genomes.
4. Understand the patterns of inheritance associated with extranuclear inheritance.

7.1 Maternal Effect

Overview

Not all genes follow a Mendelian pattern of inheritance. In this first section of the chapter you are introduced to the concept of maternal effect. Simply stated, maternal effect means that the gene products of the mother have a stronger influence on the zygote during embryonic development than the genotype of the zygote.

The section starts by introducing the principles of maternal effect using the water snail *Limnea peregra* as a model system. You should examine Figure 7.1 very carefully, noting that it is the genotype of the mother that determines the phenotype of the offspring. Recall that phenotypes are determined by gene products. In this case, the gene products released by maternal nurse cells during development of the embryo play the prime influence on the phenotype. This is illustrated in Figure 7.2.

Key Terms

Maternal effect
Nuclear genes
Reciprocal cross

Focal Points

- Experimental evidence of maternal effect (Figure 7.1). Note the relationship between genotypes and phenotypes.

Exercises and Problems

The following pertains to the water snail, *Limnea peregra*. Using the information provided in the text, complete the following table.

Mother		Father		Offspring	
Genotype	**Phenotype**	**Genotype**	**Phenotype**	**Genotype**	**Phenotype**
DD	Dextral	dd	Sinistral	Dd	(1)
(2)	Dextral	DD	Dextral	(3)	Sinistral
Dd	Sinistral	Dd	Sinistral	(4)	(5)
(6)	Sinistral	dd	Sinistral	dd	(7)

7.2 Epigenetic Inheritance

Overview

Epigenetic inheritance indicates a modification in the DNA, or a chromosome, so that the expression of the DNA or chromosome is altered. These modifications occur early in the development of the sperm or ovum, or in the early stages of embryonic development. As the text indicates, there are two primary forms of epigenetic modifications, dosage compensation and genomic imprinting.

For the most part, genes are responsible for the production of proteins, which in turn determine the phenotype of the individual. In diploid organisms, autosomal genes are inherited in pairs for all members of the species. However, for genes on the sex chromosomes, the sex of the individual determines how many copies of the gene it possesses. Since too much of a gene product can be just as detrimental as too little, there is a need to compensate for the dosage of protein being produced. Table 7.1 provides the most common mechanisms by which this occurs.

For female mammals, the X chromosome is inactivated by a mechanism called the Lyon hypothesis. Figure 7.4 outlines this process. The experimental proof is provided in Experiment 7A. Recognize that the gene being studied, glucose-6-phosphate dehydrogenase (G-6-PD) is not a mechanism of X inactivation, it is simply being used in this experiment as a method of tracking X inactivation. The next section introduces the major genes responsible for the process of X inactivation. Make sure at this point that you understand the role of Xic, Xist, and Xce.

The final portion of this section examines the process of genomic imprinting and monoallelic expression. Simply put, genomic imprinting allows the expression of one parents genes, regardless of dominance. The system presented in this chapter involves the Igf-2 gene in mice. Once again, this is an indicator of genomic imprinting, not a mechanism. Study Figure 7.10 carefully since genomic imprinting can often be a difficult topic. Finally, the process of DNA methylation is presented to explain genomic imprinting. Methylation is a powerful tool in genomic regulation, and you are going to see much more of it in upcoming chapters.

Key Terms

Barr body
Differentially methylated regions
Dosage compensation
Epigenetic inheritance
Genomic imprinting
Lyon hypothesis
Monoallelic expression
X chromosomal controlling element
X inactivation
X inactivation center

Focal Points

- Need for dosage compensation (pgs. 168-169)
- Mechanisms of dosage compensation (Table 7.1, Figure 7.4)

Exercises and Problems

Match each of the following genes, or genetic sequences, to its importance in the study of epigenetic inheritance.

_____ 1. Xic

_____ 2. G-6-PD

_____ 3. Xist

_____ 4. Igf-2

_____ 5. Xce

_____ 6. DMRs

a. Involved in the counting of the number of X chromosomes.
b. Produces an RNA that assists in the formation of a Barr body.
c. Chooses the X chromosome to be inactivated.
d. Binding sites for regulatory proteins that control gene expression.
e. An example of genomic imprinting.
f. Used to test the Lyon hypothesis of X inactivation.

For each of the following, choose as to whether the statement is an example of X inactivation (X), genomic imprinting (G), or dosage compensation (D). Some questions may have more than one answer.

_____ 7. Involves the use of methylated DNA.

_____ 8. The formation of Barr bodies in female mammals.

_____ 9. Offspring distinguish between maternal and paternal genes.

_____ 10. For some species, this involves the increase in gene expression in one sex.

_____ 11. Compensation for sex chromosomes in female mammals.

_____ 12. Calico cats.

7.3 Extranuclear Inheritance

Overview

Not all of the genetic material in the cell is located in the nucleus of eukaryotic cells. Mitochondria and chloroplasts represent the remnants of prokaryotic organisms. Their existence in the cell is due to an ancient endosymbiotic relationship, called the endosymbiotic theory of eukaryotic evolution (Figure 7.20). The genomes of the mitochondria and the chloroplast are the remains of the circular chromosomes that were initially found in these prokaryotic organisms. This section of the chapter examines the differences in the inheritance patterns of non-nuclear DNA.

Key Terms

cpDNA	Extranuclear inheritance	mtDNA
Cytoplasmic inheritance	Heterogamous	Nucleoid
Endosymbiosis	Heteroplasmy	Paternal leakage
Endosymbiosis theory	Maternal inheritance	Petites

Focal Points

- Maternal inheritance (Figure 7.16)
- Chloroplast inheritance (Figure 7.19)
- Endosymbiotic theory (Figure 7.20)

Exercises and Problems

For each of the following, match the term with its correct definition from the list provided.

_____ 1. Heteroplasmy

_____ 2. Paternal leakage

_____ 3. Heterogamous

_____ 4. cpDNA

_____ 5. mtDNA

_____ 6. Maternal inheritance

a. A species that makes two different types of gametes, such as eggs and sperm.
b. The name given to the mitochondrial genome.
c. A condition by which organelles within the cell are contributed by the mother only.
d. A cell that contains two different types of chloroplasts or mitochondria.
e. A rare event when mitochondria are contributed by the sperm.
f. The name given to the chloroplast genome.

For questions 7 to 12, indicate which of the following choices best describes the statement.

a. mitochondrial extranuclear inheritance
b. chloroplast extranuclear inheritance
c. both of the above

_____ 7. The basis of streptomycin resistance in *Chlamydomonas*.

_____ 8. A result of an endosymbiotic relationship between ancient prokaryotes.

_____ 9. Heteroplasmy producing cells that are green, even though the white pigment allele is present.

_____ 10. Paternal leakage

_____ 11. Human diseases that involve myopathy or neuropathy.

_____ 12. Tested using reciprocal crosses.

Chapter Quiz

1. Which of the following genes is responsible for promoting compaction of the X chromosome?
   a. Xic
   b. Xist
   c. Xce
   d. G-6-PD
   e. none of the above

2. Which of the following does not occur during genomic imprinting?
   a. The imprint is established in the gametes.
   b. The imprint is maintained in the offspring.
   c. The imprint is erased in the gametes of the offspring.
   d. A new imprint is established in the gametes of the offspring.
   e. All of the above occur.

3. Studies of *Chlamydomonas* were involved in the discovery of which of the following?
   a. genomic imprinting
   b. epigenetic inheritance
   c. maternal effect
   d. chloroplast inheritance
   e. mitochondrial inheritance

4. In which of the following does the genotype of the mother control the phenotype of the offspring?
   a. maternal inheritance
   b. maternal effect
   c. X inactivation
   d. gene dosage
   e. none of the above

5. A calico cat is the result of which of the following mechanisms?
   a. genomic imprinting
   b. maternal effect
   c. cloning
   d. X inactivation
   e. monoallelic expression

6. A human XXXY cell would contain how many Barr bodies?
   a. 0
   b. 1
   c. 2
   d. 3

7. The endosymbiosis theory attempts to explain which of the following?
   a. The development of Barr bodies in XX cells.
   b. The influence of the mother's genotype on the offspring's phenotype.
   c. The process of DNA methylation.
   d. The presence of mitochondrial and chloroplast genomes in eukaryotic cells.

8. With regards to inheritance of traits, DNA methylation is associated with which of the following?

a. maternal inheritance
b. maternal effect
c. genomic imprinting
d. X chromosome inactivation
e. none of the above

9. mtDNA and cpDNA is associated with the study of ______ inheritance.

a. extranuclear
b. epigenetic
c. Mendelian
d. nuclear

10. Monoallelic expression is usually associated with ______ in diploid organisms.

a. maternal effect
b. extranuclear inheritance
c. genomic imprinting
d. dosage compensation

Answer Key for Study Guide Questions

This answer key provides the answers to the exercises and chapter quiz for this chapter. Answers in parentheses ( ) represent possible alternate answers to a problem, while answers marked with an asterisk (*) indicate that the response to the question may vary.

7.1

1. dextral
2. dd
3. Dd
4. ¼ DD, ½ Dd, ¼ dd
5. dextral
6. dd
7. Sinistral

7.2

1. a
2. f
3. b
4. e
5. c
6. d
7. G
8. X, D
9. G
10. D
11. X
12. X

7.3

1. d
2. e
3. a
4. f
5. b
6. c
7. b
8. c
9. b
10. a
11. a
12. c

Quiz

1. b
2. e
3. d
4. b
5. d
6. c
7. d
8. c
9. a
10. c

Answers to Conceptual and Experimental Questions

C1. Epigenetic refers to the idea that a genetic phenomenon seems to be permanent but it really is not over the course of many generations. Examples include imprinting, X inactivation, and other forms of dosage compensation.

C2. A maternal effect gene is one in which the genotype of the mother determines the phenotype of the offspring. At the cellular level, this happens because maternal effect genes are expressed in the diploid nurse cells and then the gene products are transported into the oocyte. These gene products play key roles in the early steps of embryonic development.

C3. A. Genotypes: all *Nn*
Phenotypes: all abnormal

B. Genotypes: all *Nn*
Phenotypes: all normal

C. Genotypes: 1 *NN* : 2 *Nn* : 1 *nn*
Phenotypes: all normal (because the mother is heterozygous and *N* is dominant)

C4. The genotype of the mother must be $bic^- \; bic^-$. That is why it produces abnormal offspring. Since the mother is alive and able to produce offspring, its mother (the maternal grandmother) must have been $bic^+ \; bic^-$ and passed the bic^- allele to its daughter (the mother in this problem). The maternal grandfather also must have passed the bic^- allele to its daughter. The maternal grandfather could be either $bic^+ \; bic^-$ or $bic^- \; bic^-$.

C5. A. For an imprinted gene, you need to know whether the mutant *Igf-2* allele is inherited from the mother or the father.

B. For a maternal effect gene, you need to know the genotype of the mother.

C. For maternal inheritance, you need to know whether the mother has this trait, because the offspring will inherit her mitochondria.

C6. The mother must be heterozygous. She is phenotypically abnormal because her mother must have been homozygous for the abnormal recessive allele. However, since she produces all normal offspring, she must have inherited the normal dominant allele from her father. She produces all normal offspring because this is a maternal effect gene, and the gene product of the normal dominant allele is transferred to the oocyte.

C7. The mother is *hh.* We know this because it is a maternal effect gene and all of its offspring have small heads. The offspring are all *hh* because their mother is *hh* and their father is *hh.*

C8. Maternal effect genes exert their effects because the gene products are transferred from nurse cells to oocytes. The gene products, mRNA and proteins, do not last a very long time before they are eventually degraded. Therefore, they can only exert their effects during early stages of embryonic development.

C9. A. By the animal that donated the oocyte because the gene products of maternal effect genes are transferred to the oocyte by the nurse cells.

B. Perhaps a little of both. The egg would contain some mitochondria and so would the somatic cell. Since the egg is so much larger, however, it would probably donate many more mitochondria.

C. The "cloned" animal would have the vast majority of its genetic traits from the animal that donated the somatic cell. However, the animal that donated the oocyte would govern traits determined by maternal effect genes, and mitochondrial genes would probably come from both donors. All in all, the offspring is not quite a "clone" because it is not genetically identical to the animal that donated the somatic cell. Perhaps we should call it a "quasi-clone" or something.

C10. Dosage compensation refers to the phenomenon that the genes on the sex chromosomes are expressed at similar levels even though the two sexes have different numbers of sex chromosomes. It may not always be necessary, but in many species it seems necessary so that the balance of gene expression between the autosomes and sex chromosomes is similar between the two sexes.

C11. A Barr body is a mammalian X chromosome that is highly condensed. It is found in somatic cells that have two or more X chromosomes. Most genes on the Barr body are inactive.

C12. Mammals: X inactivation in females

Fruit flies: double transcription on the X chromosome in males

Worms: decrease transcription of the X chromosome in hermaphrodites to 50%

C13. X inactivation in heterozygous females produces a mosaic pattern of gene expression. Since it occurs randomly during embryonic development, certain patches of tissue have one X chromosome inactivated and other patches have the other X chromosome inactivated. In the case of a female that is heterozygous for a gene that affects pigmentation of the fur, this produces a variegated pattern of coat color. Since it is a random process in any given animal, two female cats will have variation where the orange and black patches occur. A variegated coat pattern could not occur in female marsupials due to X inactivation because the somatic cells preferentially inactivate the paternal X chromosome.

C14. X inactivation begins with the counting of *Xic*s. If there are two X chromosomes; one is targeted for inactivation. During embryogenesis, this inactivation begins at the *Xic* locus and spreads to both ends of the X chromosome until it becomes a highly condensed Barr body. The *TsiX* gene may play a role in the choice of the X chromosome that remains active. This may occur by its ability to bind to *Xist* mRNA. The *Xist* gene, which is located in the *Xic* region, remains transcriptionally active on the inactivated X chromosome. It is thought to play an important role in X inactivation by coating the inactive X chromosome. After the inactivation is established, it is maintained in the same X chromosome in somatic cells during subsequent cell divisions. In germ cells, however, the X chromosomes are not inactivated so that an egg can transmit either copy of an active (noncondensed) X chromosome.

C15. The male is XXY. The person is male due to the presence of the Y chromosome. Because of the counting of *Xic*s, one of the X chromosomes is inactivated to produce a Barr body.

C16. A. One

B. Zero

C. Two

D. Zero

C17. A. In females, one of the X chromosomes is inactivated. When the X chromosome that is inactivated carries the normal allele, only the defective color blindness allele will be expressed. Therefore, on average, about half of a female's eye cells are expected to express the normal allele. Depending on the relative amounts of cells expressing the normal versus the color-blind allele, the end result may be partial color blindness.

B. In this female, as a matter of chance, X inactivation occurred in the right eye to always, or nearly always, inactivate the X chromosome carrying the normal allele. The opposite occurred in the left eye. In the left eye, the chromosome carrying the color blindness allele was primarily inactivated.

C18. The offspring inherited X^B from its mother and X^O and Y from its father. It is an XXY animal, which is male (but somewhat feminized).

C19. The spreading stage is when the X chromosome is inactivated (i.e., condensed) as a wave that spreads outward from *Xic*. The condensation spreads from *Xic* to the rest of the X chromosome. The *Xist* gene is transcribed from the inactivated X chromosome. It encodes an RNA that coats the X chromosome, which subsequently attracts proteins that are responsible for the condensation.

C20. Erasure and reestablishment of the imprint occurs during gametogenesis. It is necessary to erase the imprint because each sex will either imprint or not imprint both alleles of a gene. In somatic cells, the two alleles for a gene are imprinted dissimilarly, depending on the sex from which they were inherited.

C21. The *de novo* methylase would be inactive in somatic cells and active in germ cells. Somehow, the methylase must be able to recognize particular DMRs, because we know that some genes are imprinted in males but not females, while other genes are imprinted in females but not males.

C22. A person born with paternal uniparental disomy 15 would have Angelman syndrome, because this individual would not have an active copy of the *AS* gene; the paternally inherited copies of the *AS* gene are silenced. This individual (if he/she reproduced) would have normal offspring, because he/she does not have a deletion in either copy of chromosome 15.

C23. A. Pat and Lynn's mother is abnormal. We know this because Pat and Lynn both have Angelman syndrome. The *AS* gene is inactivated in the sperm, so both children must have inherited the deletion from their mother. Therefore, they did not get the gene from their mother, and the gene from their father is normally inactivated. This causes them to have Angelman syndrome. We do not actually know if Pat and Lynn's mother has AS or PWS. We only know she has the deletion. Their mother could have either AS or PWS depending on whether their mother inherited the deletion from Pat and Lynn's grandmother or grandfather.

B. Pat is a male because he has children with PWS. He transmitted the chromosome carrying the deletion to his two children, and the mother of Pat's children normally inactivates the *PW* gene in the egg. Therefore, both children have PWS. As in the answer to part A, we know Lynn is a female because she has a child with AS.

C24. In some species, such as marsupials, X inactivation depends on the sex. This is similar to imprinting. Also, once X inactivation occurs during embryonic development, it is remembered throughout the rest of the life of the organism. Again, this is similar to imprinting. X inactivation in mammals is different from genomic imprinting in that it is not sex dependent. The X chromosome that is inactivated could be inherited from the mother or the father. There was no marking process on the X chromosome that occurred during gametogenesis. In contrast, genomic imprinting always involves a marking process during gametogenesis.

C25. Extranuclear inheritance is the transmission of genetic material (in eukaryotes) that is not located in the cell nucleus. The two most important examples are mitochondria and plastids. Less common examples are infectious particles that produce traits such as killer paramecia and the sex ratio trait in *Drosophila.*

C26. The term *reciprocal cross* refers to two parallel crosses that involve the same genotypes of the two parents, but their sexes are opposite in the two crosses. For example: female *BB* × male *bb* and a reciprocal cross in which a female *bb* × male *BB.* Autosomal inheritance gives the same result because the autosomes are transmitted from parent to offspring in the same way for both sexes. For extranuclear inheritance, the mitochondria and plastids are not transmitted via the gametes in the same way for both sexes. For maternal inheritance, the reciprocal crosses would show that the gene is always inherited from the mother.

C27. Extranuclear inheritance does not always occur via the female gamete. Sometimes it occurs via the male gamete. Even in species where maternal inheritance is prevalent, male leakage can also occur. With regard to cytoplasmic inheritance, maternal inheritance is the most common because the female gamete is relatively large and more likely to contain cell organelles.

C28. The phenotype of a petite mutant is that it forms small colonies on growth media that contain an energy source that does not require mitochondrial function. These mutants are unable to grow on an energy source that requires mitochondrial function. Since nuclear and mitochondrial genes are necessary for mitochondrial function, it is possible for a petite mutation to involve a gene in the nucleus or in the mitochondrial genome. Neutral petites lack most of their mitochondrial DNA, while suppressive petites usually lack small segments of the mitochondrial genetic material.

C29. Paternal leakage means that an organelle is inherited from the paternal parent in a small percentage of cases. If paternal leakage was 3%, then 3% of the time the offspring would inherit the organelles from their father. If the father was transmitting a dominant allele in the organellar genome (and the mother did not), then 3% of the offspring would exhibit the trait. Among a total of 200 offspring, 6 would be expected to inherit paternal mitochondria.

C30. The mitochondrial and chloroplast genomes are composed of a circular chromosome found in one or more copies. These copies are located in a region of the organelle known as the nucleoid. The number of genes per chromosome varies from species to species. Mitochondria tend to have fewer genes compared to chloroplasts. See Table 7.3 for examples of the variation among mitochondrial and chloroplast genomes.

C31. There is compelling evidence that mitochondria and chloroplasts evolved from an endosymbiotic relationship in which bacteria took up residence within a primordial eukaryotic cell. Throughout evolution, there has been a movement of genes out of the organellar genomes and into the nuclear genome. The genomes of modern mitochondria and chloroplasts only contain a fraction of genes that are necessary for organellar structure and function. Nuclear genes encode most of the proteins that function within chloroplasts and mitochondria. Long ago, these genes were originally in the mitochondrial and chloroplasts genomes but have been subsequently transferred to the nuclear genome.

C32. A. Yes.

B. Yes.

C. No, it is determined by a gene in the chloroplast genome.

D. No, it is determined by a mitochondrial gene.

C33. Superficially, the tendency to develop this form of leukemia would seem to be inherited from mother to offspring, much like the inheritance of mitochondria. To prove that it is not, one could separate newborn mice from their mothers and place them with mothers that do not carry AMLV. These offspring would not be expected to develop leukemia, even though their mother would.

C34. Biparental extranuclear inheritance would resemble Mendelian inheritance in that offspring could inherit alleles of a given gene from both parents. It differs, however, when you think about it from the perspective of heterozygotes. For a Mendelian trait, the law of segregation tells us that a heterozygote passes one allele for a given gene to an offspring, but not both. In contrast, if a parent has a mixed population of mitochondria (e.g., some carrying a mutant gene and some carrying a normal gene), that parent could pass both types of genes (mutant and normal) to a single offspring, because more than one mitochondrion could be contained within a sperm or egg cell.

C35. Most of the genes within mitochondria and chloroplasts have been transferred to the nucleus. Therefore, mitochondria and chloroplasts have lost most of the genes that would be needed for them to survive as independent organisms.

Experimental Questions

E1. The results of each succeeding generation depended on the genotypes of the mothers of the preceding generation. For example, if a mother was *dd,* the F_1 offspring were all sinistral. The genotypes of the F_2 mothers were 1 *DD* : 2 *Dd* : 1 *dd.* The *DD* and *Dd* mothers would produce dextral offspring of the F_3 generation and the *dd* mothers would produce sinistral. As expected, the ratio of dextral to sinistral was 3:1, which is derived from the genotypes of the F_2 mothers, which were 1 *DD* + 2 *Dd* : 1 *dd.*

E2. The first type of observation was cytological. The presence of the Barr body in female cells was consistent with the idea that one of the X chromosomes was highly condensed. The second type was genetic. A variegated phenotype that is found only in females is consistent with the idea that certain patches express one allele and other patches express the other allele. This variegated phenotype would occur only if the inactivation happened at an early stage of embryonic development and was inherited permanently thereafter.

E3. A haploid oocyte should only express either the 550- or 375-length mRNA, but not both (because it has only one copy of the gene). The nurse cells, however, can express both mRNAs if the female is heterozygous. Therefore, if we begin with heterozygous females, we could dissect and separate the nurse cells from the oocytes. We would then isolate mRNA from the nurse cells and (in a separate tube) isolate mRNA from oocytes. The mRNA would then be run on a gel, and subjected to Northern blotting, using a probe that is complementary to both the 550- and 375-length mRNA. According to our knowledge of maternal effect genes, we would expect the oocyte to contain both the 550 and 375 mRNAs, because it receives them from the nurse cells. Both forms of the mRNA would also be found in the nurse cells.

E4. The pattern of inheritance is consistent with imprinting. In every cross, the allele that is inherited from the father is expressed in the offspring, while the allele inherited from the mother is not.

E5. Mate the female to a *dd* male. If all the offspring coil to the left, you know the female must be *dd.* If they all coil to the right, she could be either *DD* or *Dd.* If the F_1 offspring coil to the right, you could let them mate with each other to produce an F_2 generation. If the original mother was *Dd,* then half the F_1 female offspring would be *Dd* and half would be *dd.* Therefore, half of the F_2 snails would coil to the right and half to the left. In contrast, if the original mother was *DD,* all the F_1 female offspring would be *Dd.* In this case, all the F_2 snails would coil to the right.

E6. We assume that the snails in the large colony on the second island are true-breeding, *DD.* Let the male snail from the deserted island mate with a female snail from the large colony. Then let the F_1 snails mate with each other to produce an F_2 generation. Then let the F_2 generation mate with each other to produce an F_3 generation. Here are the expected results.

Female *DD* × Male *DD*

All F_1 snails coil to the right.
All F_2 snails coil to the right.
All F_3 snails coil to the right.

Female *DD* × Male *Dd*

All F_1 snails coil to the right.
All F_2 snails coil to the right because all of the F_1 females are *DD* or *Dd.*
15/16 of F_3 snails coil to the right, 1/16 of F_3 snails coil to the left (because 1/16 of the F_2 females are *dd*).

Female *DD* × Male *dd.*

All F_1 snails coil to the right.
All F_2 snails coil to the right because all of the F_1 females are *Dd.*
3/4 of F_3 snails coil to the right, 1/4 of F_3 snails coil to the left (because 1/4 of the F_2 females are *dd*).

E7. A. All the lanes would show the same results. The form of G-6-PD inherited from the mother would be found because the paternal X chromosome would be inactivated.

B. We would find equal amounts of both the fast and slow forms of G-6-PD in all lanes. X inactivation does not occur in this species.

C. We would find equal amounts of both the fast and slow forms of G-6-PD in all lanes. X inactivation does not occur in this species.

E8. Let's first consider the genotypes of male A and male B. Male A must have two normal copies of the *Igf-2* gene. We know this because male A's mother was *Igf-2 Igf-2;* the father of male A must have been a heterozygote *Igf-2 Igf-2m* because half of the litter that contained male A also contained dwarf offspring. But since male A was not dwarf, it must have inherited the normal allele from its father. Therefore, male A must be *Igf-2 Igf-2.* We cannot be completely sure of the genotype of male B. It must have inherited the normal *Igf-2* allele from its father because male B is phenotypically normal. We do not know the genotype of male B's mother, but she could be either *Igf-2m Igf-2m* or *Igf-2 Igf-2m.* In either case, the mother of male B could pass the *Igf-2m* allele to an offspring, but we do not know for sure if she did. So, male B could be either *Igf-2 Igf-2m* or *Igf-2 Igf-2.*

For the *Igf-2* gene, we know that the maternal allele is inactivated. Therefore, the genotypes and phenotypes of females A and B are irrelevant. The phenotype of the offspring is determined only by the allele that is inherited from the father. Since we know that male A has to be *Igf-2 Igf-2,* we know that it can produce only normal offspring. Since both females A and B both produced dwarf offspring, male A cannot be the father. In contrast, male B could be either *Igf-2 Igf-2* or *Igf-2 Igf-2m.* Because both females gave birth to dwarf babies (and since male A and male B were the only two male mice in the cage), we conclude that male B must be *Igf-2 Igf-2m* and is the father of both litters.

E9. A clone only produces one type of G-6-PD enzyme because X inactivation has already occurred and it is inherited permanently in subsequent cell divisions. If a biopsy was taken in early embryonic development, prior to X inactivation, then a clone could express both copies of the *G-6-PD* alleles. The biopsy described in the experiment of Figure 7.6 must have been varied enough so that not all the tissue was derived from the same embryonic cell. Only if a small biopsy was taken, and all the tissue was derived from the same embryonic cell in which X inactivation had already occurred, would the cells from that biopsy produce only one of the two *G-6-PD* alleles.

E10. In mice, one of the two X chromosomes is inactivated; that is why females and males produce the same total amount of mRNA for most X-linked genes. In fruit flies, the expression of a male's X-linked genes is turned up twofold. In *C. elegans,* the expression of hermaphrodite X-linked genes is turned down twofold. Overall, the total amount of expression of X-linked genes is the same in males and females (or hermaphrodites) of these three species. In fruit flies and *C. elegans,* heterozygous females and hermaphrodites express 50% of each allele compared to a homozygous male, so that heterozygous females and hermaphrodites produce the same total amount of mRNA from X-linked genes compared to males. Note: In heterozygous females of mice, fruit flies, and worms, there is 50% of each gene product (compared to hemizygous males and homozygous females).

A.

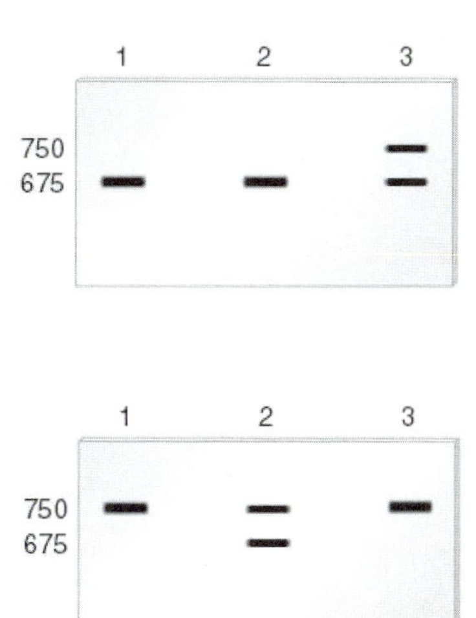

B.

C.

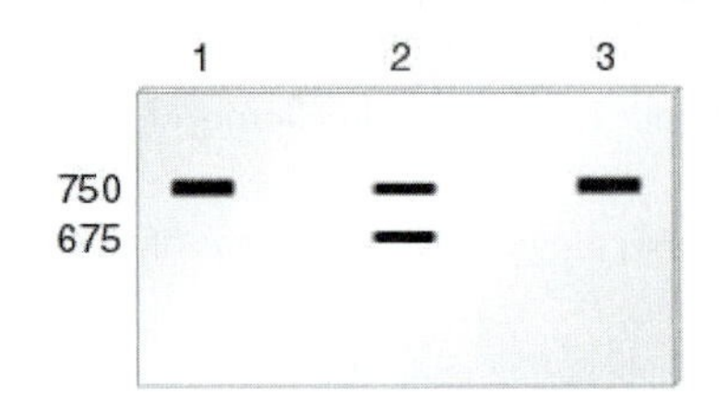

E11. Based on these results, it may be cytoplasmic inheritance involving plastids. In most cases, plastids follow a maternal inheritance pattern, but a low percentage of the time, paternal leakage does occur.

E12. In the absence of UV light, we would expect all sm^r offspring. With UV light, we would expect a greater percentage of sm^S offspring.

E13. If you examined them under the microscope, the neutral petites would lack nucleoids in their mitochondria, or their nucleoids would be very small. In contrast, suppressive petites would have fairly normal-looking nucleoids in their mitochondria. Another approach might be to isolate mitochondria from neutral and suppressive petites, extract the DNA, and determine the amount of DNA (e.g., with a spectrophotometer or by gel electrophoresis and staining with ethidium bromide).

Questions for Student Discussion/Collaboration

1. Obviously, you cannot maintain a population of flies that are homozygous for a recessive lethal allele. However, heterozygous females can produce viable offspring and these can be crossed to heterozygous or homozygous males to produce homozygous females. An experimenter would routinely have to make crosses and determine that the recessive allele was present in a population of flies by identifying homozygous females that were unable to produce any viable offspring. Maintaining a population of flies that carry a lethal recessive allele can be much easier if the recessive lethal allele is closely linked to a dominant allele that is not lethal, such as one affecting eye color or some other trait. If a fly exhibits the dominant trait, it is likely that it is also carrying the recessive lethal allele.
2. An infective particle is something in the cytoplasm that contains its own genetic material and is not an organelle. Some symbiotic infective particles, such as those found in killer paramecia, are similar to mitochondria and chloroplasts since they contain their own genomes and are known to be bacterial in origin. The observation that these endosymbiotic relationships can initiate in modern species tells us that endosymbiosis can spontaneously happen. Therefore, it is reasonable that it happened a long time ago and led to the evolution of mitochondria and plastids.

Chapter 8: Variation in Chromosome Structure and Number

Student Learning Objectives

Upon completion of this chapter you should be able to:

1. Understand the principles and terminology associated with variations in chromosome structure.
2. Understand the principles and terminology associated with variations in chromosome number.
3. Understand the processes of mitotic and meiotic nondisjunction.
4. Understand the techniques that can be used to produce changes in chromosome number, and the importance of these changes.

Introduction

8.1 Variation in Chromosome Structure

Overview

This chapter focuses on two forms of genetic variation, genome mutations and chromosome mutations. In this first section we will examine changes in the structure of a single (or sometimes two) chromosome, and how this may influence the expression of genes and the phenotype of the organism. Typically, student difficulties with this chapter rest primarily in the terminology associated with each form of variation. One of the best mechanisms of studying this material is to draw your own examples of each form of variation, noticing the loss/gain of genetic material on the chromosomes.

Before proceeding directly into the forms of chromosome variation, you need to become familiar with the terminology that describes chromosome structure. Of particular importance is the mechanism by which cytogeneticists label chromosomes (Figure 8.1).

The first two forms of chromosome variation, deletions and duplications, tend to occur simultaneously. The text provides an important experimental proof of the effects of a duplication with regards to the bar-eye phenotype in *Drosophila*. Notice in Figure 8.8 how a misaligned crossing over event can cause a duplication. While the duplication may have an effect on the phenotype of the organism, it can also have important evolutionary consequences. Gene duplications may form gene families, such as the globin gene family in humans. Gene families provide a species with a set of closely related proteins that have slight variations in function.

The next portion of the chapter examines inversions and translocations. Both of these are significantly more difficult to visualize that the duplication/deletions previously presented. In this section, focus on the key figures (8.12 and 8.15). Notice how the chromosomes align during meiosis. Pay special attention to the effects of these forms of variation on the production of gametes. Both of these may result in reduced fertility for an organism due to the loss of gametes because of chromosomal abnormalities.

Key Terms

Acentric fragment
Acrocentric
Allelic variation
Balanced translocation
Chromosome aberration
Chromosome mutation
Cytogeneticist
Deficiency
Deletion
Dicentric bridge
Dicentric chromosome
Duplication
G banding
Gene duplication
Gene family
Genetic variation
Genome mutation
Homologous
Interstitial deficiency
Inversion
Inversion heterozygote
Inversion loop
Karyotype
Metacentric
Paracentric inversion
Paralogues
Pericentric inversion
Polytene chromosomes
Position effect
Pseudodominance
Reciprocal translocation
Robertsonian translocation
Semisterility
Simple translocation
Submetacentric
Telocentric
Telomeres
Terminal deficiency
Translocation
Translocation cross
Unbalanced translocation

Focal Points

- Chromosome structure and band numbering (Figure 8.1)
- Overview of changes to chromosome structure (Figure 8.2)
- Inversion loops (Figure 8.12)
- Translocation crosses (Figure 8.15)

Exercises and Problems

For questions 1 to 5, choose which form of variation in chromosome structure the statement best applies to.

a. inversion
b. translocation
c. duplication
d. deletion

_____ 1. In some cases, the chromosomes form a cross pattern during meiosis.

_____ 2. Results in the formation of a chromosome with a dicentric bridge.

_____ 3. Usually caused by modifications to the telomere of the chromosome.

_____ 4. Responsible for the formation of gene families.

_____ 5. Causes homologous chromosomes to produce a loop formation during meiotic pairing.

_____ 6. May cause pseudodominance of a recessive allele.

For questions 7 to 11, match the condition or disease with its correct cause.

a. translocation
b. duplication
c. deficiency
d. inversion

_______ 7. Bar-eye phenotype in *Drosophila.*

_______ 8. Prader-Willi Syndrome

_______ 9. Familial Down Syndrome

_______ 10. Cri-du-chat Syndrome

_______ 11. Charcot-Marie-Tooth disease

For questions 12 to 16, assume that two non-homologous chromosomes have the following combination of genes:

```
-----------o--------------          ------------o---------------
A  B  C   D E F G                   Q  R  S       T U  V
```

For each combination indicated below, state the form of structural variation that would have to occur to produce the sequence indicated. The ----o---- indicates the location of the centromere. Note that some answers may require more than one process.

```
_____ 12. ------------- o ----------------
            A   B   C    D G F  E

_____ 13. ------------- o ----------------      ------------- o ----------------
            A   B   C      D  U   V             Q  R  S      T  E  F  G

_____ 14. ------------- o ----------------
            A   B  C     D  F  G

_____ 15. ------------- o ----------------      ------------- o ----------------
            A  B  C     D E                     Q  R  S      T  U  V G  F

_____ 16. --------------- o ----------------
           A  B  A  B C    D  E  F G
```

8.2 Variation in Chromosome Number

Overview

Changes in chromosome number also have an effect on gene expression and the phenotype of the organism. Once again, the most common problems with this section occur in the terminology. However, the terminology for variations in chromosome number are actually easy to understand if you take the time to examine the term for patterns. Before proceeding into the chapter, study the material on page 207 and Figure 8.16 carefully.

Changes in chromosome number may be represented algebraically. For a diploid organism (2n) the trisomy condition is indicated as a $2n + 1$, while monosomy is $2n–1$. Polyploids are indicated as multiples of n (2n, 3n, 6n).

One of the more important concepts of this chapter is how these conditions relate to gene expression. Previously in the text the concept of proteins being responsible for phenotypes was introduced. We have observed how many organisms are very careful to regulate levels of gene expression when the chromosome number is not the same in the sexes. Therefore, any change in chromosome number should also have an effect on the phenotype.

Key Terms

Aneuploidy	Monoploid	Tetraploid
Chromocenter	Monosomic	Triploid
Chromomere	Nondisjunction	Trisomic
Endopolyploidy	Polyploid	
Euploid	Polytene chromosome	

Focal Points

- Terminology associated with variations in chromosome number (pg. 207, Figure 8.16)
- Aneuploid conditions in humans (Table 8.1)

Exercises and Problems

For each of the following, indicate the algebraic formula for the chromosome number in the organism.

________ 1. Triploid

________ 2. Octaploid

________ 3. Trisomic

________ 4 Tetraploid

________ 5. Monosomic

________ 6. Monoploid

For each of the following conditions in humans, indicate whether the condition is due to a change in the number of autosomal or sex chromosomes, and the genotype of the individual.

7. Jacobs Syndrome

8. Klinefelter Syndrome

9. Turner Syndrome

10. Down Syndrome

11. Patau Syndrome

12. Edward Syndrome

For each of the following, indicate whether the statement is true or false. If the statement is false, correct the statement so that it is true.

_____ 13. Aneuploidy represents a change in the number of sets of chromosomes in a cell.

_____ 14. A euploid organism contains multiple sets of chromosomes for the species.

_____ 15. Polytene chromosomes are formed when chromosomes aggregate around a point called the chromomere.

_____ 16. Endopolyploidy may occur in the liver cells of humans.

_____ 17. Failure of the chromosomes to separate correctly during cell division is called nondisjunction.

_____ 18. Individuals with Down Syndrome are 3n.

_____ 19. The polyploid condition in plants frequently is a benefit to modern agriculture.

_____ 20. Polyploidy never occurs in animals.

8.3. Natural and Experimental Ways to Produce Variation in Chromosome Number

Overview

The last section of this chapter examines the mechanisms by which variations in chromosome number may occur. It is divided into two general classes, those that naturally occur as a result of cell division, and those that are used by researchers to artificially alter the chromosome number. You should recognize that regardless of which occurs, the outcome is the same since the result is a cell that produces abnormal levels of protein.

Naturally, variations in chromosome number may occur as a result of nondisjunction. While this usually occurs during meiosis (Figure 8.24), it may happen rarely during mitosis as well (Figure 8.25). Notice the difference between these forms.

This section of the chapter also presents the concept of alloploidy, or a cell that contains chromosomes from two different species. This introduces a whole new level to the terminology from the previous section, since an organism may have multiple sets of chromosomes, but not from the same species. One of the more interesting aspects of chance in chromosome number is the influence of alloploidy on sterility.

The final section of this chapter discusses some of the experimental procedures that may be utilized by researchers to create new cell lines that vary with regard to chromosome number. This is especially important in agriculture, since plants frequently exhibit a condition called hybrid vigor (or heterosis). You should familiarize yourself with the basics of these procedures.

Key Terms

Allodiploid
Alloploid
Alloploidy
Allopolyploid
Allotetraploid
Anther culture
Autopolyploid
Bilateral gynandromorphy
Cell fusion
Complete nondisjunction
Heterokaryon
Heterosis
Homeologous
Hybrid cell
Hybrid vigor
Meiotic nondisjunction
Mitotic nondisjunction
Mosaicism
Parthenogenesis
Protoplast

Focal Points

- Meiotic nondisjuntion (Figure 8.24)
- Mitotic nondisjunction (Figure 8.25)

Exercises and Problems

For questions 1 to 7, choose whether the statement applies to mitotic nondisjunction, meiotic nondisjunction, or both.

a. mitotic nondisjunction
b. meiotic nondisjunction
c. both forms of nondisjunction

_____ 1. Results in a mosaic pattern of chromosome number in the organism.

_____ 2. May result in an organism called a bilateral gynandromorph.

_____ 3. Occurs in the germ cells of the organism.

_____ 4. All of the cells of the organism will have the same chromosome number.

_____ 5. By complete nondisjunction, this may produce a gamete without chromosomes.

_____ 6. Occurs in the somatic cells of the organism.

_____ 7. Can create a cell that is an autopolyploid.

For questions 8 to 13, match the term with its correct definition.

_____ 8. Autopolyploid

_____ 9. Allodiploid

_____ 10. Allotetraploid

_____ 11. Homeologous

_____ 12. Heterokaryon

_____ 13. Alloploid

a. An organism containing chromosome sets from more than one species.
b. An organism that has two complete sets of chromosomes from two different species.
c. An increase in the number of chromosome sets in a single species.
d. Evolutionarily related chromosomes from different species.
e. A cell containing two different nuclei.
f. An organism that has one complete chromosome set from two different species.

For questions 14 to 18, identify the experimental procedure that would most likely be used to produce the results indicated in the statement.

a. anther culture
b. parthenogenesis
c. cell fusion
d. all of the above

_____ 14. In plants, starts with the creation of a protoplast so that the genetic material may be combined.

_____ 15. Production of monoploids in animals.

_____ 16. May produce diploid plants that are homozygous for all their genes.

_____ 17. The result is a hybrid cell.

_____ 18. May be used to produce true-breeding lines for the study of hybrid vigor.

Chapter Quiz

1. Which of the following represents an organism with two complete sets of chromosomes from two different species?

a. autotetraploid
b. allotetraploid
c. allodiploid
d. trisomy
e. none of the above

2. The bar-eye phenotype in *Drosophila* is a result of a(n) _______.

a. inversion
b. translocation
c. deletion
d. duplication

3. Pseudodominance is the result of a(n) ______________.

a. inversion
b. translocation
c. deletion
d. duplication

4. The chromosome number of an individual with non-familial Down's syndrome is indicated as _______.

a. $2n - 1$
b. $3n$
c. $2n + 1$
d. $n + 2$

5. Prader-Willi and Angelmans syndromes are a result of a(n)______________.
   a. inversion
   b. translocation
   c. deletion
   d. duplication

6. Liver cells in humans exhibit which of the following?
   a. alloploidy
   b. endopolyploidy
   c. monoploidy
   d. aneuploidy

7. A bilateral gynandromorph is the result of ________.
   a. mitotic nondisjunction
   b. meiotic nondisjunction
   c. cell fusion
   d. anther culture
   e. parthenogensis

8. A dicentric bridge would be caused by which of the following?
   a. pericentric inversion
   b. Robertsonian translocation
   c. mitotic nondisjunction
   d. paracentric inversion
   e. gene duplication

9. The loss of the telomere on a chromosome would cause which of the following?
   a. translocation
   b. deletion
   c. duplication
   d. inversions
   e. nondisjunction

10. The name given to the phenomenon by which a hybrid cell line is better adapted than a true-breeding line is called _______.
   a. parthenogenesis
   b. monoploidy
   c. mosaicism
   d. heterosis

Answer Key for Study Guide Questions

This answer key provides the answers to the exercises and chapter quiz for this chapter. Answers in parentheses ( ) represent possible alternate answers to a problem, while answers marked with an asterisk (*) indicate that the response to the question may vary.

8.1
1. b
2. a
3. b
4. c
5. a
6. d

7. b
8. c
9. a
10. c
11. b
12. paracentric inversion
13. reciprocal translocation
14. interstitial deficiency
15. simple translocation and paracentric inversion of F G
16. duplication

8.2

1. 3n
2. 8n
3. 2n + 1
4. 4n
5. 2n – 1
6. n
7. sex, XYY
8. sex, XXY
9. sex, XO
10. autosomal, trisomy 21
11. autosomal, trisomy 13
12. autosomal, trisomy 18
13. F, euploid
14. T
15. F, chromocenter
16. T
17. T
18. F, 2n + 1
19. T
20. F, rarely

8.3

1. a
2. a
3. b
4. b
5. b
6. a
7. c
8. c
9. f
10. b
11. d
12. e
13. a
14. c
15. b
16. a
17. c
18. a

Quiz

1. b
2. d
3. c
4. c
5. c
6. b
7. a
8. d
9. a
10. d

Answers to Conceptual and Experimental Questions

Conceptual Questions

C1. Duplications and deficiencies involve a change in the total amount of genetic material.

Duplication: a repeat of some genetic material
Deficiency: a shortage of some genetic material
Inversion: a segment of genetic material in the opposite orientation
Translocation: a segment of genetic material attached to the wrong chromosome

C2. Small deletions and duplications are less likely to affect phenotype simply because they usually involve fewer genes. If a small deletion did have a phenotypic effect, you would conclude that a gene or genes in this region are required in two functional copies in order to have a normal phenotype.

C3. It occurs when there is a misalignment during the crossing over of homologous chromosomes. One chromosome ends up with a deficiency and the other has a duplication.

C4. A gene family is a group of genes that are derived from the process of gene duplications. They have similar sequences, but the sequences have some differences due to the accumulation of mutations over many generations. The members of a gene family usually encode proteins with similar but specialized functions. The specialization may occur in different cells or at different stages of development.

C5. You would expect a_1 and a_2 to be more similar, because they have diverged more recently. Therefore, there has been less time for them to accumulate random mutations that would make their sequences different.

C6. It has a pericentric inversion.

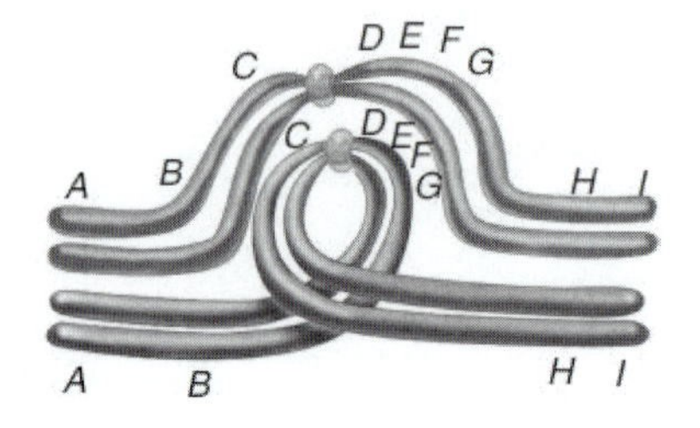

C7. There are four products from meiosis. One would be a normal chromosome and one would contain the inversion shown in the answer to conceptual question C6. The other two chromosomes would have the following order of genes:

A B C centromere *D E F G H I J B A*
M L K J I H G F E D centromere *C K L M*

C8. There are four products from meiosis. One would be a normal chromosome and one would contain the inversion shown in the drawing to conceptual question C6. The other two chromosomes would be dicentric or acentric with the following order of genes:

centromere ↓ centromere ↓

A B C D E F G H I J D C B A Dicentric
M L K J I H G F E K L M Acentric

C9. Individuals who carry inversions and reciprocal translocations have the same amount of genetic material (i.e., the same number of genes) as do normal individuals. Therefore, they commonly have a normal phenotype. In some cases, however, the breakpoint in an inversion or translocation disrupts an important gene and thereby has a phenotypic consequence. In other cases, the chromosomal rearrangement may have a position effect that alters the expression of an important gene.

C10. In the absence of crossing over, alternate segregation would yield half the cells with two normal chromosomes and half with a balanced translocation. For adjacent-1 segregation, two cells would be

ABCDE + *AIJKLM*

And the other two cells would be:

HBCDE + *HIJKLM*

C11. A terminal piece of chromosome 11 broke off and attached to the short arm of chromosome 15. A crossover occurred between the long arm of chromosome 15 and the long arm of chromosome 18.

C12. One of the parents may carry a balanced translocation between chromosomes 5 and 7. The phenotypically abnormal offspring has inherited an imbalanced translocation due to the segregration of translocated chromosomes during meiosis.

C13. It is expected to be rare because the normal driving force for segregation is the segregation of centromeres. For example, the centromeres on chromosome 2 normally align during meiosis and segregate from each other whether or not the chromosome contains a translocation. On rare occasions, a misalignment of centromeres can lead to adjacent-2 segregation in which two centromeres from one chromosome travel to one pole and two centromeres from another chromosome travel to the opposite pole.

C14. A deficiency and an unbalanced translocation are more likely to have phenotypic effects because they create genetic imbalances. For a deficiency, there are too few copies of several genes, and for an unbalanced translocation, there are too many.

C15. It is because the homologous chromosomes are trying to synapse with each other. As shown in Figure 8.15, the formation of a translocation cross allows the homologous parts of the chromosomes to line up (i.e., synapse) with each other.

C16. You should draw out the inversion loop (as is done in Figure 8.12*a*). The crossover occurred between *P* and *U*.

C17. A. 16

B. 9

C. 7

D. 12

E. 17

C18. This person has a total of 46 chromosomes. However, this person would be considered aneuploid rather than euploid. This is because one of the sets is missing a sex chromosome and one set has an extra copy of chromosome 21.

C19. One parent is probably normal while the other parent has one normal copy of chromosomes 14 and 21 and one chromosome 14, which has most of chromosome 21 attached to it.

C20. It may be related to genetic balance. In aneuploidy, there is an imbalance in gene expression between the chromosomes found in their normal copy number versus those that are either too many or too few. In polyploidy, the balance in gene expression is still maintained.

C21. Imbalances in aneuploidy, deletions, and duplications are related to the copy number of genes. For many genes, the level of gene expression is directly related to the number of genes per cell. If there are too many copies, as in trisomy, or too few, as in monosomy, the level of gene expression will be too high or too low, respectively. It is difficult to say why deletions and monosomies are more detrimental although one could speculate that having too little of a gene product causes more cellular problems than having too much of a gene product.

C22. The male offspring is the result of nondisjunction during oogenesis. The female produced an egg without any sex chromosomes. The male parent transmitted a single X chromosome carrying the red allele. This produces an X0 male offspring with red eyes.

C23. A. The F_1 offspring would probably be phenotypically normal since they would carry the correct number of genes.

B. The F_1 offspring would have lowered fertility because they are inversion heterozygotes. Since this is a large inversion, crossing over is fairly likely in the inverted region. When this occurs, it will produce deletions and duplications that will probably be lethal in the resulting F_2 offspring.

C24. Trisomies 13, 18, and 21 survive because the chromosomes are small and probably contain fewer genes compared to the larger chromosomes. Individuals with abnormal numbers of X chromosomes can survive because the extra copies are converted to transcriptionally inactive Barr bodies. The other aneuploidies are lethal because they cause a great amount of imbalance between the level of gene expression on the normal diploid chromosomes relative to the chromosomes that are trisomic or monosomic.

C25. Maybe one is diploid and the other is a closely related tetraploid species. Their offspring would be triploid, which would explain the sterility. Another possibility is that one may carry a large inversion (see answer to conceptual question C23, part B).

C26. Endopolyploidy means that a particular somatic tissue is polyploid even though the rest of the organism is not. The biological significance is not entirely understood although it has been speculated that an increase in ploidy may enable the cell to make more gene products that the cell needs.

C27. A genetic mosaic is an individual having patches of tissue that are genetically different from each other. With regard to mosaics in chromosome number, it can be due to abnormal events during mitosis, such as the movement of two homologues to the same pole, or the loss of a chromosome.

C28. In certain types of cells, such as salivary cells, the homologous chromosomes pair with each other and then replicate about nine times to produce a polytene chromosome. The centromeres from each type of chromosome associate with each other at the chromocenter. This structure has six arms that arise from one arm of two telomeric chromosomes (the X and 4) and two arms each from chromosomes 2 and 3.

C29. Polyploid plants are often more robust than their diploid counterparts. With regard to agriculture, they may produce a greater yield of fruits and vegetables. In the field, they tend to be more resistant to harsh environmental conditions. When polyploid plants have an odd number of sets, they are typically seedless. This can be a desirable trait for certain fruit-producing crops such as bananas.

C30. The turtles are two distinct species that appear phenotypically identical. The turtles with 48 chromosomes are polyploid relatives (i.e., tetraploids) of the species with 24 chromosomes. In animals, it is somewhat hard to imagine how this could occur because animals cannot self-fertilize, so there had to be two animals (i.e., one male and one female) that became tetraploids. It is easy to imagine how one animal could become a tetraploid; complete nondisjunction could occur during the first cell division of a fertilized egg, thereby creating a tetraploid cell that continued to develop into a tetraploid animal. This would have to happen independently (i.e., in two individuals of opposite sex) to create a tetraploid species. If you mated a tetraploid turtle with a diploid turtle, the offspring would be triploid and probably phenotypically normal. However, the triploid offspring would be sterile because they would make highly aneuploid gametes.

C31. Aneuploid should not be used.

C32. Polyploid, triploid, and euploid should not be used.

C33. There are 11 chromosomes per set, so there are 11 possible trisomics: trisomy 1, trisomy 2, trisomy 3, trisomy 4, trisomy 5, trisomy 6, trisomy 7, trisomy 8, trisomy 9, trisomy 10, and trisomy 11.

C34. The boy carries a translocation involving chromosome 21: probably a translocation in which nearly all of chromosome 21 is translocated to chromosome 14. He would have one normal copy of chromosome 14, one normal copy of chromosome 21, and the translocated chromosome that contains both chromosome 14 and chromosome 21. This boy is phenotypically normal because the total amount of genetic material is normal, although the total number of chromosomes is 45 (because chromosome 14 and chromosome 21 are fused into a single chromosome). His sister has Down syndrome because she has inherited the translocated chromosome, but she also must have one copy of chromosome 14 and two copies of chromosome 21. She has the equivalent of three copies of chromosome 21 (i.e., two normal copies and one copy fused with chromosome 14). This is why she has Down syndrome. One of the parents of these two children is probably normal with regard to karyotype (i.e., the parent has 46 normal chromosomes). The other parent would have a karyotype that would be like the phenotypically normal boy.

C35. The odds of producing a euploid gamete are $(1/2)^{n-1}$, which equals $(1/2)^5$ or a 1 in 32 chance. The chance of producing an aneuploid gamete is 31 out of 32 gametes will be aneuploid. We use the product rule to determine the chances of getting a euploid individual, since a euploid individual is produced from two euploid gametes: $1/32 \times 1/32 = 1/1{,}024$. In other words, if this plant self-fertilized, only 1 in 1,024 offspring would be euploid. The euploid offspring could be diploid, triploid, or tetraploid.

C36. Nondisjunction is a mechanism whereby the chromosomes do not segregate equally into the two daughter cells. This can occur during meiosis to produce cells with altered numbers of chromosomes, or it can occur during mitosis to produce a genetic mosaic individual. A third way to alter chromosome number is by interspecies crosses to produce an alloploid.

C37. It usually occurs during meiosis I when the homologues synapse to form bivalents. During meiosis II and mitosis, the homologues do not synapse. Instead, the chromosomes align randomly along the metaphase plate and then the centromeres separate.

C38. A mutation occurred during early embryonic development to create the blue patch of tissue. One possibility is a mitotic nondisjunction in which the two chromosomes carrying the *b* allele went to one cell and the two chromosomes carrying the *B* allele went to the other daughter cell. A second possibility is that the chromosome carrying the *B* allele could be lost. A third possibility is that the *B* allele could have been deleted. This would cause the recessive *b* allele to exhibit pseudodominance.

C39. An allodiploid is an organism having one set of chromosomes from two different species. Unless the two species are closely related evolutionarily, the chromosomes do not synapse during meiosis. Therefore, they do not segregate properly. This produces aneuploid gametes that are usually inviable. By comparison, allotetraploids that have two sets of chromosomes from each species are more likely to be fertile because each chromosome has a homologue to pair with during meiosis.

C40. Homeologous chromosomes are chromosomes from two species that are evolutionarily related to each other. For example, chromosome 1 in chimpanzees and gorillas is homeologous; it carries the same types of genes.

C41. For meiotic nondisjunction, the bivalents are not separating correctly during meiosis I. During mitotic nondisjunction, the sister chromatids are not separating properly.

C42. In general, Turner syndrome could be due to nondisjunction during oogenesis or spermatogenesis. However, the Turner individual with color blindness is due to nondisjunction during spermatogenesis. The sperm lacked a sex chromosome, due to nondisjunction, and the egg carried an X chromosome with the recessive color blindness allele. This X chromosome had to be inherited from the mother, because the father was not color-blind. The mother must be heterozygous for the recessive color blind allele, and the father is hemizygous for the normal allele. Therefore, the mother must have transmitted a single X chromosome carrying the color-blind allele to her offspring indicating that nondisjunction did not occur during oogenesis.

C43. Complete nondisjunction occurs during meiosis I so that one nucleus receives all the chromosomes and the other nucleus does not get any. The nucleus with all the chromosomes then proceeds through a normal meiosis II to produce two haploid sperm cells.

Experimental Questions

E1. Polytene chromosomes can be viewed in greater detail under the microscope because they are much larger. This makes it much easier to detect very small changes in chromosome structure. They are produced from the sequential replication and alignment of chromosomes. As an example, suppose a toothpick has fine lines written on it. You would probably have trouble seeing the lines. However, if you took 1,000 toothpicks with the same lines and stacked them up in a parallel manner, the lines would be much easier to see. Similarly, small changes in chromosome structure are hard to see in a single chromosome but much easier to detect in a polytene chromosome.

E2. Colchicine interferes with the spindle apparatus and thereby causes nondisjunction. At high concentrations, it can cause complete nondisjunction and produce polyploid cells.

E3. You could begin with a normal diploid strain and first use anther culture. This would create a monoploid plant. This monoploid plant could then be treated on two successive occasions with colchicine to first produce a diploid plant and then a tetraploid plant. Since this tetraploid plant would be derived from a monoploid plant, it would be homozygous for all of its genes.

E4. The primary purpose is to generate strains that are homozygous for all of their genes. Since the pollen is haploid, it has only one copy of each gene. The strain can later be made diploid (and homozygous) by treatment with colchicine.

E5. Cell fusion techniques can be used to create hybrids between strains or different species that cannot readily interbreed. Any two types of cells can be made to fuse in the laboratory, even interspecies hybrids that could never interbreed naturally. In addition, the fusion of diploid cells instantly creates allotetraploids that are usually fertile.

E6. First, you would cross the two strains together. It is difficult to predict the phenotype of the offspring. Nevertheless, you would keep crossing offspring to each other and backcrossing them to the parental strains until you obtained a great-tasting tomato strain that was resistant to heat and the viral pathogen. You could then make this strain tetraploid by treatment with colchicine. If you crossed the tetraploid strain with your great-tasting diploid strain that was resistant to heat and the viral pathogen, you may get a triploid that had these characteristics. This triploid would probably be seedless.

E7. A G band is a dark band on a chromosome that has been stained with Giemsa. The pattern of G bands on chromosomes can be used to identify chromosomes, particularly ones of similar sizes. They also make it easier to detect changes in chromosome structure.

E8. A polytene chromosome is formed when a chromosome replicates many times, and the chromatids lie side by side as shown in Figure 8.21. The homologous chromosomes also lie side by side. Therefore, if there is a deletion, there will be a loop. The loop is the segment that is not deleted from one of the two homologues.

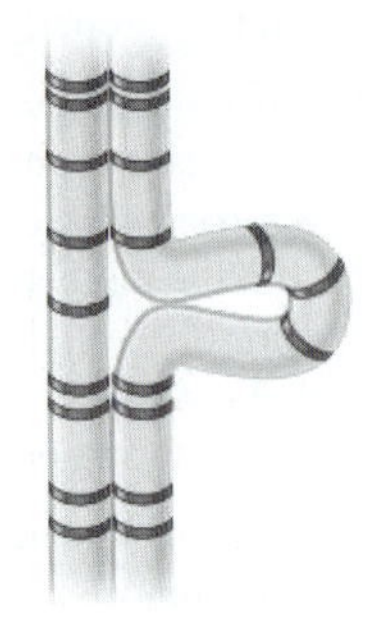

E9. The starting strain has three copies of the *bar* gene on both X chromosomes. An unequal crossover could produce chromosomes with one, two, four, or five copies, assuming that the X chromosomes are aligned over (at least) one *bar* gene. If mated to a male with three copies of the *bar* gene, the following combinations are possible:

X chromosome-1 *bar* copy + X chromosome-3 *bar* copies: *ultra-bar* heterozygote (45 facets)
X chromosome-2 *bar* copies + X chromosome-3 *bar* copies:
this fly would probably have fewer facets than the *ultra-bar* heterozygote but more facets than the *ultra-bar* homozygote, which would be somewhere between 45 and 25
X chromosome-4 *bar* copies + X chromosome-3 *bar* copies: fewer facets than the *ultra-bar* homozygote (less than 25)
X chromosome-5 *bar* copies + X chromosome-3 *bar* copies: fewer facets than the previous fly (much less than 25)

E10. The order, from most intense to least intense:

1 X chromosome-*bar,* 1 X chromosome-*ultra-bar*
1 X chromosome-*normal* and 1 X chromosome-*ultra-bar*
2 X chromosomes-*bar*
1 X chromosome-*normal* and 1 X chromosome-*bar*

It is interesting to note that the second and third genotypes of flies both have four copies of the gene. However, the fly having three copies on 1 X chromosome (1 X chromosome-*ultra-bar* and one copy on the other X chromosome) would produce more mRNA because of the position effect. That is why it has fewer facets (45) compared to a *bar* homozygote, which has 70 facets.

E11. You could cross the two species together to first create an allodiploid. The allodiploid could be treated with colchicine to make a segment of the plant an allotetraploid. As described in Figure 8.30, a cutting of the allotetraploid segment of the plant could be rooted to create a new allotetraploid plant. Alternatively, one could use cell fusion techniques as described in Figure 8.31. When you fuse two diploid cells together from two different species, you immediately create an allotetraploid.

E12. Since the plant giving the pollen is heterozygous for many genes, some of the pollen grains may be haploid for recessive alleles, which are nonbeneficial or even lethal. However, some pollen grains may inherit only the dominant (beneficial) alleles and grow quite well.

Student Discussion/Collaboration Questions

1.

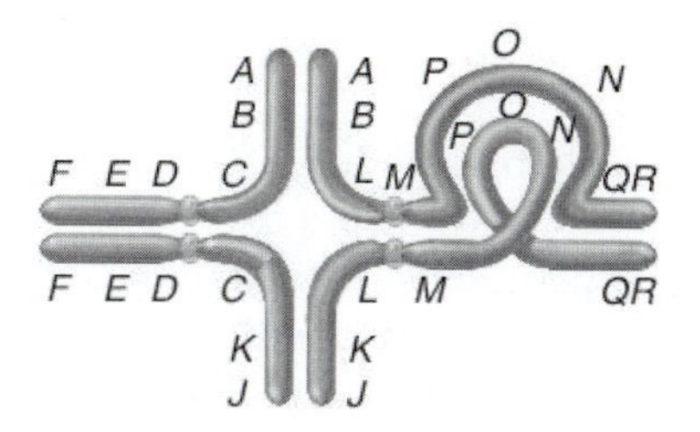

2. There are lots of possibilities. The students could look in agriculture and botany textbooks to find many examples. In the insect world, there are interesting examples of euploidy affecting sex determination. Among amphibians and reptiles, there are also several examples of closely related species that have euploid variation.
3. There are many examples: ion channels, motor proteins, transcription factors, etc. The importance usually lies in specificity. The members of the gene family are specifically used in a particular cell type, at a particular stage of development, or under a specific environmental condition.
4. 1. Polyploids are often more robust and disease resistant.
   2. Allopolyploids may have useful combinations of traits.
   3. Hybrids are often more vigorous; they can be generated from monoploids.
   4. Strains with an odd number of chromosome sets (e.g., triploids) are usually seedless.

Chapter 9: Molecular Structure of DNA and RNA

Student Learning Objectives

Upon completion of this lab you should be able to:

1. Understand the major experiments that led to the discovery of DNA as the genetic material, the experimental methods, and the investigators involved.
2. Understand the components of DNA and RNA.
3. Know the major contributors to the discovery of the structure of DNA.
4. Understand how the structure of DNA can vary, and the differences between DNA and RNA.

9.1 Identification of DNA as the Genetic Material

Overview

While today it is common knowledge that DNA is the genetic material for all living organisms, that was not always the case. Our understanding of DNA as the information storage location for cells is the result of a series of experiments conducted in the early to mid-20th century. The first part of this chapter describes those experiments that established DNA as the genetic material. As you examine these experiments, first try to place yourself in the position of the researchers and what was known at the time. Next, study how they designed their experiments to examine a specific problem. Many of these experiments are classic examples of the scientific method and still serve as model of how to design and analyze scientific experiments.

Before proceeding, examine the criteria for a genetic material on page 228. Following this, you should familiarize yourself with each of the major researchers, their experimental model, and their results. You should also be able to construct a timeline of experimental procedures that led to the discovery of DNA as the genetic material.

Key Terms

Bacteriophage
DNase
Lysis
Molecular genetics
Phage
Protease
RNase
Transformation

Focal Points

- Experimental procedures of Griffith (Figure 9.2), Avery (Figure 9.3), and Hershey and Chase (pgs. 230-234)

Exercises and Problems

For questions 1 to 7, match each statement with the researcher who contributed the information to our understanding of DNA.

a. Avery, MacLeod, and McCarty
b. Hershey and Chase
c. Griffith
d. all of the above

_____ 1. Discovered the process of transformation in bacteria.

_____ 2. Treated *Streptococcus pneumoniae* extracts with enzymes to further identify the genetic material.

_____ 3. Used *Streptococcus pneumoniae* as a model system.

_____ 4. Used radioactively labeled phages to determine if DNA or protein was the genetic material.

_____ 5. Used a bacteriophage (T2) as the model system.

_____ 6. Demonstrated that the transforming principle from the experiments with *Streptococcus pneumoniae* is DNA.

_____ 7. Demonstrated that DNA is the genetic material of bacteriophage T2.

Each of the statements below is in reference to Griffith's experiments with *Streptococcus pneumoniae*. Indicate whether each statement provides support of the genetic material having the properties of information (I), transmission (T), replication (R), or variation (V).

_____ 8. The biochemical differences in the capsule of the IIR and IIIS.

_____ 9. The copying of the genetic material within the dividing cells.

_____ 10. Instructions for type IIR and IIIS in the cell.

_____ 11. Transformation of IIR to IIIS.

For each of the following, match the chemical with its correct description.

_____ 12. Protease

_____ 13. RNase

_____ 14. ^{35}S

_____ 15. DNase

_____ 16. ^{32}P

_____ 17. RNA

a. Treating a bacterial extract with this will leave only RNA and protein.
b. Used to label DNA in the experiments of Hershey and Chase.
c. Treating a bacterial extract with this will leave only DNA and protein.
d. Used to label proteins in the experiments of Hershey and Chase.
e. An enzyme that digests proteins.
f. An alternate genetic material in some viruses.

9.2 Nucleic Acid Structure

Overview

Following the discovery of DNA as the genetic material, researchers set out to understand the structure of the molecule. This section of the chapter outlines the major contributors to this effort, as well as the structure of the DNA and RNA.

As was the case with the previous section, for each researcher you should focus on understanding not only their contribution, but also the experimental system that they utilized to make their discoveries. However, as you progress through these experiments, you need to also focus on the developing DNA molecule. As a student in a genetics class, you must comprehend the structure of DNA and the terminology that is used to describe this molecule. Discussions in later chapters of replication, transcription, and gene expression are built upon a firm understanding of the structure of DNA.

Key Terms

A DNA
Adenine
Antiparallel
B DNA
Backbone
Base pairs
Complementary
Cytosine
Deoxyribose
Directionality
Double helix
Grooves
Guanine
Major groove
Methylation
Minor groove
Nucleoside
Nucleotides
Phosphodiester linkage
Purines
Pyrimidines
Ribose
Strand
Thymine
Triplex DNA
Uracil
Z DNA

Focal Points

- Nucleotide structure (Figure 9.9)
- Structure of a DNA strand (Figure 9.11 and 9.17)
- Structure of RNA (Figure 9.22)

Exercises and Problems

The following are components of DNA structure. Match each of the following with their correct definition.

_____ 1. Nucleoside
_____ 2. Phosphodiester linkage
_____ 3. Purines
_____ 4. Deoxyribose
_____ 5. Pyrimidines
_____ 6. Nucleotides
_____ 7. Strand
_____ 8. Double helix

a. Cytosine and thymine.
b. The structural units of a DNA strand.
c. The name of the bond that connects two nucleotides in a strand.
d. Nucleotides linked by phosphodiester bonds.
e. Adenine and guanine.
f. Two interacting strands of DNA.
g. The five-carbon sugar in DNA.
h. The combination of a base and a sugar.

For questions 9 to 14, match each of the following with its correct letter from the diagram.

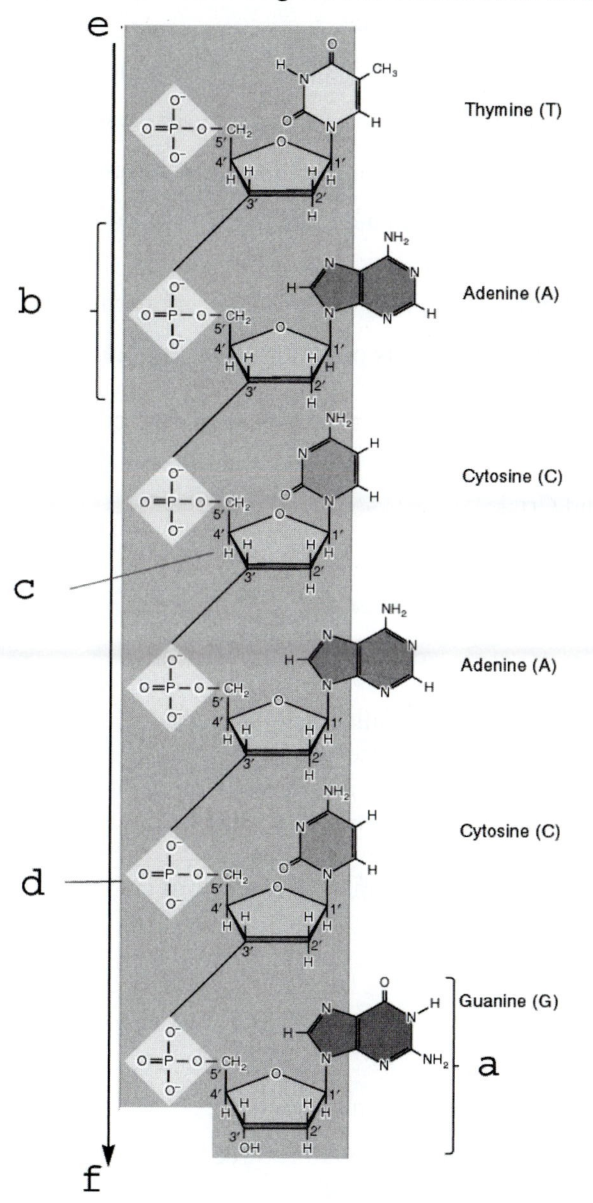

_____ 9. nucleotide

_____ 10. phosphate group

_____ 11. 5' end of the strand

_____ 12. deoxyribose sugar

_____ 13. phosphodiester linkage

_____ 14. 3' end of the strand

For each of the following, determine whether the statement is true (T) or false (F). If the statement is false, correct it so that it is true.

_____ 15. The directionality of DNA is in the 3' to 5' direction.

_____ 16. The backbone of DNA is made up of deoxyribose sugars.

_____ 17. Cytosine and thymine are examples of purine bases.

_____ 18. The two strands of DNA in a double helix are arranged in an antiparallel configuration.

_____ 19. The AT/GC rule explains the directionality properties of the DNA molecule.

_____ 20. The DNA molecule consists of major and minor grooves, to which proteins can bind.

For questions 21 to 24, match each of the researchers to their contribution in the discovery of DNA structure.

_____ 21. Watson and Crick

_____ 22. Franklin

_____ 23. Chargaff

_____ 24. Pauling

a. Studies of protein structure led to an understanding of DNA folding.
b. First to describe the AT/GC rule of base-pairing in DNA.
c. Contributed X-ray diffraction data that indicated a helical structure.
d. Developed a ball and stick model of DNA that illustrated it as a double-helix.

For questions 25 to 28, match each statement with its correct DNA form.

a. triplex DNA
b. Z DNA
c. A DNA
d. B DNA

_____ 25. The form of DNA most common in living cells.

_____ 26. Formed only in the lab and contains three DNA strands.

_____ 27. Forms under conditions of low humidity.

_____ 28. A left-handed form of DNA.

Quiz

1. The backbone of DNA is made up of ________.
   a. deoxyribose sugar
   b. purines
   c. pyrimidines
   d. phosphate groups
   e. none of the above

2. ______ used X-ray crystallography data to provide the first evidence of the three-dimensional structure of DNA.

a. Chargaff
b. Watson and Crick
c. Franklin
d. Griffith

3. In a DNA nucleotide within a DNA strand, this is the component that can vary.

a. phosphate groups
b. phosphodiester bonds
c. sugar
d. type of base
e. none of the above vary

4. This is the form of DNA most common in living cells.

a. A DNA
b. B DNA
c. triplex DNA
d. Z DNA
e. K DNA

5. ______ discovered the process of transformation in bacteria.

a. Avery and colleagues
b. Griffith
c. Chargaff
d. Pauling
e. Franklin

6. In their experiments with bacteriophage T2, Hershey and Chase labeled the DNA using ___.

a. ^{15}N
b. ^{35}S
c. ^{32}P
d. ^{3}H

7. _____ demonstrated that the genetic material in bacteriophage T2 was DNA.

a. Watson and Crick
b. Avery and colleagues
c. Franklin
d. Hershey and Chase
e. Pauling

8. The 5' to 3' configuration of DNA is called its _________ property.

a. complementary
b. AT/GC rule
c. directionality
d. antiparallel

9. These were the first individuals to demonstrate that DNA is a double-helix.
   a. Watson and Crick
   b. Pauling
   c. Franklin
   d. Chargaff

10. The structural unit of a DNA strand is the _______.
   a. nucleoside
   b. nucleotide
   c. ribose sugar
   d. purine
   e. none of the above

Answer Key for Study Guide Questions

This answer key provides the answers to the exercises and chapter quiz for this chapter. Answers in parentheses ( ) represent possible alternate answers to a problem, while answers marked with an asterisk (*) indicate that the response to the question may vary.

9.1	1. c	7. b	13. c
	2. a	8. V	14. d
	3. c	9. R	15. a
	4. b	10. I	16. b
	5. b	11. T	17. f
	6. a	12. e	

9.2	1. h	11. e	21. d
	2. c	12. c	22. c
	3. e	13. b	23. b
	4. g	14. f	24. a
	5. a	15. F, 5' to 3'	25. d
	6. b	16. F, phosphate groups	26. a
	7. d	17. F, pyrimidines	27. c
	8. f	18. T	28. b
	9. a	19. F, complementary	
	10. d	20. T	

Quiz

1. d	5. b	9. a
2. c	6. c	10. b
3. d	7. d	
4. b	8. c	

Answers to Conceptual and Experimental Questions

Conceptual Questions

C1. It is the actual substance that contains genetic information. It is usually DNA, but in some viruses it can be RNA.

C2. The transformation process is described in Chapter 6.

1. A fragment of DNA binds to the cell surface.
2. It penetrates the cell wall/cell membrane.
3. It enters the cytoplasm.
4. It recombines with the chromosome.
5. The genes within the DNA are expressed (i.e., transcription and translation).
6. The gene products create a capsule. That is, they are enzymes that synthesize a capsule using cellular molecules as building blocks.

C3. Transformation means changing from one form to another. In bacterial genetics, transformation involves the uptake of DNA into another bacterium. This may change the form (i.e., phenotype) of the bacterium. For example, transformation may change a rough bacterial strain into a smooth strain. The form or phenotype of the strain has been changed.

C4. The building blocks of a nucleotide are a sugar (ribose or deoxyribose), a nitrogenous base, and phosphate. In a nucleotide, the phosphate is already linked to the 5′ position on the sugar. When two nucleotides are hooked together, a phosphate on one nucleotide forms a covalent bond with a hydroxyl group at the 3′ position on another nucleotide.

C5. The structures can be deduced from Figures 9.8 and 9.9. Guanine is the base by itself. Guanosine is the base attached to a ribose sugar. Deoxyguanosine triphosphate is a base attached to a deoxyribose sugar with three phosphates.

C6. It is a phosphate group connecting two sugars at the 3′ and 5′ positions as shown in Figure 9.11.

C7. The bases conform to the AT/GC rule of complementarity. There are two hydrogen bonds between A and T and three hydrogen bonds between G and C. The planar rings of the bases stack on top of each other within the helical structure to provide even more stability.

C8. 3′–CCGTAATGTGATCCGGA–5′

C9. The sequence of nucleotide bases.

C10. A drawing with 10 bp per turn is like Figure 9.17 in the textbook. To make 15 bp per turn, you would have to add 5 more base pairs, but the helix should still make only one complete turn.

C11. A and B DNA are right-handed helices and the backbones are relatively helical, whereas Z DNA is left-handed and the backbone is rather zigzagged. A DNA and Z DNA have the bases tilted relative to the central axis, whereas they are perpendicular in B DNA. There are also minor differences in the number of bases per turn.

C12. The nucleotide bases occupy the major and minor grooves. Phosphate and sugar are found in the backbone. If a DNA-binding protein did not recognize a nucleotide sequence, it is probably not binding in the grooves, although it could in a nonspecific way. As an alternative, it is probably binding to the DNA backbone (i.e., sugar-phosphate sequence).

C13. DNA has deoxyribose as its sugar while RNA has ribose. DNA has the base thymine while RNA has uracil. DNA is a double helical structure. RNA is single stranded although parts of it may form double-stranded regions.

C14. The structure is shown in Figure 9.8. You begin at the carbon that is to the right of the ring oxygen and number them in a clockwise direction. Antiparallel means that the backbones are running in the opposite direction. In one strand, the sugar carbons are oriented in a 3′ to 5′ direction while in the other strand they are oriented in a 5′ to 3′ direction.

C15. Here is an example of an RNA molecule that could form a hairpin that contains 24 nucleotides in the stem and 16 nucleotides in the loop.

5′–GAUCCCUAAACGGAUCCCAGGACUCCCACGUUUAGGGAUC–3′

The complementary stem regions are underlined.

C16. Double-stranded RNA is more like A DNA than B DNA. See the text for a discussion of A-DNA structure.

C17. The sequence in part A would be more difficult to separate because it has a higher percentage of GC base pairs compared to the one in part B. GC base pairs have three hydrogen bonds compared with AT base pairs, which only have two.

C18. Its nucleotide base sequence.

C19. Complementarity is important in several ways. First, it is needed to copy genetic information. This occurs during replication, when new DNA strands are made, and during transcription, when RNA strands are made. Complementarity is also important during translation for codon/anticodon recognition. It also allows RNA molecules to form secondary structures and to recognize each other.

C20. G = 32%, C = 32%, A = 18%, T = 18%

C21. The key issue in the answer is that there are base pairing rules. Otherwise, it would not be possible to replicate the genetic material. One answer would be that the DNA is composed of double helix obeying the AT/GC rule and the third strand binds to the major groove so that X binds next to AT pairs and Y binds next to GC pairs. This would explain why the amounts of X, A, and T are approximately equal, as are the amounts of Y, G, and C. You could propose other correct scenarios.

C22. One possibility is a sequential mechanism. First, the double helix could unwind and copy itself via a semiconservative mechanism described in Chapter 11. This would produce two double helices. Next, the third strand (bound in the major groove) could copy itself via a semiconservative mechanism. This new strand could be copied to make a copy that is identical to the strand that lies in the major groove. At this point, you would have two double helices and two strands that could lie in the major groove. These could assemble to make two triple helices.

C23. The number of bases per turn is different between an RNA double helix and a DNA double helix. Also, protein binding may be affected by the structure of the sugar, which is ribose in RNA but deoxyribose in DNA.

C24. Lysines and arginines, and also polar amino acids.

C25. Both structures are helical and both are stabilized by hydrogen bonds. An α helix in proteins is a single-stranded structure; it is formed from a single polypeptide chain. A DNA double helix is formed from the interaction of two separate strands. With regard to the chemistry of the interactions that stabilize an α helix and a DNA double helix, there are some interesting similarities and differences. As already mentioned, hydrogen bonding stabilizes the α helix, but it is hydrogen bonding along the backbone; carbonyl oxygens and amide hydrogens in the polypeptide backbone interact with each other. The amino acid side chains, which project from the polypeptide backbone, may also interact favorably, but that is not a consistent feature of an α helix. In a DNA double helix, the hydrogen bonds are between bases (that project from the backbone) that are in separate strands. Base stacking also is a consistent feature that stabilizes the DNA double helix. Stacking of amino acid side chains does not occur within a single α helix.

C26. This DNA molecule contains 280 bp. There are 10 base pairs per turn, so there are 28 complete turns.

C27. They always run parallel.

C28. A hydroxyl group is at the 3′ end and a phosphate group is at the 5′ end.

C29. You would conclude that it is probably double-stranded RNA because the amount of A equals U and the amount of G equals C. Therefore, this molecule could be double stranded and obey the AU/GC rule. However, it is also possible that it is merely a coincidence that A happens to equal U and G happens to equal C and the genetic material is really single stranded.

C30. Not necessarily. The AT/GC rule is required only of double-stranded DNA molecules.

C31. There are 10^8 base pairs in this chromosome. In a double helix, a single base pair traverses about 0.34 nm, which equals 0.34×10^{-9} meters. If we multiply the two values together:

$$10^8 (0.34 \times 10^{-9}) = 0.34 \times 10^{-1} \text{ m, or } 0.034 \text{ m, or } 3.4 \text{ cm.}$$

The answer is 3.4 cm, which equals 1.3 inches! That is enormously long considering that a typical human cell is only 10 to 100 μm in diameter. As described in Chapter 10, the DNA has to be greatly compacted to fit into a living cell.

C32. The first thing we need to do is to determine how many base pairs are in this DNA molecule. The linear length of 1 base pair is 0.34 nm, which equals 0.34×10^{-9} m. One centimeter equals 10^{-2} meters.

$$\frac{10^{-2}}{0.34 \times 10^{-9}} = 2.9 \times 10^{7} \text{ bp}$$

There are approximately 2.9×10^7 bp in this DNA molecule, which equals 5.8×10^7 nucleotides. If 15% are adenine, then 15% must also be thymine. This leaves 70% for cytosine and guanine. Since cytosine and guanine bind to each other, there must be 35% cytosine and 35% guanine. If we multiply 5.8×10^7 times 0.35, we get

$$(5.8 \times 10^7)(0.35) = 2.0 \times 10^7 \text{ cytosines, or about 20 million cytosines}$$

C33. Yes, as long as there are sequences that are complementary and antiparallel to each other. It would be similar to the complementary double-stranded regions observed in RNA molecules (e.g., see Figures 9.23 and 9.24).

C34. The methyl group is not attached to one of the atoms that hydrogen bonds with guanine, so methylation would not directly affect hydrogen bonding. It could indirectly affect hydrogen bonding if it perturbed the structure of DNA. Methylation may affect gene expression because it could alter the ability of proteins to recognize DNA sequences. For example, a protein might bind into the major groove by interacting with a sequence of bases that includes one or more cytosines. If the cytosines are methylated, this may prevent a protein from binding into the major groove properly. Alternatively, methylation could enhance protein binding. In chapter 7, we considered DNA-binding proteins that were influenced by the methylation of DMRs (differentially methylated regions) that occur during genomic imprinting.

C35. Region 1 cannot form a stem-loop with region 2 and region 3 at the same time. Complementary regions of RNA form base pairs, not base triplets. The region 1/region 2 interaction would be slightly more stable than the region 1/region 3 interaction because it is one nucleotide longer, and it has a higher amount of GC base pairs. Remember that GC base pairs form three hydrogen bonds compared to AU base pairs, which form two hydrogen bonds. Therefore, helices with a higher GC content are more stable.

Experimental Questions

E1. A trait of pneumococci is the ability to synthesize a capsule. There needs to be a blueprint for this ability. The blueprint for capsule formation was being transferred from the type IIIS to the type IIR bacteria. (Note: At the molecular level, the blueprint is a group of genes that encode enzymes that can synthesize a capsule.)

E2. A. There are different possible reasons why most of the cells were not transformed.

1. Most of the cells did not take up any of the type IIIS DNA.
2. The type IIIS DNA was usually degraded after it entered the type IIR bacteria.
3. The type IIIS DNA was usually not expressed in the type IIR bacteria.

B. The antibody/centrifugation steps were used to remove the bacteria that had not been transformed. It enabled the researchers to determine the phenotype of the bacteria that had been transformed. If this step was omitted, there would have been so many colonies on the plate it would have been difficult to identify any transformed bacterial colonies, since they would have represented a very small proportion of the total number of bacterial colonies.

C. They were trying to demonstrate that it was really the DNA in their DNA extract that was the genetic material. It was possible that the extract was not entirely pure and could contain contaminating RNA or protein. However, treatment with RNase and protease did not prevent transformation, indicating that RNA and protein were not the genetic material. In contrast, treatment with DNase blocked transformation, confirming that DNA is the genetic material.

E3. 1. Isolate and purify DNA from resistant bacteria.

2. In three separate tubes, add DNase, RNase, or protease.
3. Add sensitive bacteria to each tube. A small percentage may be transformed.
4. Plate on petri plates containing tetracycline.

Expected results: Tetracycline-resistant colonies should only grow when the DNA has been exposed to RNase and protease, but not to DNase.

E4. A. There are several possible explanations why 30% of the DNA is in the supernatant. One possibility is that not all of the DNA was injected into the bacterial cells. Alternatively, some of the cells may have been broken during the shearing procedure, thereby releasing the DNA.

B. If the radioactivity in the pellet had been counted instead of the supernatant, the following figure would be produced:

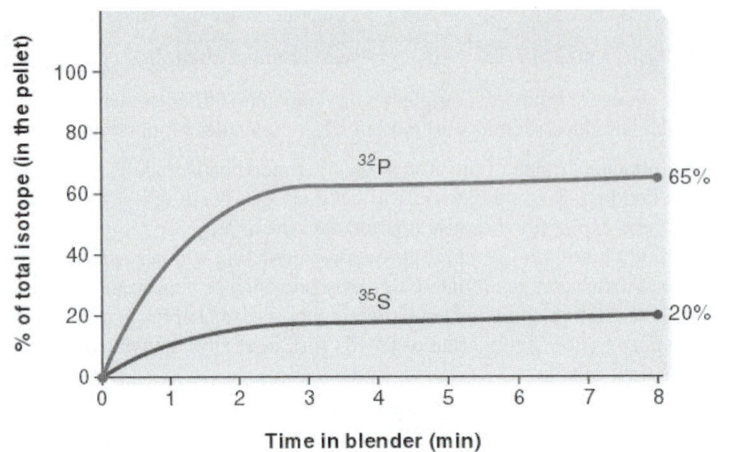

C. ^{32}P and ^{35}S were chosen as radioisotopes to label the phages because phosphorous is found in nucleic acids, while sulfur is found only in proteins.

D. There are multiple reasons why less than 100% of the phage protein is removed from the bacterial cells during the shearing process. Perhaps the shearing just is not strong enough to remove all of the phages. Perhaps the tail fibers remain embedded in the bacterium and only the head region is sheared off.

E5. It does not rule out the possibility that RNA is the genetic material because RNA and DNA both contain phosphorus. One way to distinguish RNA and DNA is to provide bacteria with radiolabeled uracil in order to label RNA or provide bacteria with radiolabeled thymine to label DNA. (Note: Uracil is found only in RNA and thymine is found only in DNA.) If they had propagated T2 phage in *E. coli* cells exposed to radiolabeled uracil, the phages would not be radiolabeled. However, if they had propagated phage in *E. coli* cells exposed to radiolabeled thymine, the T2 phages would be radiolabeled. This would indicate that T2 phages contain DNA and not RNA because radiolabeled uracil would not label the genetic material of T2 bacteriophage.

E6. This is really a matter of opinion. The Avery, MacLeod, and McCarty experiment seems to indicate directly that DNA is the genetic material since DNase prevented transformation while RNase and protease did not. However, one could argue that the DNA is required for the rough bacteria to take up some other contaminant in the DNA preparation. It would seem that the other contaminant would not be RNA or protein. The Hershey and Chase experiments indicate that DNA is being injected into bacteria, although quantitatively the results are not entirely convincing. Some ^{35}S-labeled protein was not sheared off, so the results do not definitely rule out the possibility that protein could be the genetic material. But the results do indicate that DNA is the more likely candidate.

E7. 1. You can make lots of different shapes.

2. You can move things around very quickly with a mouse.

3. You can use mathematical formula to fit things together in a systematic way.

4. Computers are very fast.

5. You can store the information you have obtained from model building in a computer file.

E8. A. The purpose of chromatography was to separate the different types of bases.

B. It was necessary to separate the bases and determine the total amount of each type of base. In a DNA strand, all the bases are found within a single molecule, so it is difficult to measure the total amount of each type of base. When the bases are removed from the strand, each type can be purified, and then the total amount of each type of base can be measured with a spectrophotometer.

C. His results would probably not be very convincing if done on a single species. The strength of his data was that all species appeared to conform to the AT/GC rule, suggesting that this is a consistent feature of DNA structure. In a single species, the observation that A = T and G = C could occur as a matter of chance.

E9. If the RNA was treated with RNase, the plants would not be expected to develop lesions. If treated with DNase or protease, lesions would still occur because RNA is the genetic material, and DNase and protease do not destroy RNA.

Questions for Student Discussion/Collaboration

1. There are lots of possibilities. The main points are that there needs to be some way for the structure of a genetic material to store information, like a sequence of some building blocks, and there needs to be some way for it to be replicated.
2. Again, there are lots of possibilities. You could use a DNA-specific chemical and show that it causes heritable mutations. Perhaps you could inject an oocyte with a piece of DNA and produce a mouse with a new trait.

Chapter 10: Chromosome Organization and Molecular Structure

Student Learning Objectives

Upon completion of the chapter you should be able to:

1. Recognize the differences between viral, bacterial, and eukaryotic genomes.
2. Understand the structure of bacterial and eukaryotic chromosomes.
3. Know the levels of organization of a eukaryotic chromosome.

10.1 Viral Genomes

Overview

Although viruses are not considered to be living organisms, they do contain genetic material. This material may consist either of DNA or RNA, and is organized into a long stretch of nucleotides that is sometimes called the viral chromosome.

The packaging of the viral genomes occurs within the host cell, and may occur by one of two mechanisms. Some viral genomes can self-assemble, while others undergo directed assembly using the assistance of noncapsid proteins.

Key Terms

Bacteriophages
Chromosomes
Directed assembly
Genome
Host cells
Host range
Self-assemble
Viral chromosome
Viral genome
Viruses

Focal Points

- Viral structure (Figure 10.1)

Exercises and Problems

For questions 1 to 6, match the term with its correct definition.

_____ 1. Host cells
_____ 2. Directed assembly
_____ 3. Host range
_____ 4. Viral chromosome
_____ 5. Self-assemble
_____ 6. Bacteriophages

a. Involves the use of non-capsid proteins.
b. The site where replication of the virus occurs.
c. A form of virus that infects bacteria.
d. A spontaneous interaction between the nucleic acids and capsid proteins.
e. The specific types of cells that a certain virus may infect.
f. May the either DNA or RNA in a virus.

10.2 Bacterial Chromosomes

Overview

Bacterial chromosomes are larger than viral chromosomes, and thus have the need for additional levels of organization and packaging. In addition to structural genes, the bacterial chromosome contains intergenic regions of DNA that contain sequences of nucleotides that regulate DNA folding, gene expression and recombination.

With regards to the packaging of genetic material in bacteria, there are two primary processes to understand. This first is the formation of loop domains (Figure 10.5) and the second is the process of supercoiling (Figure 10.7). Supercoiling exists in two forms, negative supercoiling and positive supercoiling. You should become familiar with the physical differences in these processes, as well as the influence of each on the function of the chromosome.

Key Terms

DNA gyrase	Nucleoid	Topoisomerase I
DNA supercoiling	Origin of replication	Topoisomers
Intergenic regions	Repetitive sequences	
Loop domains	Structural gene sequences	

Focal Points

- Difference between negative and positive supercoiling (Figure 10.7)

Exercises and Problems

For questions 1 to 6, match each of the following definitions with its correct term.

_____ 1. Short DNA sequences found in multiple copies in the intergenic regions.

_____ 2. The region of a bacterial cell in which compact DNA is found.

_____ 3. Nontranscribed sequences of DNA.

_____ 4. An intergenic site that serves as an initiation point for copying the chromosome.

_____ 5. The first level of DNA packaging in bacteria.

_____ 6. The second level of DNA packaging in bacteria.

a. origin of replication
b. nucleoid
c. loop domains
d. DNA supercoiling
e. intergenic regions
f. repetitive sequences

For questions 7 to 10, provide the name of the enzyme that is responsible for the process indicated.

7. The _________ enzyme introduces negative supercoiling into the bacterial chromosome.
8. The tension created by negative supercoils may be released by the __________ enzyme.
9. The effects of positive supercoils may be reduced by the ________ enzyme.
10. Coumarin and quinolones act by inhibiting the action of the ______ and other bacterial topoisomerases.

10.3 Eukaryotic Chromosomes

Overview

The final section of this chapter examines the structure of eukaryotic chromosomes. The first thing that you will notice is that the level of complexity in the eukaryotic chromosomes is significantly higher than that found in the bacterial or viral chromosomes. Eukaryotic genomes are large and frequently contain multiple chromosomes. However, this genetic material must be packaged into a relatively small area of the cell, the nucleus. The levels of compaction are a major focus of this section.

The section starts out by outlining the structure of eukaryotic chromosomes. While telomeres and centromeres have previously been introduced, a new concept is that of repetitive sequences and satellite DNA. Repetitive DNA comes in several forms, all of which are of importance to the eukaryotic genome. You should understand how researchers determine the percent of repetitive sequences in a eukaryotic genome using the C_0t curve analysis (page 264).

The chapter then moves to a discussion of how the genetic material is compacted into chromosome form in eukaryotes. Use Figure 10.21 as a guide for this section. As you proceed, frequently refer back to this figure so that you can develop a big picture perspective on how the different levels of organization interrelate.

Key Terms

30 nm fiber
Barr body
Catenanes
Centromere
Chromatin
Chromatin remodeling
Cohesion
Condensation
Condensin
Constitutive chromatin
C_0t curve
Euchromatin
Facultative chromatin
Heterochromatic
Heterochromatin
Highly repetitive sequences
Histone proteins
Kinetochore
Matrix-attachment regions
Nonrepetitive sequences
Nuclear matrix
Nucleosome
Nucleus
Radial loop domains
Satellite DNA
Scaffold
Scaffold-attachment regions
Sequence complexity
SMC proteins
Telomeres
Transposable elements
Unique sequences

Focal Points

- Understanding C_0t curves (Figure 10.13)
- Compacting the eukaryotic chromosome (Figure 10.21)

Exercises and Problems

Complete each of the following statements regarding chromosome structure.

1. The ends of eukaryotic chromosomes contain regions called the ________.
2. During mitosis and meiosis, the _______ region of the chromosome are associated with segregation.
3. The centromere acts as an attachment site for a protein complex called the ________.
4. Highly repetitive sequences that separate from other DNA during equilibrium density centrifugation is called ______ DNA.
5. The centromeres and teleomeres of a chromosome represent ________ DNA.

For questions 6 to 12, use the illustration below.

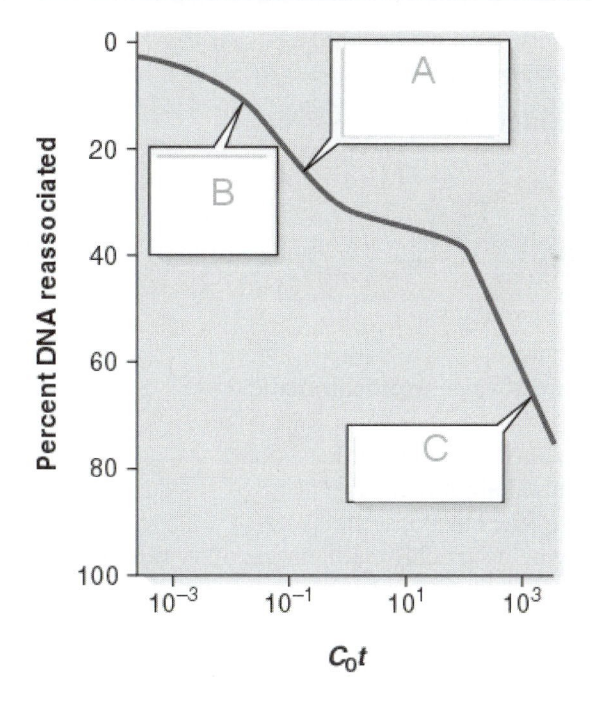

_____ 6. This area represents unique DNA.

_____ 7. This area represents highly repetitive DNA.

_____ 8. This area represents moderately repetitive DNA.

_____ 9. The sequences for transposable elements would be found in this region.

_____ 10. The sequences for rRNA would be found in this region.

_____ 11. The majority of gene sequences would be found in this region.

_____ 12. The *Alu* sequence in humans would be found here.

For questions 13 to 19, match each statement with its appropriate level of DNA organization.

a. 30 nm fiber
b. nucleosome
c. radial loop domains
d. double-helix

_____ 13. A combination of DNA and histone proteins.

_____ 14. The zig-zag model of this level is now accepted as correct.

_____ 15. These are connected to the nuclear matrix.

_____ 16. Involves the use of MARs and SARs.

_____ 17. These are connected by linker regions.

_____ 18. These are assembled into radial loop domains.

_____ 19. The only level that does not involve the use of proteins.

20. Using these same answers, arrange the terms in increasing levels of DNA compaction.

For questions 21 to 26, match each term with its correct definition.

_____ 21. Constitutive heterochromatin

_____ 22. Facultative heterochromatin

_____ 23. Euchromatin

_____ 24. Condensin

_____ 25. Cohesion

_____ 26. Catenanes

a. Chromosomal regions that are always heterochromatic
b. Intertwined DNA molecules
c. A protein that is involved in chromosome condensation
d. Responsible for sister chromatid alignment
e. The less condensed regions of the chromosome
f. Chromatin that at times may be heterochromatin and at other times euchromatin

Chapter Quiz

1. The concept of directed assembly and self assembly is associated with this type of chromosome.

a. bacterial
b. viral
c. eukaryotic
d. all of the above

2. The lowest level of DNA- protein interactions in the chromosome.
   a. 30 nm fiber
   b. radial loop domains
   c. nucleosome
   d. MARs and SARs

3. This is not used in the compaction of bacterial chromosomes.
   a. loop domains
   b. DNA supercoiling
   c. nucleosomes
   d. all of the above are used

4. In a C_0t curve, the sequences of DNA that correspond to structural genes would be found where?
   a. at the start of the curve
   b. in the middle of the curve
   c. at the end of the curve
   d. dispersed throughout the curve

5. The centromeres and telomeres are examples of ______.
   a. euchromatin
   b. constitutive chromatin
   c. facultative chromatin
   d. nucleosomes

6. Which of the following interacts directly with the nuclear matrix?
   a. nucleosomes
   b. 30 nm fibers
   c. radial loop domains
   d. histone proteins
   e. the double-helix

7. Barr bodies are an example of _________.
   a. euchromatin
   b. constitutive chromatin
   c. facultative chromatin
   d. MARs and SARs

8. The interaction of sister chromatids is the result of the action of ______.
   a. cohesion
   b. condension
   c. kinetochores
   d. telomeres
   e. nucleosomes

9. The enzyme that introduces negative supercoils into bacterial chromosomes is called ____.
   a. topoisomerase I
   b. condensin
   c. transposable elements
   d. DNA gyrase

10. Which of the following contains the largest genome?
 a. viruses
 b. bacteria
 c. eukaryotes
 d. can't be determined based on this information

Answer Key for Study Guide Questions

This answer key provides the answers to the exercises and chapter quiz for this chapter. Answers in parentheses ( ) represent possible alternate answers to a problem, while answers marked with an asterisk (*) indicate that the response to the question may vary.

10.1	1. b	3. e	5. d
	2. a	4. f	6. c
10.2	1. f	5. c	9. DNA gyrase
	2. b	6. d	10. DNA gyrase
	3. e	7. DNA gyrase	
	4. a	8. topoisomerase I	
10.3	1. telomeres	10. a	19. d
	2. centromeres	11. c	20. d-b-a-c
	3. kinetochore	12. b	21. a
	4. satellite	13. b	22. b
	5. heterochromatin	14. a	23. e
	6. c	15. c	24. c
	7. b	16. c	25. d
	8. a	17. b	26. b
	9. a	18. a	

Quiz

1. b	5. b	9. d
2. c	6. c	10. c
3. c	7. c	
4. c	8. a	

Answers to Conceptual and Experimental Questions

Conceptual Questions

C1. Self-assembly occurs spontaneously, without the aid of other proteins. Directed assembly involves the aid of proteins that are not found in the mature viral coat.

C2. Viruses also need sequences that enable them to be replicated. These sequences are equivalent to the origins of replication found in bacterial and eukaryotic chromosomes.

C3. The bacterial nucleoid is a region in a bacterial cell that contains a compacted circular chromosome. Unlike eukaryotic nuclei, a nucleoid is not surrounded by a membrane.

C4. A bacterium with two nucleoids is similar to a diploid eukaryotic cell because it would have two copies of each gene. The bacterium is different, however, with regard to alleles. A eukaryotic cell can have two different alleles for the same gene. For example, a cell from a pea plant could be heterozygous, *Tt,* for the gene that affects height. By comparison, a bacterium with two nucleoids has two identical chromosomes. Therefore, a bacterium with two nucleoids is homozygous for its chromosomal genes.

Note: As we will learn in Chapter 14, on rare occasions a bacterium can contain another piece of DNA, called an F′ factor, that can carry a few genes. The alleles on an F′ factor can be different from the alleles on the bacterial chromosome.

C5. One mechanism is DNA looping. Loops of DNA are anchored to DNA-binding proteins. Secondly, the DNA double helix is twisted further to make it more compact, much like twisting a rubber band.

C6. A. One loop is 40,000 bp. One base pair is 0.34 nm, which equals 0.34×10^{-3} μm. If we multiply the two together:

$$(40{,}000)(0.34 \times 10^{-3}) = 13.6\ \mu m$$

B. Circumference= πD

$13.6\ \mu m = \pi D$

$D = 4.3\ \mu m$

C. No, it is too big to fit inside of *E. coli.* Supercoiling is needed to make the loops more compact.

C7. DNA is a double helix. The helix is a coiled structure. Supercoiling involves additional coiling to a structure that is already a coil. Positive supercoiling is called overwinding because it adds additional twists in the same direction as the DNA double helix; it is in a right-handed direction. Negative supercoiling is in the opposite direction. Z DNA is a left-handed helix. Positive supercoiling in Z DNA is in a left-handed direction while negative supercoiling is in the right-handed direction. This is opposite to the meaning of positive and negative supercoiling in B DNA.

C8. These drugs would diminish the amount of negative supercoiling. Negative supercoiling is needed to compact the chromosomal DNA, and it also aids in strand separation. Bacteria might not be able to survive and/or transmit their chromosomes to daughter cells if their DNA was not compacted properly. Also, since negative supercoiling aids in strand separation, these drugs would make it more difficult for the DNA strands to separate. Therefore, the bacteria would have a difficult time transcribing their genes and replicating their DNA, since both processes require strand separation. As discussed in Chapter 11, DNA replication is needed to make new copies of the genetic material to transmit from mother to daughter cells. If DNA replication was inhibited, the bacteria could not grow and divide into new daughter cells. As discussed in Chapters 12–15, gene transcription is necessary for cells to make proteins. If gene transcription was inhibited, the bacteria could not make many proteins that are necessary for survival.

C9. A. The three turns would create either three fewer or three more turns for a total of seven or thirteen, respectively.

B. If the helix now has seven turns, it was left-handed. The three right-handed turns you made would cause three fewer turns in a left-handed helix. If the helix now has thirteen turns, it was right-handed. The three right-handed turns you made would add three more turns to a right-handed helix (compare Figures 10.7a and e).

C. The turning motion would probably not make supercoils because the two strings are not tightly interacting with each other. It's easy for the two strings to change the number of coils.

D. If you glued the two strings together with rubber cement, the three additional turns would probably make supercoils. A glued pair of strings is more like the DNA double helix. In a double helix, the two strands are hydrogen bonding to each other. The hydrogen bonding is like the glue. Additional turning motions tend to create supercoils rather than alter the number of coils.

C10.

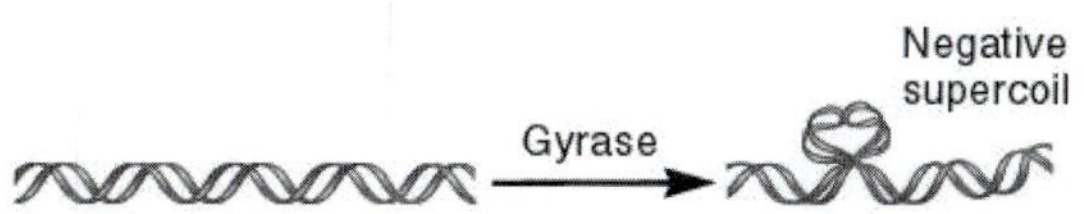

Gyrase

Negative supercoil

C11. Topoisomers are different with regard to the number of supercoils they contain. They are identical with regard to the number of base pairs in the double helix.

C12. The centromere is the attachment site for the kinetochore, which attaches to the spindle. If a chromosome is not attached to the spindle, it is free to "float around" within the cell, and it may not be near a pole when the nuclear membrane re-forms during telophase. If a chromosome is left outside of the nucleus, it is degraded during interphase. That is why the chromosome without a centromere is not likely to be found in daughter cells.

C13. Centromeres are found in eukaryotic chromosomes. They provide an attachment site for kinetochore proteins so that the chromosomes are sorted (i.e., segregated) during mitosis and meiosis. They are most important during M phase.

C14. Highly repetitive DNA, as its name suggests, is a DNA sequence that is repeated many times. It can be tandemly repeated or interspersed. Tandemly repeated DNA often has a base content that is significantly different from the rest of the chromosomal DNA so it sediments as a satellite band. In DNA renaturation studies, highly repetitive DNA renatures at a much faster rate because it is found at a higher concentration.

C15. A nucleosome is composed of double-stranded DNA wrapped 1.65 times around an octamer of histones. In the 30 nm fiber, histone H1 helps to compact the nucleosomes. The three-dimensional zigzag model is a current model that describes how this compaction occurs. It looks like a somewhat random (zigzagging) of the nucleosomes within the 30 nm fiber.

C16. During interphase (i.e., G_1, S, and G_2), the euchromatin is found primarily as a 30 nm fiber in a radial loop configuration. Most interphase chromosomes also have some heterochromatic regions where the radial loops are more highly compacted. During M phase, each chromosome becomes entirely heterochromatic. This is needed for the proper sorting of the chromosomes during nuclear division.

C17. Assuming a size of 3 billion bp, and if we assume that 146 bp wrap around a histone octamer, with 50 bp in the intervening region:

$$3,000,000,000/196 = 15,306,122, \text{ or about } 15.3 \text{ million}$$

C18.

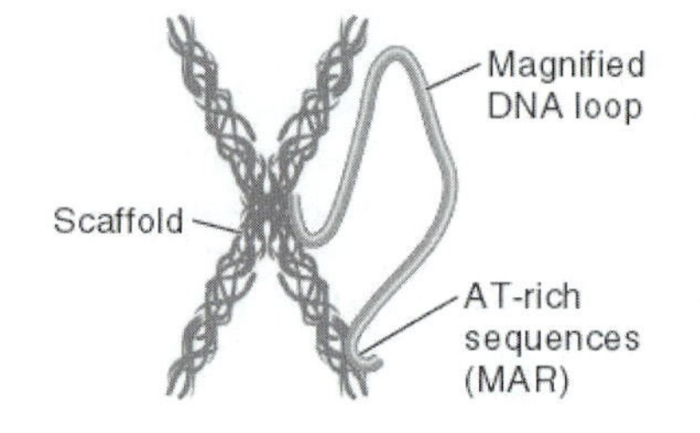

C19. Heterochromatin is more tightly packed. This is due to a greater compaction of the radial loop domains. Functionally, euchromatin can be transcribed into RNA, while heterochromatin is inactive.

C20. During interphase, the chromosomes are found within the cell nucleus. They are less tightly packed and are transcriptionally active. Segments of chromosomes are anchored to the nuclear matrix. During M phase, the chromosomes become highly condensed and the nuclear membrane is fragmented into vesicles. The chromosomal DNA remains anchored to a scaffold, formed from the nuclear matrix. The chromosomes eventually become attached to the spindle apparatus via microtubules that attach to the kinetochore, which is attached to the centromere.

C21. The main activities that can occur during interphase are transcription and DNA replication. For these activities to occur, the DNA must be in a relatively loose conformation. During M phase, there is relatively little genetic activity, although there is evidence that a few genes are transcribed. However, most genes are transcriptionally inactive during M phase.

C22. There are 146 bp around the core histones. If the linker region is 54 bp, we expect 200 bp of DNA (i.e., 146 + 54) for each nucleosome and linker region. If we divide 46,000 bp by 200 bp we get 230. Since there are two molecules of H2A for each nucleosome, there would be 460 molecules of H2A in a 46,000 bp sample of DNA.

C23. We are looking at a 30 nm fiber. This is the predominant form of DNA found in the radial loops of a cell that is in interphase.

C24. The role of the core histones is to form the nucleosomes. In a nucleosome, the DNA is wrapped 1.65 times around the core histones. Histone H1 binds to the linker region. It may play a role in compacting the DNA into a 30 nm fiber.

C25. A. There are 10^8 bp in this chromosome. In a double helix, a single nucleotide traverses about 0.34 nm, which equals 0.34×10^{-3} μm. If we multiply the two values together:

$$10^8 (0.34 \times 10^{-3}) = 0.34 \times 10^5 \ \mu\text{m, or } 34{,}000 \ \mu\text{m.}$$

B. The 30 nm fiber is about 49 times shorter than a linear double helix. (Note: The 11 nm fiber compacts the DNA about seven times. The 30 nm structure compacts the DNA an additional seven times compared to the 11 nm fiber. Therefore, compared to linear DNA, the 30 nm fiber is $7 \times 7 = 49$ times more compact.)

If we divide 34,000 μm by 49 we get 694 μm.

C. This would not fit inside the nucleus if it were stretched out in a linear manner because a typical nucleus is much smaller. However, the 30 nm fiber is very thin and is compacted into many radial loop domains.

C26. The answer is B and E. A Barr body is composed of a type of chromatin called heterochromatin. Heterochromatin is highly compacted. Euchromatin is not so compacted. A Barr body is not composed of euchromatin. The term *genome* refers to all the types of chromosomes that make up the genetic composition of an individual. A Barr body is just one chromosome, the X chromosome.

C27. During interphase, much of the chromosomal DNA is in the form of the 30 nm fiber, and some of it is more highly compacted heterochromatin. During metaphase, all of the DNA is highly compacted, as shown in Figure 10.21*d*. A high level of compaction prevents gene transcription and DNA replication from taking place. Therefore, these events occur during interphase.

Experimental Questions

E1. The second possibility in which molecule A has +4 supercoils and molecule B has –1 supercoils fits these data. Molecule A would be more compacted because it has more supercoils. Also, molecule B would be more transcriptionally active since it is more negatively supercoiled. The first possibility does not fit the data because both molecules have the same level of supercoiling so molecule A and molecule B would have the same level of compaction. The third possibility does not fit the data because molecule B would be more compact.

E2. This type of experiment gives the relative proportions of highly repetitive, moderately repetitive, and unique DNA sequences within the genome. The highly repetitive sequences renature at a fast rate, the moderately repetitive sequences renature at an intermediate rate, and the unique sequences renature at a slow rate.

E3. It affects only the rate of renaturation. Denaturation occurs because the heat breaks the hydrogen bonds between the two strands. The rate of denaturation depends on the hydrogen bonding, not on the number of copies of a sequence. The rate of renaturation, however, depends on the two complementary strands "finding" each other. The rate at which two complementary strands find each other will be faster if the concentration of the two strands is higher. That is why highly repetitive sequences renature at a faster rate.

E4. The amphibian probably has fewer structural genes than the mammal. The extra DNA is due to highly repetitive DNA sequences, which do not encode genes.

E5. Supercoiled DNA would look all curled up into a relatively compact structure. You could add different purified topoisomerases and see how they affect the structure via microscopy. For example, gyrase relaxes positive supercoils while topoisomerase I relaxes negative supercoils. If we added topoisomerase I to a DNA preparation and it became less compacted, then the DNA was negatively supercoiled.

E6.
1. The repeating nucleosome structure was revealed from DNase I digestion studies.
2. Purification studies showed that the biochemical composition is an octamer of histones.
3. More recently, crystallography has shown the precise structure of the nucleosome.
4. Microscopy has revealed information about the 30 nm fiber and the attachment of chromatin to the nuclear matrix.

In general, it is easier to understand the molecular structure of something when it forms a regular repeating pattern. The eukaryotic chromosome has a repeating pattern of nucleosomes. The bacterial chromosome seems to be more irregular in its biochemical composition.

E7. Yes, if its base composition is similar to the main chromosomal DNA.

E8.

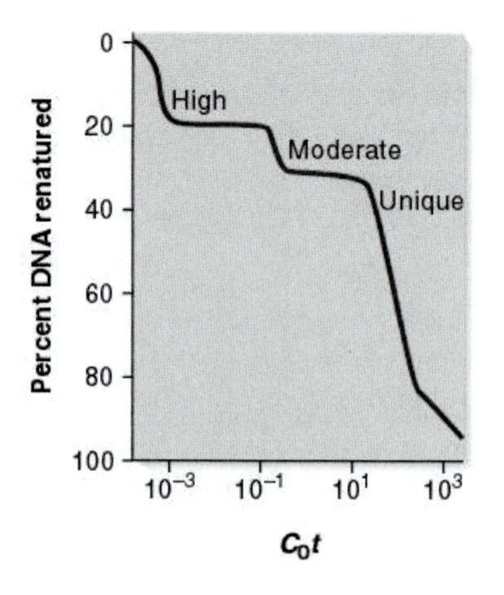

E9. A. One way is to do this by hand. You could make a series of solutions: 70% CsCl, 65% CsCl, 60% CsCl, 55% CsCl, 50% CsCl, 45% CsCl, 40% CsCl, 35% CsCl, and 30% CsCl. You could then add 1 ml of the 70% solution to the bottom, then gently layer 1 ml of the 65% solution on top, and then gently layer 1 ml of the 60% solution on top, and so on, until you finally add 1 ml of the 30% solution to the very top. This makes a step gradient; in this case, the gradient is found in 5% steps. Alternatively, some laboratories are equipped with a gradient maker. This is a machine that makes a continuous gradient. The experimenter would make a 70% CsCl solution and a 30% CsCl solution. The machine draws the 70% solution and 30% solution into a mixing chamber. After mixing, the solution is dripped into a centrifuge tube. At first, mostly the 70% solution is drawn into the mixing chamber, so the concentration of CsCl is greater near the bottom of the tube. Over time, more and more of the 30% solution (and less and less of the 70% solution) is drawn into the mixing chamber. Therefore, the solution dripped into the centrifuge tube gradually changes from a 70% solution to a 30% solution.

B. The gradient does not last forever, because diffusion eventually causes it to dissipate. Nevertheless, the gradient lasts long enough (at least many hours) so that the DNA sample can reach its equilibrium density.

C. Actually, you could add the sample anywhere because the DNA moves in the direction where the density of the DNA matches the density of the CsCl. It is usually easier to add it to the top. If you tried to add it in the middle, you might disrupt the CsCl gradient.

E10. You would expect all of the DNA in the sample to renature at a fast rate because it is a purified sample of highly repetitive DNA.

E11. With a moderate salt concentration, the nucleosome structure is still preserved so the same pattern of results would be observed. DNase-I would cut the linker region and produce fragments of DNA that would be in multiples of 200 bp. However, if a high salt concentration was used, the core histones would be lost, and DNase-I could cut anywhere. On the gel, you would see fragments of almost any size. Since there would be a continuum of fragments of many different sizes, the lane on the gel would probably look like a smear, rather than have a few prominent bands of DNA.

E12. You would get DNA fragments of about 446 to 496 bp (i.e., 146 bp plus 300 to 350).

E13. Lots of possibilities. You could digest it with DNase-I and see if it gives multiples of 200 bp or so. You could try to purify proteins from the sample and see if eukaryotic proteins or bacterial proteins are present.

E14. Histones are positively charged and DNA is negatively charged. They bind to each other by these ionic interactions. Salt is composed of positively charged ions and negatively charged ions. For example, when dissolved in water, NaCl becomes individual ions of Na^+ and Cl^-. When chromatin is exposed to a salt such as NaCl, the positively charged Na^+ ions could bind to the DNA and the negatively charged Cl^- ions could bind to the histones. This would prevent the histones and DNA from binding to each other.

E15. A. Since the *Alu* sequence is interspersed throughout all of the chromosomes, there would be many brightly colored spots along all chromosomes.

B. Only the centromeric region of the X chromosome would be brightly colored.

Questions for Student Discussion/Collaboration

1. You need to have the DNA compact to fit in the cell. Compaction allows a cell to store more information. Without this ability, organisms would not have been able to evolve to the high levels they have achieved. The downside is that cells need to copy and access this information. Therefore, cells need to make the structure looser. To achieve both goals, the structure of chromosomes is extremely dynamic.
2. This is a matter of opinion. It seems strange to have so much DNA that seems to have no obvious function. It is a waste of energy. Perhaps it has a function that we do not know about yet. On the other hand, evolution does occasionally cause bad things to accumulate within genomes, such as genes that cause diseases in the homozygous condition, etc. Perhaps this is just another example of the negative consequences of evolution.
3. The DNA structure is the same, except the degree of supercoiling. The protein compositions are very different. Bacteria compact their chromosomes by supercoiling and looping. Eukaryotes wrap their DNA around histones to form nucleosomes. The nucleosomes associate with one another to form fibers, and these fibers are supported on a scaffold.

Chapter 11: DNA Replication

Student Learning Objectives

Upon completion of this chapter you should be able to:

1. Understand the models of DNA replication and the experimental evidence that supports the semiconservative model.
2. Understand the process of baceterial DNA replication, including the function of the major molecules and regions of the DNA.
3. Know the differences between bacterial and eukaryotic DNA replication.
4. Understand the process of eukaryotic DNA replication.
5. Understand how eukaryotic cells handle the problem of replication involving the telomeres and the histone proteins.

11.1 Structural Overview of DNA Replication

Overview

Before proceeding into the mechanics of DNA replication, it is first necessary to examine some of the underlying principles that are common to both prokaryotic and eukaryotic systems. Figure 11.1 provides an excellent overview of the process, without interference from the various molecules that will be introduced in the following sections. Notice that DNA replication is semi-conservative, meaning that the daughter strands contain one original DNA strand, and one newly synthesized strand. The experimental proof of semiconservative replication was first provided by Meselson and Stahl. The experimental procedure is outlined in Figure 11.2 and Experiment 11A.

Key Terms

Conservative
Daughter strands
Dispersive
DNA replication
Parental strands
Semiconservative
Template strands

Focal Points

- The structural basis of DNA replication (Figure 11.1)
- Possible models of DNA replication (Figure 11.2)

Exercises and Problems

Match each of the following terms to its correct definition.

_____ 1. The model of replication that has segments of newly synthesized DNA and parental DNA randomly scattered throughout the new double-helices.

_____ 2. The end result of DNA replication.

_____ 3. A form of replication in which the two parental DNA strands end up in the same double-helix following replication.

_____ 4. The correct model of DNA replication.

_____ 5. The starting material for DNA replication.

a. conservative
b. semi conservative
c. template strands
d. dispersive
e. daughter strands

11.2 Bacterial DNA Replication

Overview

A significant amount of research has been conducted on the mechanism of DNA replication in bacteria, specifically the bacteria *E. coli*. The basic process is the same in both the bacterial and eukaryotic systems. The bacterial system utilizes fewer enzymes and proteins, making it the ideal model to develop an overview of the process. The differences between the two will be covered more in the next section.

A key point to focus on in this section is the interaction of the various proteins in bacterial replication. Each of the proteins has a specific role (Table 11.1) and location (Figure 11.7) during the process. If you can visualize the location of these proteins, it will be much easier to understand the process.

In addition to the discussion of DNA replication, this section introduces proof-reading and control of the replication process. One of the key aspects of control is DNA methylation, or an addition of the methyl ($-CH_3$) functional group to the DNA molecule. We have previously discussed DNA methylation in genomic imprinting and have discovered that methylation acts as a type of "on-off" switch. The same is true in the case of DNA replication. Finally, this section examines the methods by which researchers have been able to study replication in prokaryotic systems through the use of temperature sensitive mutations.

Key Terms

Bidirectional replication
Bidirectionally
Catenanes
Conditional mutants
Dam methylase
Dimeric DNA polymerase
DNA gyrase
DNA helicase
DNA primase
DnaA box sequences
DnaA proteins
Exonuclease
Fidelity
Lagging strand
Leading strand
Okazaki fragments
Origin of replication
Primase
Primosome
Processive enzyme
Proofreading function
Replisome
RNA primer
Single-strand binding protein
Temperature-sensitive mutant
Termination sequences
Topoisomerase type II

Focal Points

- The enzyme of bacterial replication (Figure 11.7)
- Proteins involved in bacterial replication (Table 11.1)

Exercises and Problems

For questions 1 to 6, indicate whether the statement applies to the initiation of replication (I), synthesis of the new strand (S), or termination of replication (T).

_____ 1. Following this, the DNA strands may form a catenane structure.

_____ 2. Polymerases covalently attach nucleotides together.

_____ 3. DnaA proteins bind to the *oriC* region.

_____ 4. DNA helicases begin strand separation.

_____ 5. The polymerase reaches the *ter* sequence opposite of *oriC.*

_____ 6. Okazaki fragments allow synthesis of the lagging strands.

For questions 7 to 15, match each of the following to its correct letter on the diagram. Some answers may be used more than once.

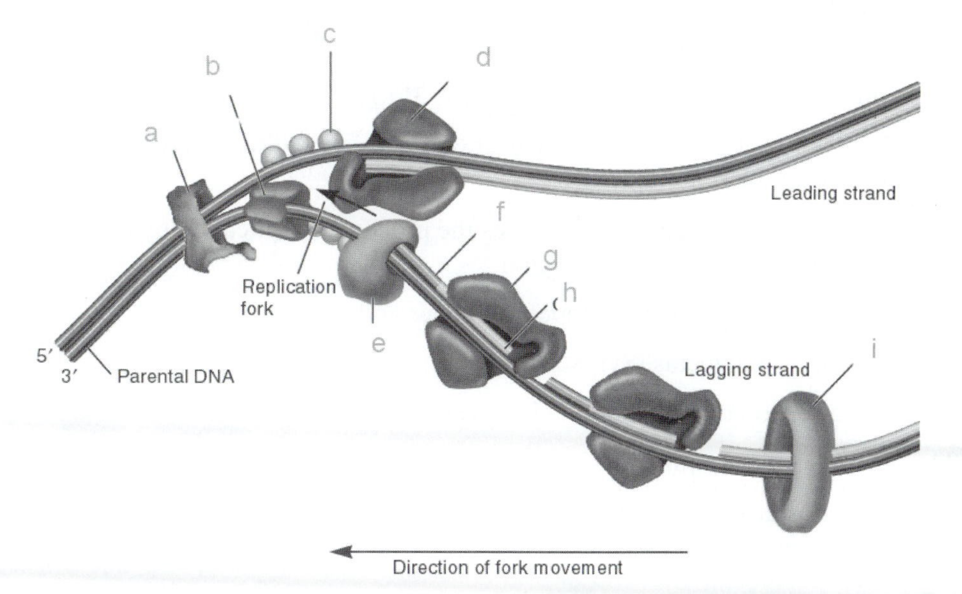

_____ 7. Catalyzes a bond between DNA fragments on the lagging strand.

_____ 8. These prevent the DNA from recoiling after the action of the helicase.

_____ 9. This molecule removes the supercoiling ahead of the replication fork.

_____ 10. This molecule forms short RNA primers that start replication.

_____ 11. Unwraps the double-helix ahead of the replication fork.

_____ 12. The protein that is synthesizing the new daughter strand from the leading strand.

_____ 13. DNA fragments that allow replication of the lagging strand.

_____ 14. The RNA primers formed by the primase enzyme.

_____ 15. This molecule is using the Okazaki fragments to replicate the lagging strand.

Match the definition in questions 16 to 25 to their correct term.

_____ 16. A processive enzyme.

_____ 17. A combination of the helicase and primase enzymes.

_____ 18. This corrects errors in replication in the 3' to 5' direction.

_____ 19. The DNA polymerase synthesizes this strand in the direction opposite of the replication fork.

_____ 20. The action of this enzyme regulates the process of replication by adding a $-CH_3$ funtional group to the *oriC* region.

_____ 21. This was used to identify important genes and sequences in DNA replication.

_____ 22. The DNA polymerase synthesizes this strand in the direction of the replication fork.

_____ 23. The direction of replication in bacteria.

_____ 24. The site where replication begins in bacteria.

_____ 25. A primosome that has bound to two polymerases.

a. lagging strand
b. leading strand
c. temperature-sensitive mutations
d. DNA polymerase
e. proofreading function
f. Dam methylase
g. primosome
h. bidirectional replication
i. origin of replication
j. replisome

11.3 Eukaryotic DNA Replication

Overview

As noted in the previous section, more is known of bacterial replication than eukaryotic replication. While there are some important differences between eukaryotic and prokaryotic replication, the process is fundamentally the same. This section examines how the size of the eukaryotic genome results in multiple replication initiation sites, the problems facing compaction of DNA into nucleosomes and the different forms of polymerases. The last section of the chapter examines how the ends of the chromosomes, or the telomeres, are replicated. Currently there is an interest in this area due to the possible relationship of the telomeres to aging and cancer.

Key Terms

ARS elements
Base excision repair
Lesion replicating polymerase
Origin recognition complex
Polymerase switch
Telomerase

Focal Points

- Action of the telomerase (Figures 11.24 and 11.25)

Exercises and Problems

For each of the following, match each definition to its correct term.

_____ 1. The enzyme that prevents shortening of the chromosomes during replication.

_____ 2. A special form of polymerase that allows replication to occur over damaged regions.

_____ 3. These regions have a high percentage of A and T, are present in multiple copies in the genome, and are involved in the initiation of eukaryotic replication.

_____ 4. This allows the initiation of replication to be conducted by one polymerase, while elongation of the same strand is completed by a second polymerase.

_____ 5. The initiator of eukaryotic replication.

a. ARS elements
b. polymerase switching
c. telomerase
d. lesion-replicating polymerases
e. origin recognition complex

Chapter Quiz.

1. Which of the following molecules is responsible for synthesizing the daughter DNA strands during bacterial replication?
   a. DNA ligase
   b. DNA primase
   c. DNA helicase
   d. the primosome
   e. none of the above

2. Replication in bacterial occurs in what direction along the chromosome?
   a. in a clockwise direction
   b. in a counter-clockwise direction
   c. bidirectionally
   d. randomly

3. DNA is replicated according to which of the following models?
   a. dispersive
   b. conservative
   c. destructing
   d. semi-conservative

4. The name of the site of replication initiation in bacteria is _____.

a. ARS
b. oriC
c. ORC
d. Okazaki-1
e. none of the above

5. During bacterial replication, what prevents the newly separated DNA strands from reforming a helix?

a. the topoisomerase
b. the ligase
c. the helicase
d. the polymerase
e. the single-stranded binding proteins

6. Which of the following does not occur in bacterial replication?

a. initiation of replication at multiple sites
b. formation of new nucleosomes
c. replication of the telomeres
d. a polymerase switch
e. all of the above are correct

7. Dam methylase allows for what process?

a. initiation of replication
b. synthesis of RNA primers
c. regulation of replication
d. proofreading
e. all of the above

8. The _____ of the DNA indicates that relatively few errors are made during replication.

a. accuracy
b. integrity
c. fidelity
d. honestly
e. faithfulness

9. The leading strand of DNA is replicated in the direction of movement of the replication fork.

a. True
b. False

10. The primosome contains which of the following?

a. helicase and primase
b. primase and polymerase
c. polymerase and ligase
d. helicase and topoisomerase
e. topoisomerase and Okazaki fragments

Answer Key for Study Guide Questions

This answer key provides the answers to the exercises and chapter quiz for this chapter. Answers in parentheses ( ) represent possible alternate answers to a problem, while answers marked with an asterisk (*) indicate that the response to the question may vary.

11.1
1. d
2. e
3. a
4. b
5. c

11.2
1. T
2. S
3. I
4. I
5. T
6. S
7. i
8. c
9. a
10. e
11. b
12. d
13. h
14. f
15. g
16. d
17. g
18. e
19. a
20. f
21. c
22. b
23. h
24. i
25. j

11.3
1. c
2. d
3. a
4. b
5. e

Quiz

1. e
2. c
3. d
4. b
5. e
6. e
7. c
8. c
9. a
10. a

Answers to Conceptual and Experimental Questions

Conceptual Questions

C1. It is a double-stranded structure that follows the AT/GC rule.

C2. Bidirectionality refers to the idea that two replication forks emanate from one origin of replication. There are four DNA strands being made in two directions radiating outward from the origin.

C3. Statement C is not true. A new strand is always made from a preexisting template strand. Therefore, a double helix always contains one strand that is older than the other.

C4. A.

TTGGHTGUTGG
HHUUTHUGHUU

B.

TTGGHTGUTGG
HHUUTHUGHUU

↓

TTGGHTGUTGG CCAAACACCAA
AACCCACAACC HHUUTHUGHUU

↓

TTGGHTGUTGG TTGGGTGTTGG CCAAACACCAA CCAAACACCAA
AACCCACAACC AACCCACAACC GGTTTGTGGTT HHUUTHUGHUU

C5. No. In a conservative mechanism, one double helix would always be fully methylated so the cell would not have any way to delay the next round of DNA replication via a methylation mechanism.

C6. If we assume there are 4,600,000 bp of DNA, and that DNA replication is bidirectional at a rate of 750 nucleotides per second:

If there were just a single replication fork:

$$4{,}600{,}000/750 = 6{,}133 \text{ seconds, or } 102.2 \text{ minutes}$$

Because replication is bidirectional: 102.2/2 = 51.1 minutes

Actually, this is an average value based on a variety of growth conditions. Under optimal growth conditions, replication can occur substantially faster.

With regard to errors, if we assume an error rate of one mistake per 100,000,000 nucleotides:

$$4{,}600{,}000 \times 1{,}000 \text{ bacteria} = 4{,}600{,}000{,}000 \text{ nucleotides of replicated DNA}$$

$$4{,}600{,}000{,}000/100{,}000{,}000 = 46 \text{ mistakes}$$

When you think about it, this is pretty amazing. In this population, DNA polymerase would cause only 46 single mistakes in a total of 1,000 bacteria, each containing 4.6 million bp of DNA.

C7.

```
5′—————————————DNA POLYMERASE———>3′
3′———————————————————————————————————————————————————————5′
                      Template strand
```

C8. DNA polymerase would slide from right to left. The new strand would be

3′–CTAGGGCTAGGCGTATGTAAATGGTCTAGTGGTGG–5′

C9. 1. DnaA boxes—binding sites for the DnaA protein

2. Methylation sites—sites of adenine methylation that are important for regulating DNA replication

3. AT-rich region—site where the DNA initially denatures to form a replication bubble

C10. A. When looking at Figure 11.5, the first and third DnaA boxes are running in the same direction and the second and fourth DnaA boxes are running in the opposite direction. Once you realize that, you can see that the sequences are very similar to each other.

B. According to the direction of the first DnaA box, the consensus sequence is

TGTGGATAA
ACACCTATT

C. This sequence is nine nucleotides long. Because there are four kinds of nucleotides (i.e., A, T, G, and C), the chance of this sequence occurring by random chance is 4^{-9}, which equals once every 262,144 nucleotides. Because the *E. coli* chromosome is more than 10 times longer than this, it is fairly likely that this consensus sequence occurs elsewhere. The reason why there are not multiple origins, however, is because the origin has four copies of the consensus sequence very close together. The chances of having four copies of this consensus sequence occurring close together (as a matter of random chance) are very small.

C11. A. Your hand should be sliding along the white string. The free end of the white string is the 5′ end, and DNA helicase travels in the 5′ to 3′ direction.

B. The white string should be looped by your right hand. The black string is the template strand for the synthesis of the leading strand. DNA polymerase moves directly toward the replication fork to synthesize the leading strand. The white string is the template for the lagging strand. The template DNA for the lagging strand must be looped around DNA polymerase III so it can move toward the replication fork.

C12. First, according to the AT/GC rule, a pyrimidine always hydrogen bonds with a purine. A transition still involves a pyrimidine hydrogen bonding to a purine, but a transversion causes a purine to hydrogen bond with a purine or a pyrimidine to hydrogen bond with a pyrimidine. The structure of the double helix makes it much more difficult for this later type of hydrogen bonding to occur. Second, the induced-fit phenomenon of the catalytic site of DNA polymerase makes it unlikely for DNA polymerase to catalyze covalent bond formation if the wrong nucleotide is bound to the template strand. A transition mutation creates a somewhat bad interaction between the bases in opposite strands, but it is not as bad as the fit caused by a transversion mutation. In a transversion, a purine is opposite another purine or a pyrimidine is opposite a pyrimidine. This is a very bad fit. And finally, the proofreading function of DNA polymerase is able to detect and remove an incorrect nucleotide that has been incorporated into the growing strand. A transversion is going to cause a larger distortion in the structure of the double helix and make it more likely to be detected by the proofreading function of DNA polymerase.

C13. A primer is needed to make each Okazaki fragment. The average length of an Okazaki fragment is 1,000 to 2,000 bp. If we use an average value of 1,500 bp for each Okazaki fragment, then there needs to be approximately

$$\frac{4,600,000}{1,500} = 3,067 \text{ copies}$$

C14. Primase and DNA polymerase are able to knock the single-strand binding proteins off the template DNA.

C15. A. The removal of RNA primers occurs in the 5′ to 3′ direction, while the proofreading function occurs in the 3′ to 5′ direction.

B. No. The removal of RNA primers occurs from the 5′ end of the strand.

C16. A. The right Okazaki fragment was made first. It is farthest away from the replication fork. The fork (not seen in this diagram) would be to the left of the three Okazaki fragments, and moving from right to left.

B. The RNA primer in the right Okazaki fragment would be removed first. DNA polymerase would begin by elongating the DNA strand of the middle Okazaki fragment and remove the right RNA primer with its 5′ to 3′ exonuclease activity. DNA polymerase I would use the 3′ end of the DNA of the middle Okazaki fragment as a primer to synthesize DNA in the region where the right RNA primer is removed. If the middle fragment was not present, DNA polymerase could not fill in this DNA (because it needs a primer).

C. You only need DNA ligase at the right arrow. DNA polymerase I begins at the end of the left Okazaki fragment and synthesizes DNA to fill in the region as it removes the middle RNA primer. At the left arrow, DNA polymerase I is simply extending the length of the left Okazaki fragment. No ligase is needed here. When DNA polymerase I has extended the left Okazaki fragment through the entire region where the RNA primer has been removed, it hits the DNA of the middle Okazaki fragment. This occurs at the right arrow. At this point, the DNA of the middle Okazaki fragment has a 5′ end that is a monophosphate. DNA ligase is needed to connect this monophosphate with the 3′ end of the region where the middle RNA primer has been removed.

D. As mentioned in the answer to part C, the 5′ end of the DNA in the middle Okazaki fragment is a monophosphate. It is a monophosphate because it was previously connected to the RNA primer by a phosphoester bond. At the location of the right arrow, there was only one phosphate connecting this deoxyribonucleotide to the last ribonucleotide in the RNA primer. For DNA polymerase to function, the energy to connect two nucleotides comes from the hydrolysis of the incoming triphosphate. In this location shown at the right arrow, however, the nucleotide is already present at the 5′ end of the DNA, and it is a monophosphate. DNA ligase needs energy to connect this nucleotide with the left Okazaki fragment. It obtains energy from the hydrolysis of ATP.

C17. DNA methylation is the covalent attachment of methyl groups to bases in the DNA. Immediately after replication, there has not been sufficient time to attach methyl groups to the bases in the newly made daughter strand. The time delay of DNA methylation helps to prevent premature DNA replication immediately after cell division.

C18. 1. It recognizes the origin of replication.

2. It initiates the formation of a replication bubble.

3. It recruits helicase to the region.

C19. It would be difficult to delay DNA replication after cell division because the dilution of the DnaA protein is one mechanism that regulates replication. One might expect that such a strain would have more copies of the bacterial chromosome per cell compared to a normal strain.

C20. The picture would depict a ring of helicase proteins traveling along a DNA strand and breaking the hydrogen bonding between the two helices, as shown in Figure 11.6.

C21. An Okazaki fragment is a short segment of newly made DNA in the lagging strand. It is necessary to make short fragments because the fork is exposing the lagging strand in a 5′ to 3′ direction but DNA polymerase can slide along a template strand in a 3′ to 5′ direction. Therefore, the newly made lagging strand is synthesized in short pieces in the direction away from the replication fork.

C22. The leading strand is primed once, at the origin, and then DNA polymerase III makes it continuously in the direction of the replication fork. In the lagging strand, many short pieces of DNA are made. This requires many RNA primers and DNA polIII. The primers are removed by polI, which then fills in the gaps with DNA. DNA ligase then covalently connects the Okazaki fragments together. Having the enzymes within a complex provides coordination among the different steps in the replication process and thereby allows it to proceed more efficiently and faster.

C23. The active site of DNA polymerase has the ability to recognize a distortion in the newly made strand and remove it. This occurs by a 3′ to 5′ exonuclease activity. After the mistake is removed, DNA polymerase resumes DNA synthesis.

C24. A processive enzyme is one that remains clamped to one of its substrates. In the case of DNA polymerase, it remains clamped to the template strand as it makes a new daughter strand. This is important to ensure a fast rate of DNA synthesis.

C25. Nucleosomes are made during the S phase. The original histones become bound to each of the two DNA double helices in a random manner. The newly made histones then bind to regions where histones are missing. Following DNA replication, the nucleosomes in both double helices are a random mixture of newly made histone octamers and original histone octamers.

C26. It is necessary to have the correct number of chromosomes per cell. If DNA replication is too slow, daughter cells may not receive any chromosomes. If it is too fast, they may receive too many chromosomes.

C27. The inability to synthesize DNA in the 3′ to 5′ direction and the need for a primer necessitate the action of telomerase. Telomerase is different in that it uses a short RNA sequence, which is part of its structure, as a template for DNA synthesis. Since it uses this sequence many times in row, it produces a tandemly repeated sequence in the telomere.

C28. The opposite strand is made in the conventional way by DNA polymerase using the "telomerase added strand" as a template.

C29. Fifty, because two replication forks emanate from each origin of replication. DNA replication is bidirectional.

C30. The ends labeled *B* and *C* could not be replicated by DNA polymerase. DNA polymerase makes a strand in the 5′ to 3′ direction using a template strand that is running in the 3′ to 5′ direction. Also, DNA polymerase requires a primer. At the ends labeled *B* and *C*, there is no place (upstream) for a primer to be made.

C31. A. Both reverse transcriptase and telomerase use an RNA template to make a complementary strand of DNA.

B. Since reverse transcriptase does not have a proofreading function, it makes it more likely for mistakes to occur. This creates many mutant strains of the virus. Some mutations might prevent the virus from proliferating. However, other mutations might prevent the immune system from battling the virus. These kinds of mutations would enhance the proliferation of the virus.

C32. As shown in Figure 11.23, the first step involves a binding of telomerase to the telomere. The 3′ overhang binds to the complementary RNA in telomerase. For this reason, a 3′ overhang is necessary for telomerase to replicate the telomere.

Experimental Questions

E1. A. Four generations: 7/8 light, 1/8 half-heavy
Five generations: 15/16 light, 1/16 half-heavy

B. All of the DNA double helices would be 1/8 heavy.

C. The CsCl gradient separates molecules according to their densities. ^{14}N-containing compounds have a lighter density compared to ^{15}N-containing compounds. The bases of DNA contain nitrogen. If the bases contain only ^{15}N, the DNA will be heavy; it will sediment at a higher density. If the bases contain only ^{14}N, the DNA will be light; it will sediment at a lower density. If the bases in one DNA strand contain ^{14}N and the bases in the opposite strand contain ^{15}N, the DNA will be half-heavy; it will sediment at an intermediate density.

E2. A. You would probably still see a band of DNA, but you would only see a heavy band.

B. You would probably not see a band because the DNA would not be released from the bacteria. The bacteria would sediment to the bottom of the tube.

C. You would not see a band. Ethidium bromide stains the DNA and you are actually seeing the ethidium bromide when the gradient is exposed to UV light.

E3. You might be able to determine the number of replication forks and their approximate locations. For chromosomes with a single origin, you can determine that replication is bidirectional. However, you do not get any molecular information about the DNA replication process.

E4. You would need to add a primer (or primase), dNTPs, and DNA polymerase. If the DNA were double stranded, you would also need helicase. Adding single-strand binding protein and topoisomerase may also help.

E5. You would need to add a DNA template with a 3′ overhang that was complementary to the telomere sequence. You would also have to add telomerase and dNTPs.

E6. This is a critical step because you need to separate the radioactivity in the free nucleotides from the radioactivity in the newly made DNA strands. If you used an acid that precipitated free nucleotides and DNA strands, all of the radioactivity would be in the pellet. You would get the same amount of radioactivity in the pellet no matter how much DNA was synthesized into newly made strands. For this reason, the perchloric step is very critical. It separates radioactivity in the free nucleotides from radioactivity in the newly made strands.

E7. No, because the hydrolysis of the deoxynucleoside triphosphates provides the energy for the synthesis of new strands.

E8. A. The left end is the 5′ end. If you flip the sequence of the first primer around, you will notice that it is complementary to the right end of the template DNA. The 5′ end of the first primer binds to the 3′ end of the template DNA.

B. The sequence would be 3′–CGGGGCCATG–5′. It could not be used because the 3′ end of the primer is at the end of the template DNA. There wouldn't be any place for nucleotides to be added to the 3′ end of the primer and bind to the template DNA strand.

E9. For a DNA strand to grow, a phosphoester bond is formed between the 3′ —OH group on one nucleotide and phosphate group on the incoming nucleotide (see Figure 11.10). If the —OH group is missing, a phosphoester bond cannot form.

E10. A. Heat is used to separate the DNA strands, so you do not need helicase.

B. Each primer must be a sequence that is complementary to one of the DNA strands. There are two types of primers, and each type binds to one of the two complementary strands. The arrowheads are at the 3' end of the primers.

C. A termophilic DNA polymerase is used because DNA polymerases isolated from nonthermophilic species would be permanently inactivated during the heating phase of the PCR cycle. Remember that DNA polymerase is a protein, and most proteins are denaturated by heating. However, proteins from thermophilic organisms have evolved to withstand heat. That is how thermophilic organisms survive at high temperatures.

D. With each cycle, the amount of DNA is doubled. Since there are initially 10 copies of the DNA, there will be 10×2^{27} copies after 27 cycles. $10 \times 2^{27} = 1.34 \times 10^{9}$ = 1.34 billion copies of DNA. As you can see, PCR can amplify the amount of DNA by a staggering amount!

Questions for Student Discussion/Collaboration

1. Lots of possibilities. The main point is the idea of chemical recognition. There must be strict rules about how two structures recognize each other.
2. Basically, the idea is to add certain combinations of enzymes and see what happens. You could add helicase and primase to double-stranded DNA along with radiolabeled ribonucleotides. Under these conditions, you would make short (radiolabeled) primers, and if you added DNA polymerase and deoxyribonucleotides, you would also make DNA strands. Alternatively, if you added DNA polymerase plus an RNA primer, you would not need to add primase. Or if you used single-stranded DNA as a template, rather than double-stranded DNA, you would not need to add helicase.
3. This is a matter of opinion. It needs to be fast so that life can occur on the time scale that we are accustomed to. The error-free issue is very important. Since the DNA stores information, it is important that the information be maintained with high fidelity. Otherwise, its storage ability is greatly comprised. Finally, the regulation of DNA replication is necessary to ensure the proper of amount of DNA per cell.

Chapter 12: Gene Transcription and RNA Modification

Student Learning Objectives

Upon completion of this lab you should be able to:

1. Understand the role of transcription in gene expression.
2. Know the process of initiation, elongation, and termination in bacterial cells.
3. Understand how eukaryotic transcription differs from that of bacterial transcription.
4. Understand the importance of regulatory elements and transcription factors to eukaryotic transcription.
5. Understand the processes of RNA modification and why each of these processes is needed in a eukaryotic cell.

12.1 Overview of Transcription

Overview

The next two chapters in the text will examine the process by which the genetic information is interpreted to form a functional protein. This is a two stage process. In this chapter we will examine the biochemical process of transcription. The basic purpose of transcription is to make a copy of the protein-building instructions of a gene. The next chapter will examine the process of translation, in which the instructions contained within the copy are used to construct a functional protein.

This initial section of the chapter serves as an important overview to transcription. The terms and processes described within this section are used for both prokaryotic and eukaryotic transcription. You should master these terms before proceeding to the next sections.

Key Terms

Closed promoter complex
Coding Strand
Codons
Elongation
Initiation
Messenger RNA
Open complex
Open promoter complex
Promoter
Regulatory sequences
Ribosomal-binding site
Start codon
Stop codon
Structural gene
Synthesis
Template strand
Termination
Terminator
Transcription
Transcriptional unit

Focal Points

- Organization of a bacterial gene (Figure 12.1)
- Overview of transcription (Figure 12.2)
- Forms of RNA (Table 12.1)

Exercises and Problems

For questions 1 to 7, match the term to its correct definition.

_____ 1. Terminator

_____ 2. Codons

_____ 3. Regulatory sequences

_____ 4. Promoter

_____ 5. Template stand

_____ 6. Gene

_____ 7. Structural genes

a. Binding sites for proteins that control transcription.
b. Contains genetic information that is complementary to the RNA transcript.
c. Groups of three nucleotides within the mRNA.
d. The site that controls the start of transcription.
e. The site that controls the stop of transcription.
f. A transcriptional unit that can be transcribed into RNA.
g. Produces mRNA when transcribed.

For questions 8 to 11, indicate whether the statement is associated with the initiation (I), elongation (E), or termination of transcription (T).

_____ 8. The RNA polymerase synthesizes a mRNA transcript.

_____ 9. The RNA polymerase interacts with the promoter.

_____ 10. The DNA forms an open complex.

_____ 11. The RNA polymerase and mRNA dissociate from the DNA.

For each of the following statements, indicate the type of RNA from Table 12.1 that is best matches the statement.

12. Targets the proteins to the endoplasmic reticulum of eukaryotes.

13. Small nuclear RNA that is involved in the splicing of eukaryotic mRNA.

14. Contains the instructions for a protein sequence.

15. Cytoplasmic RNA of prokaryotes that is used in protein secretion.

16. A component of the ribosome.

17. Transfer RNA.

12.2 Transcription in Bacteria

Overview

The section of the chapter outlines the process of transcription in bacteria. As was the case with replication, it is better to understand the simpler prokaryotic transcription system before proceeding to the eukaryotes. Before entering into the discussion of initiation, elongation, and termination, the chapter takes a good look at the structure of a prokaryotic promoter. A promoter is a regulatory sequence that plays a central role in transcriptional regulation. There are two things that you should focus on with regards to the promoter. The first is the numbering of nucleotides (Figure 12.3) and the second is the concept of a consensus sequence (Figure 12.4). Notice that not all promoters have the same sequence of nucleotides, although they are very similar.

The best way to study initiation, elongation and termination is to first read the sections of the text, and then study the appropriate diagram. Remember that these are three-dimensional systems, and it is a complex interaction of proteins and nucleic acids (both DNA and RNA) that makes the process possible.

Key Terms

Closed complex
Coding strand
Consensus sequence
Core enzyme
Helix-turn-helix structure
Holoenzyme
Intrinsic terminators
Noncoding strand
Pribnow box
Rho
Rho-dependent termination
Rho-independent termination
RNA polymerase
Sigma factor
Template strand
Transcription factors

Focal Points

- Numbering of nucleotides (Figure 12.3)
- Synthesis of the RNA transcript (Figure 12.7)
- Termination (Figures 12.8 and 12.9)

Exercises and Problems

For questions 1 to 6, match the term with its correct definition.

_____ 1. Helix-turn-helix

_____ 2. Transcription factors

_____ 3. RNA polymerase

_____ 4. Holoenzyme

_____ 5. Closed complex

_____ 6. Promoter

a. An RNA polymerase that has sigma factor attached.
b. The region of a gene between -10 and -35 that contains the Pribnow box.
c. Occurs when the holoenzyme is bound to the promoter region.
d. Proteins that influence the relationship of the holoenzyme with the promoter.
e. A four subunit molecule that catalyzes the synthesis of RNA.
f. A region of the sigma factor that facilitates binding to the DNA.

For questions 7 to 14 indicate whether the statement is associated with initiation (I), elongation (E), or termination of bacterial transcription.

_____ 7. Disrupts the hydrogen bonding between the DNA and RNA.

_____ 8. Starts with the formation of a closed complex.

_____ 9. May occur either dependently or independently of the rho-protein.

_____ 10. Actively uses the template strand for RNA synthesis.

_____ 11. Unraveling of the DNA occurs in AT-rich regions

_____ 12. The RNA polymerase moves along the DNA.

_____ 13. RNA is synthesized in a 5' to 3' direction.

_____ 14. The sigma factor within the holoenzyme binds to the DNA.

12.3 Transcription in Eukaryotes

Overview

Transcription in eukaryotes is slightly more complicated than in prokaryotes, although the principles remain the same. During initiation, the RNA polymerase must still bind to the promoter and form an open complex. However, unlike the bacterial systems, eukaryotic systems utilize a variety of proteins, called transcription factors, which regulate the binding of the polymerase to the promoter. The interaction of the transcription factors can be complex (Figure 12.12) and allows for a fine tuning of the transcription process. In addition to transcription factors, the structure of the chromatin (Chapter 11) also has an influence on the transcription process. As you proceed through this section, focus on these interactions.

Key Terms

ATP-dependent chromatin remodeling
Basal transcription
Basal transcription apparatus
cis-acting elements
Core promoter
Enhancers
General transcription factors
Histone acetyltransferases
Histone deactylases
Mediator
Preinitiation complex
Regulatory elements
Regulatory transcription factors
Silencers
SWI/SNF family
TATA box
trans-acting elements
Transcriptional start site

Focal Points

- Promoter regions of a eukaryotic gene (Figure 12.11)
- Interactions of transcription factors and RNA polymerase in initiation (Figure 12.12)

Exercises and Problems

For questions 1 to 7, match each of the components of a eukaryotic structural gene to its correct definition.

_____ 1. *trans*-acting factors

_____ 2. TATA box

_____ 3. enhancers

_____ 4. core promoter

_____ 5. regulatory elements

_____ 6. *cis*-acting factors

_____ 7. silencers

a. Any sequence that effects the ability of the RNA polymerase to bind the promoter.
b. A regulatory sequence that inhibit the level of transcription.
c. A general term for regulatory elements that influence the activity of a nearby gene.
d. A regulatory sequence that increase the level of transcription.
e. This will produce a low level of transcription even in the absence of regulatory elements.
f. This determines the precise start site of transcription.
g. A gene that controls a second gene from a distance.

For each of the following components of the transcription preinitiation complex, match the protein with its correct function.

_____ 8. TFIIB

_____ 9. TFIID

_____ 10. TFIIF

_____ 11. Mediator

_____ 12. TFIIH

_____ 13. TFIIE

a. Interacts with the transcription factors and the RNA polymerase.
b. Serves as a junction between the RNA polymerase, TFIIB and the promoter.
c. This protein binds to the TATA box in the promoter.
d. Binds to TFIID, allows the RNA polymerase to bind to the core promoter.
e. Plays the major role in the formation of the open complex.
f. Maintains the open complex.

Questions 14 to 17 relate to the influence of chromatin on transcription. For each, complete the sentence.

14. The use of energy in the form to change the structure of the nucleosomes is called _______.

15. The _____ enzyme loosens the interaction between the DNA and the histone proteins, allowing a higher level of transcription.

16. The ____ enzyme tightens the interaction between the histones and the DNA.

17. During interphase, most chromatin is in the form of _______ organized into radial loop domains.

12.4 RNA Modification

Overview

In eukaryotic organisms, transcription does not directly produce a functional mRNA. Instead, following transcription, the pre-mRNA molecule is modified by a number of difference processes that can drastically alter the size and function of the pre-mRNA produced by the RNA polymerase. These processes can be divided into two general categories (Table 12.3).

The first form of modification is a chemical change to the RNA transcript. This is usually in the form of 5' capping or 3'polyA tailing. In addition, the length of the RNA may be altered to remove any non-coding or intervening sequences. These sequences, called introns, do not contain information that is necessary for the production of the functional protein. They are removed by RNA splicing (Figure 12.18).

In this section you need to familiarize yourself with the forms of modification and the terminology associated with the modification processes.

Key Terms

Alternative splicing
Capping
Colinearity
Complementary DNA
Exons
Genomic clone
Group I and II introns
Heterogenous nuclear RNA
Intervening sequences
Introns
Maturases
Nucleolus
PolyA tail
Polyadenylation sequence
Pre-mRNA
R loop
Ribozyme
RNA splicing
RNaseP
Self-splicing
snRNPs
Spliceosome

Focal Points

- Overview of RNA modifications (Table 12.3)
- Mechanisms of RNA splicing (Figure 12.18)

Exercises and Problems

For questions 1 to 9, use the following answers. Some answers may be used more than once.

a. 5' capping
b. 3' polyA tailing
c. Splicing

_____ 1. May involve the use of snRNPs.
_____ 2. Usually serves to shorten the length of the RNA transcript.
_____ 3. This process may occur while the transcript is being made by the RNA polymerase.
_____ 4. This plays an important role in the early stages of translation.
_____ 5. This provides stability to the mRNA molecule.
_____ 6. Removes the introns from the pre-mRNA.
_____ 7. May produce an mRNA with different combinations of introns.
_____ 8. Identifies the consensus sequence AAUAAA in the pre-mRNA.
_____ 9. Attaches a guanosine monophosphate to the pre-mRNA.

For questions 10 to 14, identify the type of splicing that is involved in the statement. Some answers may be used more than once.

a. group II
b. spliceosome
c. group I
d. all of the above

_____ 10. Involves the use of snRNPs.

_____ 11. Uses a free guanosine to enable splicing.

_____ 12. Produces a conformational change in the pre-mRNA to enable splicing.

_____ 13. Identifies consensus sequences in the pre-mRNA.

_____ 14. Utilizes an adenine nucleotide present within the intron.

For questions 15 to 18, provide the correct term that completes the statement.

15. An enzyme that is composed of RNA, not protein, is called a _________.

16. R loops were first used to discover the existence of ________.

17. The coding information for a functional protein is contained within the ____ of a gene and its pre-mRNA.

18. The fact that one gene may produce many similar proteins is due to ___________.

Chapter Quiz

1. Which of the following would reduce the overall level of transcription?
    a. transcription factors
    b. enhancers
    c. silencers
    d. core promoter
    e. none of the above

2. Sigma factor is associated with transcription in _____.
    a. prokaryotes
    b. eukaryotes
    c. viruses
    d. all organisms

3. This process increases the stability of the mRNA.
    a. 5' capping
    b. alternative splicing
    c. 3'polyA tailing
    d. spliceosomes
    e. none of the above

4. The transcription factor that binds to the TATA sequence in eukaryotes is _____.
   a. sigma
   b. rho
   c. TFIIB
   d. TFIID
   e. mediator

5. This process involves the use of snRNPs.
   a. 3' polyA tailing
   b. initiation of transcription
   c. spiceosome activity
   d. transcription termination

6. The Pribnow box is located where?
   a. the prokaryotic promoter
   b. the core promoter of eukaryotes
   c. eukaryotic introns
   d. the end of the coding sequence
   e. none of the above

7. Which of the following is not associated with the initiation of transcription in eukaryotes?
   a. rho
   b. mediator
   c. formation of an open complex
   d. transcription factors
   e. RNA polymerase II

8. The movement of nucleosomes to facilitate transcription is called ______.
   a. RNA splicing
   b. ATP-dependent chromatin remodeling
   c. capping
   d. termination

9. Transcription factors are found in ________ .
   a. prokaryotes
   b. eukaryotes
   c. viruses
   d. all organisms

10. The information for producing a functional protein is contained within the _____ of a eukaryotic gene.
   a. introns
   b. exons
   c. polyA tail
   d. 5'cap
   e. group I introns

Answer Key for Study Guide Questions

This answer key provides the answers to the exercises and chapter quiz for this chapter. Answers in parentheses ( ) represent possible alternate answers to a problem, while answers marked with an asterisk (*) indicate that the response to the question may vary.

12.1

1. e
2. c
3. a
4. d
5. b
6. f
7. g
8. E
9. I
10. I
11. T
12. 7S RNA
13. snRNA
14. mRNA
15. scRNA
16. rRNA
17. tRNA

12.2

1. f
2. d
3. e
4. a
5. c
6. b
7. T
8. I
9. T
10. E
11. I
12. E
13. E
14. I

12.3

1. g
2. f
3. d
4. e
5. a
6. c
7. b
8. d
9. c
10. b
11. a
12. e
13. f
14. ATP-dependent chromatin remodeling
15. histone acetyltransferase
16. histone deacetylases
17. 30 nm fibers

12.4

1. c
2. c
3. a
4. a
5. b
6. c
7. c
8. b
9. a
10. b
11. c
12. d
13. d
14. a
15. ribozyme
16. introns
17. exons
18. alternative splicing

Quiz

1. c
2. a
3. c
4. d
5. c
6. a
7. a
8. b
9. b
10. b

Answers to Conceptual and Experimental Questions

Conceptual Questions

C1. A. tRNA genes encode tRNA molecules, and rRNA genes encode the rRNAs found in ribosomes. There are also genes for the RNAs found in snRNPs, etc.

B. The term *template strand* is still appropriate because one of the DNA strands is used as a template to make the RNA. The term *coding strand* is not appropriate because the RNA made from nonstructural genes does not code for a polypeptide sequence.

C. Yes.

C2. The formation of the open complex, and the release of sigma factor.

C3. A consensus sequence is the most common nucleotide sequence that is found within a group of related sequences. An example is the –35 and –10 consensus sequence found in bacterial promoters. At –35, it is TTGACA, but it can differ by one or two nucleotides and still function efficiently as a promoter. In the consensus sequences within bacterial promoters, the –35 site is primarily for recognition by sigma factor. The –10 site, also known as the Pribnow box, is the site where the DNA will begin to unwind to allow transcription to occur.

C4. GGCATTGTCA

C5. Mutations that make a sequence more like the consensus sequence are likely to be up promoter mutations while mutations that cause the promoter to deviate from the consensus sequence are likely to be down promoter mutations. Also, in the –10 region, AT pairs are favored over GC pairs, because the role of this region is to form the open complex. AT pairs are more easily separated because they form only two hydrogen bonds compared to GC pairs, which form three hydrogen bonds.

A. Up promoter

B. Down promoter

C. Up promoter

C6. The most highly conserved positions are the first, second, and sixth. In general, when promoter sequences are conserved, they are more likely to be important for binding. That explains why changes are not found at these positions; if a mutation altered a conserved position, the promoter would probably not work very well. By comparison, changes are occasionally tolerated at the fourth position and frequently at the third and fifth positions. The positions that tolerate changes are less important for binding by sigma factor.

C7. In Figure 12.5, each α–helix (labeled *2* and *3*) appears to occupy a region that is approximately one-half of a complete turn of the double helix. Since a complete turn of a DNA double helix involves 10 bp, each *a* helix appears to be bound to about 5 bp in the major groove of the DNA; 5 bp would span a linear distance of approximately 1.7 nm (as described in Chapter 9, Figure 9.17). If we divide 1.7 nm by 0.15 nm/amino acid, we obtain a value of 11.3 amino acids per helix. This would create about 3.15 turns of an *a* helix; there are three complete turns per helix, for a total of six complete turns for both helices 2 and 3.

C8. This will not affect transcription. However, it will affect translation by preventing the initiation of polypeptide synthesis.

C9. RNA polymerase holoenzyme consists of sigma factor plus the core enzyme, which is a tetramer, $a_2\beta\beta'$. The role of sigma factor is to recognize the promoter sequence. The *a* subunits are necessary for the assembly of the core enzyme and for loose DNA binding. The β and β' subunits are the portion that catalyze the covalent linkages between adjacent ribonucleotides.

C10. Sigma factor can slide along the major groove of the DNA. In this way, it is able to recognize base sequences that are exposed in the groove. When it encounters a promoter sequence, hydrogen bonding between the bases and the sigma factor protein can promote a tight and specific interaction.

C11. The mutation would alter the bases so that hydrogen bonding with the sigma factor would not occur, or would not occur in the correct way. From Figure 12.4, if we look at the –35 region, we may expect that changing the first two nucleotides to anything besides T might inhibit transcription, changing the third base to an A (perhaps), changing the fourth base to a G or T, or changing the last base to anything besides an A. We would think these substitutions may have an effect, because they would deviate from known functional bacterial promoters.

C12. DNA-G/RNA-C
DNA-C/RNA-G
DNA-A/RNA-U
DNA-T/RNA-A

The template strand is 3′–CCGTACGTAATGCCGTAGTGTGATCCCTAG–5′ and the coding strand is 5′–GGCATGCATTACGGCATCACACTAGGGATC–3′. The promoter would be to the left (in the 3′ direction) of the template strand.

C13. RNA polymerase slides down the DNA and forms an open complex as it goes. The open complex is a DNA bubble of about 17 nucleotide bp. Within the open complex, one of the DNA strands, the one running in the 3′ to 5′ direction, is used as template for RNA synthesis. This occurs as individual nucleotides hydrogen bond to the DNA template strand according to the rules described in conceptual question C12. As the RNA polymerase slides along, the DNA behind the open complex rewinds back into a double helix.

C14. Transcriptional termination occurs when the hydrogen bonding is broken between the DNA and the part of the newly made RNA transcript that is located in the open complex.

C15. In ρ-dependent termination, the p protein binds to the RNA transcript after the p site has been transcribed. Eventually, RNA polymerase will transcribe a termination stem-loop that will cause it to pause in the transcription process. As it is pausing, the p protein, which functions as a helicase, will catch up to RNA polymerase and knock it off the DNA. In p-independent transcription, there is no p protein. Again, the RNA polymerase transcribes a termination hairpin that causes it to pause. However, when it pauses, an AU-rich region is left base pairing in the open complex. Since this is holding on by fewer hydrogen bonds, it is rather unstable. Therefore, it tends to dissociate from the open complex and thereby end transcription.

C16. Helicase and p protein bind to a nucleic acid strand and travel in the 5′ to 3′ direction. When they encounter a double-stranded region, they break the hydrogen bonds between complementary strands. p protein is different from DNA helicase in that it moves along an RNA strand, while DNA helicase moves along a DNA strand. The purpose of DNA helicase function is to promote DNA replication while the purpose of p protein function is to promote transcriptional termination.

C17. RNA and DNA polymerase are similar in the following ways:

1. They both use a template strand.
2. They both synthesize in the 5′ to 3′ direction.
3. The chemistry of synthesis is very similar in that they use incoming triphosphates and make a phosphoester bond between the previous nucleotide and the incoming nucleotide.
4. They are both processive enzymes that slide along a template strand of DNA.

RNA and DNA polymerase are different in the following ways:

1. RNA polymerase makes an RNA strand while DNA polymerase makes a DNA strand.
2. RNA polymerase uses only one DNA strand as a template.
3. RNA polymerase recognizes a particular DNA sequence (i.e., promoters) and synthesizes only a defined region of RNA from the promoter to the terminator.
4. RNA polymerase cannot proofread.
5. RNA polymerase does not need a primer.

C18. A. Mutations that alter the AU-rich region by introducing Gs and Cs, and mutations that prevent the formation of the stem-loop (hairpin) structure.

B. Mutations that alter the termination sequence and mutations that alter the p recognition site.

C. Eventually, somewhere downstream from the gene, there would be found another transcriptional termination sequence and transcription would terminate there. This second termination sequence might be found randomly or it might be at the end of an adjacent gene.

C19. A. Ribosomal RNA (5.8S, 18S, and 28S)

B. All of the mRNA and certain snRNA genes

C. All of the tRNAs and the 5S rRNA

C20. Eukaryotic promoters are somewhat variable with regard to the pattern of sequence elements that may be found. In the case of structural genes that are transcribed by RNA polymerase II, it is common to have a TATA box, which is about 25 bp upstream from a transcriptional start site. The TATA box is important in the identification of the transcriptional start site and the assembly of RNA polymerase and various transcription factors. The transcriptional start site defines where transcription actually begins.

C21. A. RNA polymerase would not be bound to the TATA box.

B. TTIID contains the TATA-binding protein. If it were missing, RNA polymerase would not bind to the TATA box either.

C. The formation of the open complex would not take place.

C22. It is primarily an accessibility problem. When the DNA is tightly wound around histones, it becomes difficult for large proteins, like RNA polymerase and transcription factors, to recognize the correct base sequence in the DNA and to catalyze the movement of the open complex. It is thought that a partial or complete disruption of the nucleosome structure is necessary for transcription to occur.

C23. TFIID and TFIIB would play analogous roles. Sigma factor does two things. It recognizes the promoter (as does TFIID) and it recruits RNA polymerase to the promoter (as does TFIIB).

C24. Hydrogen bonding is usually the predominant type of interaction when proteins and DNA follow an assembly and disassembly process. In addition, ionic bonding and hydrophobic interactions could occur. Covalent interactions would not occur. High temperature and high salt concentrations tend to break hydrogen bonds. Therefore, high temperature and high salt would inhibit assembly and stimulate disassembly.

C25. Only the first intron would be spliced out. The mature RNA would be: exon 1–exon 2–intron 2–exon 3.

C26. In bacteria, the 5′ end of the tRNA is cleaved by RNaseP. The 3′ end is cleaved by a different endonuclease, and then a few nucleotides are digested away by an exonuclease that removes nucleotides one at a time until it reaches a CCA sequence.

C27. The spliceosome is composed of multiple protein subunits and some RNAs. The function of the spliceosome is to cut the RNA in two places, hold the ends together, and then catalyze covalent bond formation between the two ends. The role of small snRNAs during this process could be involved in binding the pre-mRNA and/or it could be catalytic, like the RNA in RNaseP.

C28. A ribozyme is an enzyme whose catalytic part is composed of RNA. Examples are RNaseP and self-splicing group I and II introns. It is thought that the spliceosome may have catalytic RNAs as well.

C29. A gene is colinear when the sequence of bases in the sense strand of the DNA (i.e., the DNA strand that is complementary to the template strand for RNA synthesis) corresponds to the sequence of bases in the mRNA. Most prokaryotic genes and many eukaryotic genes are colinear. Therefore, you can look at the gene sequence in the DNA and predict the amino acid sequence in the polypeptide. Many eukaryotic genes, however, are not colinear. They contain introns that are spliced out of the pre-mRNA.

C30. Self-splicing means that an RNA molecule can splice itself without the aid of a protein. Group I and II introns can be self-splicing, although proteins can also facilitate the process.

C31. In eukaryotes, pre-mRNA can be capped, tailed, and spliced and then exported out of the nucleus.

C32. A transesterification reaction involves the breakage of one ester bond and formation of another ester bond.

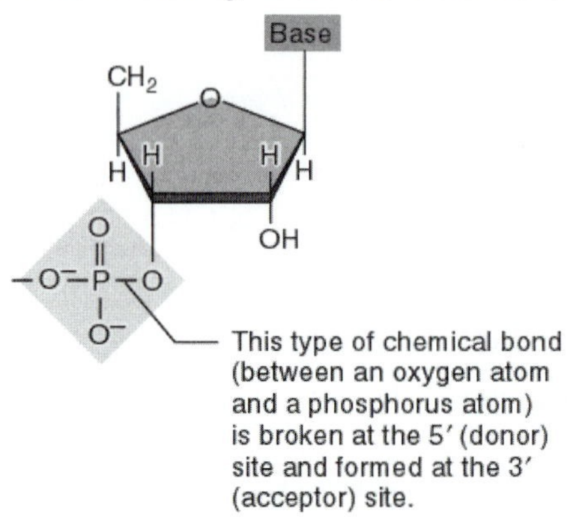

C33. Alternative splicing occurs when exons are spliced out or alternative splice sites are used at intron-exon boundaries. The biological significance is that two or more polypeptide sequences can be derived from a single gene. This is a more efficient use of the genetic material. In multicellular organisms, alternative splicing is often used in a cell-specific manner.

C34. This is called trimming or processing rather than splicing because none of the mature rRNA fragments are ever ligated (i.e., connected) to each other after the cleavage of the primary rRNA transcript.

C35. As shown at the *left* side of Figure 12.18, the guanosine, which binds to the guanosine-binding site, does not have a phosphate group attached to it. This guanosine is the nucleoside that winds up at the 5′ end of the intron. Therefore, the intron does not have a phosphate group at its 5′ end.

C36. A. U1 and U4/U6

B. U5

C. U2

C37. U5

C38. The 60-nucleotide sequence would be found in the closed loop. The closed loop is the region between the 5′ splice site and the branch site.

Experimental Questions

E1. The location of the intron within the cDNA is shown below:

cDNA:

5′-ATTGCATCCAGCGTATACTATCTCGGGCCCAATTAATGCCAGC
GGCCAGACTATCACCCAACTCG...**INTRON**...GTTACCTACTAGTATATCCCATATACTAGCATATATTTT
ACCCATAATTTGTGTGTGGGTATACAGTATAATCATATA–3′

You can figure this out by finding where the sequence of the genomic DNA begins to differ from the sequence of the cDNA. The genomic DNA has the normal splice sites that are described in Figure 12.21.

Genomic DNA:

5′–
ATTGCATCCAGCGTATACTATCTCGGGCCCAATTAATGCCAGCGGCCAGACTATCACCCAACTCGGCC
CACCCCCCAGGTTTACACAGTCATACCATACATACAAAAATCGCAGTTACTT**A**TCCCAAAAAAACCTAG
ATACCCCACATACTATTAACTCTTTCTTTCTAGTTACCTACTAGTATATCCCATATACTAGCATATATTTT
ACCCATAATTTGTGTGTGGGTATACAGTATAATCATATA–3′

The splice donor and acceptor sites are underlined. The space indicates where the strands in the corresponding RNA would be cut. The branch site is also underlined. The large A is the adenine that participates in the transesterification reaction.

E2. An R loop is a loop of DNA that occurs when RNA is hybridized to double-stranded DNA. While the RNA is hydrogen bonding to one of the DNA strands, the other strand does not have a partner to hydrogen bond with so it bubbles out as a loop. RNA is complementary to the template strand, so that is the strand it binds to.

E3.

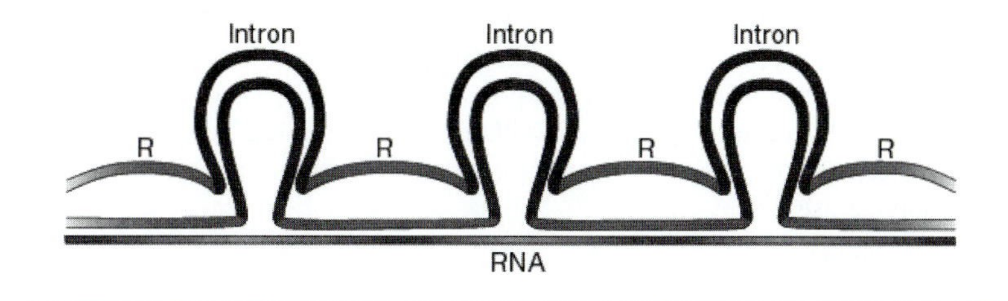

E4. The 1,100-nucleotide band would be observed from a normal individual (lane 1). A deletion that removed the –50 to –100 region would greatly diminish transcription, so the homozygote would produce hardly any of the transcript (just a faint amount as shown in lane 2) and the heterozygote would produce roughly half as much of the 1,100-nucleotide transcript (lane 3) compared to a normal individual. A nonsense codon would not have an effect on transcription; it affects only translation. So the individual with this mutation would produce a normal amount of the 1,100-nucleotide transcript (lane 4). A mutation that removed the splice acceptor site would prevent splicing. Therefore, this individual would produce a 1,550-nucleotide transcript (actually, 1,547 to be precise, 1,550 minus 3). The Northern blot is shown here:

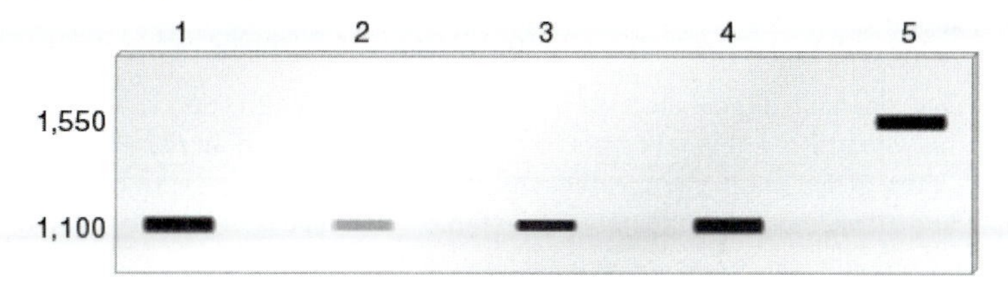

E5. When the 900 bp fragment is mixed with TFIID (lane 1), it would be retarded because TFIID would bind. When mixed with TFIIB (lane 2), it would not be retarded because TFIIB cannot bind without TFIID. Compared to lane 1, the 900 bp fragment would be retarded even more when mixed with TFIID and TFIIB (lane 3), because both transcription factors could bind. It would not be retarded when mixed with TFIIB and RNA polymerase (lane 4) because you do not have TFIID, which is needed for the binding of TFIIB and RNA polymerase. Finally, when mixed with TFIID, TFIIB, and RNA polymerase/TFIIF, the 900 bp fragment would be retarded a great deal because all four could bind (lane 5).

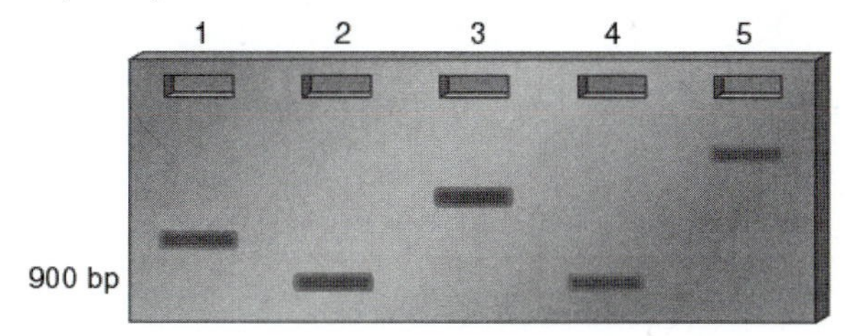

E6. A. It would not be retarded because ρ protein would not bind to the mRNA that is encoded by a gene that is terminated in a ρ -independent manner. The mRNA from such genes does not contain the sequence near the 3′ end that acts as a recognition site for the binding of ρ protein.

B. It would be retarded because ρ -protein would bind to the mRNA.

C. It would be retarded because U1 would bind to the pre-mRNA.

D. It would not be retarded because U1 would not bind to mRNA that has already had its introns removed. U1 binds only to pre-mRNA.

E7. A. The region of the gel from about 250 bp to 75 bp does not contain any bands. This is the region being covered up; it is about 175 base pairs long.

B. In a nucleosome, the DNA is wrapped twice around the core histones; a nucleosome contains 146 bp of DNA. The region bound by RNA polymerase II plus TFIID and TFIIB would be slightly greater than this length. Therefore, if the DNA was in a nucleosome structure, these proteins would have to be surrounding a nucleosome. It is a little hard to imagine how large proteins such as TFIID, TFIIB, and RNA polymerase II could all be wrapped around a single nucleosome (although it is possible). Therefore, the type of results shown here makes it more likely that the DNA is released from the core histones during the binding of transcription factors and RNA polymerase II.

E8. A.mRNA molecules would bind to this column because they have a polyA tail. A polyA tail is complementary to poly-dT, so the two would hydrogen bond to each other. To purify mRNAs, one begins with a sample of cells; the cells need to be broken open by some technique such as homogenization or sonication. This would release the RNAs and other cellular macromolecules. The large cellular structures (e.g., organelles, membranes, etc.) could be removed from the cell extract by a centrifugation step. The large cellular structures would be found in the pellet, while soluble molecules such as RNA and proteins would stay in the supernatant. At this point, you would want the supernatant to contain a high salt concentration and neutral pH. The supernatant would then be poured over the poly-dT column. The mRNAs would bind to the poly-dT column and other molecules (i.e., other types of RNAs and proteins) would flow through the column. The mRNAs would bind to the poly-dT column via hydrogen bonds. To break the hydrogen bonds between the mRNAs and poly-dT, you could add a solution that contains a low salt concentration and/or a high pH. This would release the mRNAs, which would then be collected in a low salt/high pH solution as it dripped from the column.

B. The basic strategy is to attach an oligonucleotide (i.e., a short sequence of DNA) to the column matrix that is complementary to the type of RNA that you want to purify. For example, if an rRNA contained a sequence 5′–AUUCCUCCA–3′, a researcher could chemically synthesize oligonucleotides with the sequence 3′–TAAGGAGGT–5′ and attach these oligonucleotides to the column matrix. To purify rRNA, one would use this 3′–TAAGGAGGT–5′ column and follow the general strategy described in part A.

Questions for Student Discussion/Collaboration

1. It would seem like a sufficient reason. It would seem that natural selection wants to create organisms that are very efficient with regard to their DNA storage abilities. Therefore, having a single gene that encodes several different polypeptides makes an efficient use of the genetic material. Other people may not see it that way. For example, in Chapter 10, we discussed how some organisms, like amphibians, have a lot of unused DNA. Therefore, it does not seem like a small genome size is necessarily an important attribute.

2. RNA transcripts come in two basic types: those that function as RNA (e.g., tRNA, rRNA, etc.) versus those that are translated (i.e., mRNA). As described in Chapter 12, they play a myriad of functional roles. RNAs that form complexes with proteins carry out some interesting roles. In some cases, the role is to bind other types of RNA molecules. For example, rRNA in bacteria plays a role in binding mRNA. In other cases, the RNA plays a catalytic role. An example is RNaseP.

Chapter 13: Translation of mRNA

Student Learning Objectives

Upon completion of this chapter you should be able to:

1. Understand the function of translation in gene expression.
2. Understand the genetic basis for protein synthesis and the key experiments which led to the discovery of gene function.
3. Understand the structure and function of the genetic code.
4. Understand the structure of proteins.
5. Understand the role of tRNA and ribosomes in translation.
6. Understand the process of translation and the role of the initiation, elongation, and termination stages.

13.1 Genetic Basis for Protein Synthesis

Overview

The final stage in the information flow from the genome to the cell is the manufacturing of a functional protein. The first section of this chapter provides an important foundation for understanding the process of translation and protein synthesis. It starts with a historical perspective on the relationship between genes and metabolic pathways (Archibald) and proteins (Beadle and Tatum).

The section then outlines the relationship between transcription (Chapter 12) and the functional protein. This is outlined in Figure 13.2. You should use this figure as a reference for the remainder of the chapter. In order for the genetic information, which is written as nucleic acids, to be converted to a functional protein, a code must be present so that groups of nucleotides dictate specific amino acids. This code is presented in Table 13.2, and the discoveries leading to its discovery are outline in Experiment 13A. It is important to note that the genetic code is a chemical code, and actually represents the chemical relationships of several molecules during translation. These will be covered in the next sections.

The section concludes with a discussion of amino acid structure and the linking of amino acids to form three-dimensional proteins. The figure on protein structure (13.6) is especially important since variations in the genetic information may produce alterations in the protein structure. This may change the phenotype of the organism, alter a metabolic pathway, or in some cases cause lethality. You need to firmly understand that the DNA and functional proteins are linked before leaving this section.

Key Terms

α-helix
β-sheet
Alkaptonuria
Codons
Degenerate
Enzymes
Genetic code
Inborn error of metabolism
Messenger RNA
One gene-One enzyme theory
Polypeptide
Primary structure
Protein
Quaternary structure
R group
Secondary structures
Side chain
Start codon
Stop codons
Structural genes
Subunits
Termination
Tertiary structure
Translation
Universal
Wobble base

Focal Points

- The genetic code (Table 13.2)
- The relationship of the DNA, mRNA and amino acids (Figure 13.2)
- Levels of protein structure (Figure 13.6)

Exercises and Problems

For questions 1 to 5, complete the following sentences regarding the early experiments on the relationship between genes and traits.

1. The first researcher to suggest that a relationship exists between the function of a gene and the production of an enzyme was ________.
2. The one-gene one-enzyme theory was introduced by ________.
3. The disease that Archibald described as an inborn error of metabolism was _________.
4. Beadle and Tatum studied the inheritance of defects in the metabolic pathways of _____, a common bread mold.
5. The term ______ denotes the physical structure of a protein.

For questions 6 to 11, match each of the following terms that are associated with the genetic code to its correct definition. You may need to refer to Table 12.2 in the text for information.

_____ 6. Universal

_____ 7. Codons

_____ 8. Start codon

_____ 9. Stop codons

_____ 10. Degenerate

_____ 11. Wobble base

a. There are very few exceptions to the genetic code in living organisms.
b. A group of three nucleotides that specify a single amino acid.
c. The codon AUG.
d. The fact that the codons UAU and UAC code for the same amino acid.
e. The codons UAA, UGA, and UAG.
f. There are more codons in the genetic code than there are amino acids.

For questions 12 to 17, select the level of protein structure that each statement refers to.

a. amino acid structure
b. primary structure
c. secondary structure
d. tertiary structure
e. quaternary structure

_____ 12. α-helices and β-sheets

_____ 13. The linear sequence of amino acids in a polypeptide

_____ 14. R groups

_____ 15. Hydrophobic and ionic interactions

_____ 16. A carboxyl and amino functional group

_____ 17. The interaction of two or more polypeptides

13.2 Structure and Function of tRNA

Overview

The next step in understanding the process of translation involves examining the structure and function of the transfer RNA (tRNA). The tRNA plays an important role in the translation of the mRNA codon into a specific amino acid. It is the structure of the tRNA (Figure 13.10) that makes this possible. Another important molecule in this process is the aminoacyl-tRNA synthetase (Figure 13.11) which catalyzes the interaction of a tRNA with its corresponding amino acid.

Key Terms

Adaptor hypothesis
Aminoacyl-tRNA synthetases
Anticodon
Charged tRNA
Wobble hypothesis
Isoacceptor tRNAs

Focal Points

- Structure of tRNA (Figure 13.10)
- Action of aminoacyl-tRNA synthetase (Figure 13.11)

Exercises and Problems

For questions 1 to 6, refer to the labels in the figure below.

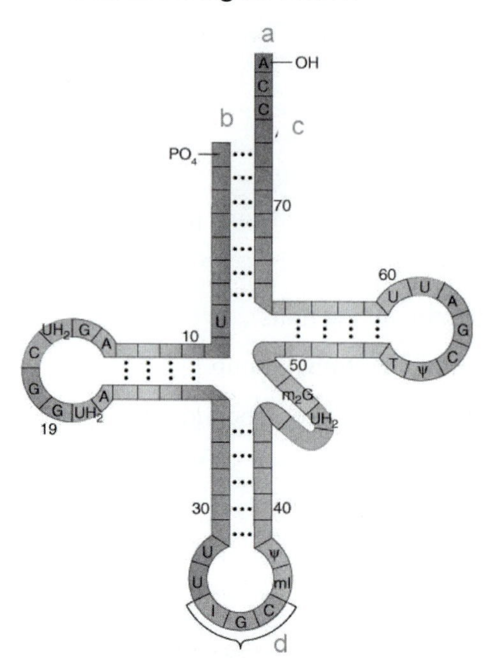

______ 1. The label that indicates the location of the anticodon.

______ 2. The label that indicates the acceptor stem.

______ 3. The 5' end of the tRNA.

______ 4. The site that will directly interact with the mRNA.

______ 5. The 3' end of the tRNA.

______ 6. A charged tRNA will have an amino acid attached at this point.

For questions 7 to 11, match the definition to its correct term.

______ 7. The enzyme that attaches the amino acids to the tRNA.

______ 8. According to the AT/GC rule, the complementary sequence to the codon.

______ 9. The idea that tRNAs recognize specific regions of the mRNA.

______ 10. The third base of the anticodon can tolerate mismatches.

______ 11. Two or more tRNAs that are recognized by the same codon.

a. aminoacyl-tRNA synthetase
b. adaptor hypothesis
c. wobble hypothesis
d. anticodon
e. isoacceptor tRNAs

13.3 Ribosome Structure and Assembly

Overview

Before proceeding to the stages of translation it is first necessary to study the site of translation, the ribosome. The ribosome is a composite of protein and ribosomal RNA (rRNA) arranged into structures called the large and small subunits. The purpose of the ribosome is to provide a workspace for the interaction of the mRNA and tRNA, and to ensure that the mRNA is being read in the correct sequence.

Key Terms

Aminoacyl site
Exit site
Nucleolus
Peptidyl site
Ribosome

Focal Points

- Model of ribosome structure (Figure 13.14c)

Exercises and Problems

For questions 1 to 5, match the term with its correct definition.

_______ 1. Ribosome

_______ 2. P site

_______ 3. E site

_______ 4. A site

_______ 5. Nucleolus

a. A site where the ribosomal subunits are assembled.
b. A site on the ribosome where uncharged tRNAs bind.
c. The site of translation.
d. The aminoacyl site.
e. The peptidyacyl site.

13.4 Stages of Translation

Overview

After introducing the genetic code, tRNAs and the ribosome, we are prepared to discuss the process of translation. Note that translation involves three stages – initiation, elongation, and termination. Although this is the same terminology as was used in transcription, the processes are very different.

Before entering into the details of translation, take a moment to review Figure 13.15. Notice the interaction of the mRNA, ribosome and tRNAs. Once you are familiar with the overview, you can proceed into the more detailed examinations of initiation, elongation, and termination.

Following the discussion of translation is a section that relates the directionality of the mRNA to the structure of the polypeptide. Notice that the 5' end of the mRNA corresponds to the amino (N-terminal) end of the polypeptide, while the 3' end of the mRNA is associated with the C-terminal end.

The final topic in this chapter examines the sorting of polypeptides. This is illustrated in Figure 13.22. Note the difference between cotranslational sorting and posttranslational sorting with regards to the organelles involved.

Key Terms

Amino terminal end
C-terminal
Carboxyl terminal end
Cotranslational import
Cotranslational sorting
Decoding function
Elongation
Elongation stage
Initiation
Initiator tRNA
Kozak's rules
N-terminal
Nonsense codons
Peptide bond
Peptidyl transfer
Peptidyltransferase
Polyribosome
Polysome
Posttranslational sorting
Release factors
Ribosomal binding site
Shine-Dalgarno sequence
Sorting signals
Termination
Traffic signals

Focal Points

- Overview of translation (Figure 13.15)
- Directionality in polypeptide synthesis (Figure 13.20)
- Protein sorting (Figure 13.22)

Exercises and Problems

For questions 1 to 7, indicate whether the statement is associated with prokaryotic (P) or eukaryotic (E) translation.

______ 1. Initiation factors

______ 2. A 7-methylguanosine cap on the mRNA enhances initiation

______ 3. The mRNA binds to the ribosomal subunit at the Shine-Dalgarno sequence.

______ 4. Sorting of the proteins occurs either during or after translation.

______ 5. Termination involves a single release factor.

______ 6. Transcription and translation occur simultaneously.

______ 7. Polyribosomes may occur on the mRNA.

For each of the following, indicate whether the statement is associated with initiation (I), elongation (E), or termination (T) of translation.

_______ 8. IF factors stabilize the mRNA and ribosomal subunits.

_______ 9. Nonsense codons enter into the A site.

_______ 10. Release factors interact with stop codons.

_______ 11. The ribosomal subunits associate with the mRNA.

_______ 12. The ribosomal subunits dissociate.

_______ 13. The peptidyl transfer reaction occurs between the amino acids of the P and A sites.

_______ 14. The decoding function of translation occurs.

_______ 15. Translocation of the ribosome occurs.

For each of the following, indicate whether cotranslational (C) or posttranslational (P) is involved.

_______ 16. Translation occurs in the cytosol.

_______ 17. The protein is moved into the Golgi apparatus.

_______ 18. The protein remains in the cytosol.

_______ 19. Translation occurs along the endoplasmic reticulum.

_______ 20. The protein is sent to the mitochondria or chloroplast.

Chapter Quiz

1. Hydrophobic and ionic interactions between the R groups of amino acids occurs at the ____ level of protein structure.

a. primary
b. secondary
c. tertiary
d. quaternary

2. The anticodon is located on the ______.

a. ribosome
b. mRNA
c. tRNA
d. rRNA

3. The molecule that attaches the amino acid to the tRNA is called ______.

a. peptidyltransferase
b. release factor
c. aminoacyl-tRNA transferase
d. polysome

4. Simultaneous transcription and translation occurs in _______ organisms.
   a. prokaryotic
   b. eukaryotic
   c. both a and b
   d. none of the above

5. Translocation of the ribosome occurs during ______.
   a. elongation
   b. termination
   c. initiation
   d. transcription

6. The Shine-Dalgarno sequence is involved in prokaryotic ______.
   a. initiation
   b. termination
   c. elongation
   d. posttranslational sorting
   e. none of the above

7. Which of the following is not a stop codon in most species?
   a. UAA
   b. AUG
   c. UGA
   d. UAG

8. Which of the following is true regarding the genetic code?
   a. It is universal.
   b. It is degenerate.
   c. It allows wobble in the third base.
   d. All of the above are correct.

9. The decoding function of translation occurs during _______.
   a. initiation
   b. elongation
   c. translocation
   d. cotranslational sorting

10. In which of the following ribosomal site receives the polypeptide chain during the peptidyl transfer reaction?
   a. A site
   b. P site
   c. E site
   d. K site

Answer Key for Study Guide Questions

This answer key provides the answers to the exercises and chapter quiz for this chapter. Answers in parentheses ( ) represent possible alternate answers to a problem, while answers marked with an asterisk (*) indicate that the response to the question may vary.

13.1	1. Archibald	6. a	12. c
	2. Beadle and Tatum	7. b	13. b
		8. c	14. a
	3. alkaptonuria	9. e	15. d
	4. Neurospora	10. f	16. a
	5. polypeptide	11. d	17. e
13.2	1. d	5. a	9. b
	2. c	6. a	10. c
	3. b	7. a	11. e
	4. d	8. d	
13.3	1. a	3. b	5. a
	2. e	4. d	
13.4	1. E	8. I	15. E
	2. E	9. T	16. P
	3. P	10. T	17. C
	4. E	11. I	18. P
	5. E	12. T	19. C
	6. P	13. E	20. P
	7. P	14. E	

Quiz

1. c	5. a	9. b
2. c	6. a	10. a
3. c	7. b	
4. a	8. d	

Answers to Conceptual and Experimental Questions

Conceptual Questions

C1. The start codon begins at the fifth nucleotide. The amino acid sequence would be Met Gly Asn Lys Pro Gly Gln STOP.

C2. When we say the genetic code is degenerate, it means that more than one codon can specify the same amino acid. For example, GGG, GGC, GGA, and GGU all specify glycine.

In general, the genetic code is nearly universal, because it is used in the same way by viruses, prokaryotes, fungi, plants, and animals. As discussed in Table 13.3, there are a few exceptions, which occur primarily in protozoa and organellar genetic codes.

C3. A. true

B. false

C. false

C4. A. This mutant tRNA would recognize glycine codons in the mRNA but would put in tryptophan amino acids where glycine amino acids are supposed to be in the polypeptide chain.

B. This mutation tells us that the aminoacyl-tRNA synthetase is primarily recognizing other regions of the tRNA molecule besides the anticodon region. In other words, tryptophanyl-tRNA synthetase (i.e., the aminoacyl-tRNA synthetase that attaches tryptophan) primarily recognizes other regions of the $tRNA^{trp}$ sequence (i.e., other than the anticodon region), such as the T- and D-loops. If aminoacyl-tRNA synthetases recognized only the anticodon region, we would expect glycyl-tRNA synthetase to recognize this mutant tRNA and attach glycine. That is not what happens.

C5. As shown in Figure 13.11, the energy comes from ATP. It is this energy conversion that explains the term *charged* tRNA.

C6. A. The answer is three. There are six leucine codons: UUA, UUG, CUU, CUC, CUA, and CUG. The anticodon AAU would recognize UUA and UUG. You would need two other tRNAs to *efficiently* recognize the other four leucine codons. These could be GAG and GAU or GAA and GAU.

B. The answer is one. There is only one codon, AUG, so you need only one tRNA with the anticodon UAC.

C. The answer is three. There are six serine codons: AGU, AGC, UCU, UCC, UCA, and UCG. You would need only one tRNA to recognize AGU and AGC. This tRNA could have the anticodon UCG or UCA. You would need two tRNAs to efficiently recognize the other four tRNAs. These could be AGG and AGU or AGA and AGU.

C7. There are four proline codons, four glycine codons, one methionine codon, and six serine codons. We apply the product rule to solve this problem.

$$4 \times 4 \times 1 \times 6 = 96$$

C8. 3′–CUU–5′ or 3′–CUC–5′

C9. The codon is 5′–CCA–3′, which specifies proline.

C10. It can recognize 5′–GGU–3′, 5′–GGC–3′, and 5′–GGA–3′. All of these specify glycine.

C11. An anticodon that was 3′–UUG–5′ would recognize the two codons. To recognize 5′–AAA–3′, it would have to be modified to 3′–UUI–5′.

C12. All tRNA molecules have some basic features in common. They all have a cloverleaf structure with three stem-loop structures. The second stem-loop contains the anticodon sequence that recognizes the codon sequence in mRNA. At the 3′ end, there is an acceptor site, with the sequence CCA, that serves as an attachment site for an amino acid. Most tRNAs also have base modifications that occur within their nucleotide sequences.

C13. They are very far apart, at opposite ends of the molecule.

C14. The role of aminoacyl-tRNA synthetase enzymes is to specifically recognize tRNA molecules and attach the correct amino acid to them. They are sometimes described as the second genetic code because the specificity of their attachment is a critical step in deciphering the genetic code. For example, if a tRNA has a 3′–GGG–5′ anticodon, it will recognize a 5′–CCC–3′ codon, which should specify proline. It is essential that the prolyl-tRNA-synthetase recognizes this tRNA and attaches proline to the 3′ end. The other aminoacyl-tRNA synthetases should not recognize this tRNA.

C15. In the context of translation, an activated amino acid has had AMP attached to it. This provides necessary energy so that the amino acid can be attached to the correct tRNA.

C16. Bases that have been chemically modified can occur at various locations throughout the tRNA molecule. The significance of all of these modifications is not entirely known. However, within the anticodon region, base modification alters base pairing to allow the anticodon to recognize two or more different bases within the codon.

C17. A formyl group is covalently attached to methionine after the methionine has been attached to the tRNA containing a UAC anticodon.

C18. No, it is not. Due to the wobble rules, the 5′ base in the anticodon of a tRNA can sometimes recognize two or more bases in the third (3′) position of the mRNA. Therefore, any given cell type synthesizes far fewer than 61 types of tRNAs.

C19. Translation requires mRNA, tRNAs, ribosomes, many proteins such as initiation, elongation, and termination factors, and many small molecules. ATP and GTP are small molecules that contain high-energy bonds. The mRNA, tRNAs, and proteins are macromolecules. The ribosomes are a large complex of macromolecules.

C20. The assembly process is very complex at the molecular level. In eukaryotes, 33 proteins and one rRNA assemble to form a 40S subunit, and 49 proteins and three rRNAs assemble to form a 60S subunit. This assembly occurs within the nucleolus.

C21. A protein subunit is a polypeptide. A ribosomal subunit is a much larger complex that is composed of RNA and many proteins. A ribosomal subunit is a much larger structure compared to a protein subunit.

C22. A. On the surface of the 30S subunit and at the interface

B. Within the 50S subunit

C. From the 50S subunit

D. To the 30S subunit

C23. *Initiation:* The mRNA, initiator tRNA, and initiation factors associate with the small ribosomal subunit; then the large subunit associates.

Elongation: The ribosome moves one codon at a time down the mRNA, adding one amino acid at a time to the growing polypeptide chain. There are three sites on the ribosome, the A, P, and E sites, that are important in this process. The A site is where the tRNA (except for the initiator tRNA) binds to the ribosome and recognizes the codon in the mRNA. The growing polypeptide chain is then transferred to the amino acid attached to this tRNA. The ribosome then translocates so that this tRNA is now moved to the P site. The empty tRNA that was in the P site is moved into the E site. This empty tRNA in the E site is then expelled, and now the next charged tRNA can bind to the A site.

Termination: A stop codon is reached and a termination factor binds to the A site. The hydrolysis of GTP initiates a series of events that leads to the disassembly of the ribosomal subunits and the release of the completed polypeptide chain.

C24. Most bacterial mRNAs contain a Shine-Dalgarno sequence, which is necessary for the binding of the mRNA to the small ribosomal subunit. This sequence, UAGGAGGU, is complementary to a sequence in the 16S rRNA. Due to this complementarity, these sequences will hydrogen bond to each other during the initiation stage of translation.

C25. A. The initiator tRNA would not bind to the small ribosomal subunit.

B. It may prevent the mRNA from binding to the small ribosomal subunit, and/or it may prevent the start codon from being recognized due to secondary structure in the mRNA.

C. The large ribosomal subunit would not assemble after the start codon had been identified by the small ribosomal subunit.

C26. The ribosome binds at the 5′ end of the mRNA and then scans in the 3′ direction in search of an AUG start codon. If it finds one that reasonably obeys the Kozak's rules, it will begin translation at that site. Aside from an AUG start codon, the two other features are a purine at the –3 position and a guanosine at the +4 position.

C27. 1. GCCACCAUGG

2. GACGCCAUGG

3. GCCUCCAUGC

4. GCCAUCAAGG

The last one does not have a start codon, so it would not work. The third one may be translated, but very poorly.

C28. The A site is the acceptor site. It is the location where a tRNA initially "floats in" and recognizes a codon in the mRNA. The only exception is the initiator tRNA that binds to the P site. The P site is the next location where the tRNA moves. When it first moves to the P site, it carries with it the polypeptide chain. In each round of elongation, the polypeptide chain is transferred from the tRNA in the P site to the amino acid attached to the tRNA in the A site. The third site is the E site. During translocation, the uncharged tRNA in the P site is transferred to the E site. It exits or is released from this site.

C29. The amino acid sequence is methionine tyrosine tyrosine glycine alanine. Methionine is at the amino terminus, alanine at the carboxyl terminus. The peptide bonds should be drawn as shown in Figure 13.20.

C30. Sorting signals provide an address that sends the protein to the correct location (i.e., compartment) within the cell. Proteins destined for the ER, Golgi, lysosomes, plasma membrane, or secretion have an SRP sorting signal at their amino terminal end. Nuclear proteins have an NLS sequence, etc. These sorting signals are recognized by cellular proteins/complexes that then act in a way to traffic the proteins to their correct destination.

C31. Cotranslational sorting occurs via the SRP signal. This form of sorting is necessary for proteins that are destined for the ER, Golgi, lysosomes, plasma membrane, or secretion. The SRP recognizes the signal in the amino terminus of the protein as it is being translated. It then directs the ribosome to proteins within the ER membrane so that the protein is synthesized into the ER. By comparison, posttranslational sorting occurs after the protein has been completely made. In this case, the protein usually contains a different amino acid sequence that acts as a traffic signal for its direction to a particular cellular organelle. Examples of these types of sorting signals include mitochondrial, chloroplast, and nuclear sorting signals.

C32. The initiation phase involves the binding of the Shine-Dalgarno sequence to the rRNA in the small ribosomal subunit. The elongation phase involves the binding of anticodons in tRNA to codons in mRNA.

C33. The nucleolus is a region inside the eukaryotic nucleus where the assembly of ribosomal subunits occurs.

C34. A. E site and P site (Note: A tRNA without an amino acid attached is only briefly found in the P site, just before translocation occurs.)

B. P site and A site (Note: A tRNA with a polypeptide chain attached is only briefly found in the A site, just before translocation occurs.)

C. Usually the A site, except the initiator tRNA can be found in the P site.

C35. A polysome is an mRNA molecule with many ribosomes attached to it.

C36. The tRNAs bind to the mRNA because their anticodon and codon sequences are complementary. When the ribosome translocates in the 5′ to 3′ direction, the tRNAs remain bound to their complementary codons, and the two tRNAs shift from the A site and P site to the P site and E site. If the ribosome tried to move in the 3′ direction, it would have to dislodge the tRNAs and drag them to a new position where they would not (necessarily) be complementary to the mRNA.

C37. A. False.

B. True.

C38. 52

C39. This means that translation can begin before transcription of the mRNA is completed. This cannot occur in eukaryotic cells because transcription and translation occur in different cellular compartments. Transcription occurs in the nucleus, while translation occurs in the cytosol.

Experimental Questions

E1. A codon contains three nucleotides. Since G and C are present at 50% each, if we multiply $0.5 \times 0.5 \times 0.5$, we get a value of 0.125, or 12.5%, of each of the eight possible codons. If we look the codons up in Table 13.2, we would expect 25% glycine (GGG and GGC), 25% alanine (GCC and GCG), 25% proline (CCC and CCG), and 25% arginine (CGG and CGC).

E2. A. There could have been other choices, but this template would be predicted to contain a cysteine codon, UGU, but would not contain any alanine codons.

B. You do not want to use ^{35}S because the radiolabel would be removed during the Raney nickel treatment.

C. There would not be a significant amount of radioactivity incorporated into newly made polypeptides with or without Raney nickel treatment. The only radiolabeled amino acid in this experiment was cysteine, which became attached to $tRNA^{cys}$. When exposed to Raney nickel, these cysteines were converted to alanine but only after they were already attached to $tRNA^{cys}$. If there were not any cysteine codons in the mRNA template, the $tRNA^{cys}$ would not recognize this mRNA. Therefore, we would not expect to see much radioactivity in the newly made polypeptides.

E3. The threonine has been changed to serine. Based on their structures, a demethylation of threonine has occurred. In other words, the methyl group has been replaced with hydrogen.

E4. The initiation phase of translation is very different between bacteria and eukaryotes, so they would not be translated very efficiently. A bacterial mRNA would not be translated very efficiently in a eukaryotic translation system because it does not have a cap structure. A eukaryotic mRNA would probably not have a Shine-Dalgarno sequence near its 5′ end, so it would not be translated very efficiently in a bacterial translation system.

E5. The overall function of the ribosome is to facilitate translation of mRNA. That is already clear, and an understanding of the molecular structure of the ribosome would not affect that knowledge. However, molecular biologists are interested in the steps that occur so that the process of translation can be elucidated. In this regard, an understanding of the structure of ribosomes can be quite helpful. For example, the sequencing of bacterial 16S rRNA showed that a sequence within this rRNA was complementary to the Shine-Dalgarno sequence. This helped molecular biologists understand how mRNA initially binds to the small ribosomal subunit. Electron microscopy studies revealed how the mRNA, ribosomal subunits, and tRNAs "fit together" during the translation process. Overall, the studies of molecular biologists provide bits and pieces to the puzzle of how a large macromolecular complex can catalyze the synthesis of a polypeptide with a defined amino acid sequence based upon an mRNA with a defined nucleotide sequence.

E6. Looking at the figure, the 5′ end of the template DNA strand is toward the right side. The 5′ ends of the mRNAs are farthest from the DNA, and the 3′ ends of the mRNAs are closest to the DNA. The start codons are slightly downstream from the 5′ ends of the RNAs.

E7.

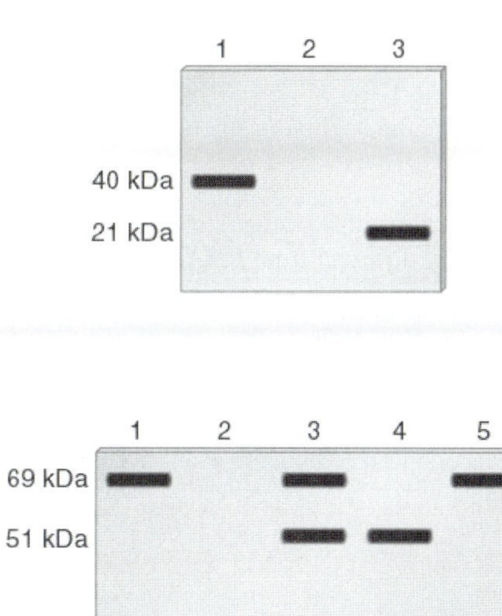

E8.

E9. The sample in lane 2 came from an individual who is homozygous for a mutation that introduces an early stop codon into the coding sequence. As seen in lane 2, the protein is shorter than normal. The sample in lane 3 came from an individual who was homozygous for a mutation that prevented the expression of this polypeptide. It could be a down promoter mutation (i.e., a mutation in the promoter that decreases the rate of transcription), or it could be a mutation in the coding sequence that causes the protein to be degraded very rapidly. The sample in lane 4 came from an individual who is homozygous for a mutation that changed one amino acid to another amino acid. This type of mutation, termed a missense mutation, may not be detectable on a gel. However, a single amino acid substitution within a polypeptide may block protein function. This would explain the albino phenotype.

E10. A. If codon usage were significantly different between kangaroo and yeast cells, this would inhibit the translation process. For example, if the preferred leucine codon in kangaroos was CUU, translation would probably be slow in a yeast translation system. We would expect the yeast translation system to primarily contain leucine tRNAs with an anticodon sequence that is AAC because this $tRNA^{leu}$ would match the preferred yeast leucine codon, which is UUG. In a yeast translation system, there probably would not be a large amount of tRNA with an anticodon of GAA, which would match the preferred leucine codon, CUU, of kangaroos. For this reason, kangaroo mRNA would not be translated very well in a yeast translation system, but it probably would be translated to some degree.

B. The advantage of codon bias is that a cell can rely on a smaller population of tRNA molecules to efficiently translate its proteins. A disadvantage is that mutations, which do not change the amino acid sequence but do change a codon (e.g., UUG to UUA), may inhibit the production of a polypeptide if a preferred codon is changed to a nonpreferred codon.

Questions for Student Discussion/Collaboration

1. In some ways, it seems like a waste of energy, but the process of translation is pretty amazing and complex. It starts with a sequence of nucleotides and uses that sequence as a "director" to make a sequence of amino acids within a polypeptide. As a metaphor, it is like taking a string of pearls and using them to copy them into a string of colored glass beads. To do this reproducibly the same way every time, you would need to examine each pearl and have some "rules" that would correlate each pearl with a particular colored bead. You would need tools to carefully examine each pearl and then another set of tools to actually make the string of beads. When you think about it in this context, it makes sense that translation at the molecular level would involve a large amount of cellular proteins and other molecules.
2. This could be a very long list. There are similarities along several lines.
    1. There is a lot of molecular recognition going on, either between two nucleic acid molecules or between proteins and nucleic acid molecules. Students may see these as similarities or differences, depending on their point of view.
    2. There is biosynthesis going on in both processes. Small building blocks are being connected. This requires an input of energy.
    3. There are genetic signals that determine the beginning and ending of these processes.

    There are also many differences.

    1. Transcription produces an RNA molecule with a similar structure to the DNA, while translation produces a polypeptide with a structure that is very different from RNA.
    2. Depending on your point of view, it seems that translation is more biochemically complex, requiring more proteins and RNA molecules to accomplish the task.
3. The recognition between the Shine-Dalgarno sequence and 16S rRNA is an RNA-RNA recognition. Likewise, the binding of codons and anticodons is RNA-RNA. The scanning of the start codon in eukaryotes is an RNA recognition process. Protein-protein recognition is also important in ribosome assembly. Protein-RNA recognition occurs during the termination process.

Chapter 14: Gene Regulation in Bacteria and Bacteriophages

Student Learning Objectives

Upon completion of this chapter you should be able to:

1. Understand the various mechanisms of transcriptional regulation.
2. Understand the regulation of the *lac*, *ara* and *trp* operons.
3. Understand the mechanisms of translational and posttranslational regulation.
4. Understand the regulation of genes in the viral life cycles.

14.1 Transcriptional Regulation

Overview

Gene regulation can occur at a number of levels, but the most common is at the transcriptional level. The first section of this chapter examines the variety of means by which genes may be transcriptionally regulated in bacteria. It is important that you understand the primary mechanisms of regulation and the terminology associated with each. An overview of these are provided in Figure 14.2.

The chapter presents three different mechanisms of transcriptional regulation. These are the *lac* operon, the *ara* operon, and the *trp* operon. Each is similar in that it utilizes a regulatory protein as the mechanism of gene regulation. As you proceed through these units, focus on the mechanism of regulation and whether it is a positively or negatively regulated system.

Key Terms

Activators
Attenuation
Attenuator
CAP site
Catabolite activator protein
Catabolite repression
cis-acting element
cis-effect
Constitutive genes
Corepressor
Cyclic AMP
Diauxic growth
Enzyme adaptation
Gene regulation
Induced
Inducer
Inducible
Inhibitor
lac repressor
Merozygote
Negative control
Operator
Operator site
Operon
Polycistronic mRNA
Positive control
Promoter
Regulated
Repressible
Repressor
Structural genes
Terminator
trans-effect
trp repressor

Focal Points

- Overview of regulatory proteins (Figure 14.2)
- *lac* operon (Figure 14.5a, b; 14.8)
- *ara* operon (Figure 14.11; 14.12)
- *trp* operon (Figure 14.13)

Exercises and Problems

For questions 1 to 7, match the following regulatory terms with its correct definition.

_____ 1. Activator

_____ 2. Inhibitor

_____ 3. Repressor

_____ 4. Inducer

_____ 5. Negative control

_____ 6. Positive control

_____ 7. Corepressor

a. Gene regulation by repressor proteins.
b. Binds to an activator protein and inhibits it from binding to the DNA.
c. Gene regulation by activator proteins.
d. A protein the binds to the DNA and inhibits transcription.
e. A protein that binds to the repressor and causes it to bind to the DNA.
f. An effector molecule that increases transcription.
g. Proteins that increase transcription.

For questions 8 and 9, complete the statement regarding the structure of the *lac* operon.

8. The operator of the lac operon is the location where a _____ protein binds.

9. The lac operon is _________, meaning that it encodes more than one structural gene.

For questions 10 to 17, choose whether the statement refers to the lac operon (L), ara operon (A), or trp operon (T). Some statements may have more than one answer.

_____ 10. An example of a polycistronic operon.

_____ 11. An example of a repressible system.

_____ 12. Uses a process called catabolite repression as a form of regulation.

_____ 13. Regulated by attenuation.

_____ 14. An example of an inducible system.

_____ 15. The use of cAMP as an effector molecule.

_____ 16. The operon is turned off by a repressor protein.

_____ 17. Uses the same protein as both an activator and repressor.

For questions 18 to 21, indicate whether the operon will be on (+) or off (-) under the described conditions.

_____ 18. The trp operon in the presence of tryptophan.

_____ 19. The lac operon in the presence of allolactose.

_____ 20. The lac operon in the presence of glucose.

_____ 21. The ara operon in the presence of arabinose.

14.2 Translational and Posttranslational Regulation

Overview

Gene regulation at the translational and posttranslational levels is relatively rare, but it does occur in a number of systems. Translational regulation typically involves interactions with the machinery of initiation. In posttranslational regulation, the target of regulation is the functional protein. In these cases, some mechanism interferes with the activity of a formed protein.

Key Terms

Allosteric enzyme
Antisense RNA
Feedback inhibition
Posttranslational
Translational regulatory protein
Translational repressors

Focal Points

- Feedback inhibition (Figure 14.17)

Exercises and Problems

For the following questions, match the definition with the appropriate term.

_____ 1. Any regulatory mechanism that acts on the functional protein.

_____ 2. The final product in a metabolic pathway influences an earlier enzyme in the pathway.

_____ 3. An RNA strand that is complementary to the mRNA strand.

_____ 4. Binds to the mRNA and prevents it from interacting with the ribosome.

_____ 5. An enzyme with two different binding site, one catalytic and one regulatory.

_____ 6. A regulatory mechanism that interacts with the initiation, elongation, or termination of translation.

a. antisense RNA
b. translational repressors
c. allosteric enzymes
d. feedback inhibition
e. translational regulation
f. posttranslational regulation.

14.3 Gene Regulation in the Bacteriophage Life Cycle

Overview

Viral genomes are small and may only contain a few genes. Thus, the regulatory mechanisms demonstrated in bacteria are rare in viral systems. However, many viruses, specifically temperate viruses that must switch between a lytic and lysogenic cycle, possess regulatory systems that are governed by proteins. This section introduces the genetic switch for bacteriophage λ.

Key Terms

Antitermination
Lysogenic cycle
phage
Lytic cycle
Prophage
Temperate

Focal Points

- The pathways of the lytic and lysogenic cycles (Figure 14.19)

Exercises and Problems

For questions 1 to 6, choose the correct function for the following bacteriophage λ regulatory and structural proteins.

_____ 1. λ repressor

_____ 2. cro

_____ 3. Integrase

_____ 4. cII-cIII

_____ 5. recA

_____ 6. Proteases

a. During starvation, these enzymes are inactive, allowing for the accumulation of cII.
b. This protein detects damage to the bacteria. When present it inactivates λ repressor.
c. The enzyme that incorporates the phage into the bacterial chromosome.
d. Accumulation of this protein favors the lysogenic cycle.
e. Accumulation of this protein favors the lytic cycle.
f. Promotes movement into the lysogenic cycle by binding to the lytic operator (O_R).

Chapter Quiz

1. Which of the following operons in bacteria represents a repressible system?
   a. ara
   b. lac
   c. trp
   d. cro
   e. none of the above

2. The ____ operon utilizes attenuation as a regulatory mechanism.
   a. trp
   b. lac
   c. ara
   d. all of the above

3. Genes that are turned on by regulatory proteins are called _______.
   a. repressible
   b. reversible
   c. inducible
   d. inductable

4. The lac operon is turned off in the presence of what?
   a. lactose
   b. glucose
   c. arabinose
   d. any sugar

5. The presence of tryptophan does what to the trp repressor?
   a. inactivates it
   b. activates it
   c. degrades it
   d. initiates transcription of
   e. none of the above

6. In bacteriophage λ, an accumulation of the cro protein signal what cycle?
   a. the lysogenic cycle
   b. the lytic cycle
   c. the transcription cycle
   d. self-destruct

7. An enzyme that contains both a catalytic site and a regulatory site is called _____.
   a. antisense
   b. inducible
   c. cis-acting
   d. allosteric

8. The name of the site in the bacterial genome that the repressor protein binds to is called the _____.

a. CAP site
b. terminator
c. promoter
d. operator
e. none of the above

9. What is the effect of the λ repressor on bacteriophage λ?

a. It activates the lytic cycle.
b. It activates the lysogenic cycle.
c. It integrates the phage into the bacterial genome.
d. It causes damage to the host DNA.

10. In the presence of tryptophan, the *trp* operon is _____.

a. turned on
b. turned off

Answer Key for Study Guide Questions

This answer key provides the answers to the exercises and chapter quiz for this chapter. Answers in parentheses ( ) represent possible alternate answers to a problem, while answers marked with an asterisk (*) indicate that the response to the question may vary.

14.1

1. g
2. b
3. d
4. f
5. a
6. c
7. e
8. repressor
9. polycistronic
10. L A T
11. T
12. L
13. T
14. L A
15. L
16. L A T
17. A
18. –
19. +
20. –
21. +

14.2

1. f
2. d
3. a
4. b
5. c
6. e

14.3

1. f
2. e
3. c
4. d
5. b
6. a

Quiz

1. c
2.a
3. c
4. b
5. b
6. b
7. d
8. d
9. b
10. b

Answers to Conceptual and Experimental Questions

Conceptual Questions

C1. A constitutive gene is unregulated, which means that its expression level is relatively constant. The expression of a regulated gene varies under different conditions. In bacteria, the regulation of genes oftentimes occurs at the level of transcription by combinations of regulatory proteins and small effector molecules. In addition, gene expression can be regulated at the level of translation or the function of a protein can be regulated after translation is completed.

C2. In bacteria, gene regulation greatly enhances the efficiency of cell growth. It takes a lot of energy to transcribe and translate genes. Therefore, a cell is much more efficient and better at competing in its environment if it expresses genes only when the gene product is needed. For example, a bacterium will express only the genes that are necessary for lactose metabolism when a bacterium is exposed to lactose. When the environment is missing lactose, these genes are turned off. Similarly, when tryptophan levels are high within the cytoplasm, the genes that are required for tryptophan biosynthesis are repressed.

C3. In this case, an activator protein and inhibitor molecule are involved. The binding of the inhibitor molecule to the activator protein would prevent it from binding to the DNA and thereby inhibit its ability to activate transcription.

C4. A. Regulatory protein

B. Effector molecule

C. DNA segment

D. Effector molecule

E. Regulatory protein

F. DNA segment

G. Effector molecule

C5. Choices B and C are correct. In both of these cases, the presence of the small effector molecule will turn off transcription. In contrast, the presence of an inducer turns on transcription.

C6. A *cis*-mutation is within a genetic regulatory sequence, such as an operator site, that affects the binding of a genetic regulatory protein. A *cis*-mutation affects only the adjacent genes that the genetic regulatory sequence controls. A *trans*-mutation is usually in a gene that encodes a genetic regulatory protein. A *trans*-mutation can be complemented in a merozygote experiment by the introduction of a normal gene that encodes the regulatory protein.

C7. The term *enzyme adaptation* means that a particular enzyme is made only when a cell is exposed to the substrate for that enzyme. It occurs because the gene that encodes the enzyme that is involved in the metabolism of the substrate is expressed only when the cells have been exposed to the substrate.

C8. A. No transcription would take place. The *lac* operon could not be expressed.

B. No regulation would take place. The operon would be continuously turned on.

C. The rest of the operon would function normally but none of the transacetylase would be made.

C9. It would be impossible to turn the *lac* operon on even in the presence of lactose because the repressor protein would remain bound to the operator site.

C10. Diauxic growth refers to the phenomenon whereby a cell first uses up one type of sugar (e.g., glucose) before it begins to metabolize a second sugar (e.g., lactose). In this case, it is caused by gene regulation. When a bacterial cell is exposed to both sugars, the uptake of glucose causes the cAMP levels in the cell to fall. When this occurs, the catabolite activator protein is removed from the *lac* operon so that it is not able to be (maximally) activated.

C11. A. *Cis*-effect. It would affect only the genes that are in the adjacent operon.

B. *Trans*-effect. This is a mutation that affects a protein that can move throughout the cell.

C. *Trans*-effect. This is a mutation that affects a protein that can move throughout the cell.

D. *Cis*-effect. It would affect only the genes that are in the adjacent operon.

C12. A mutation that prevented the *lac* repressor from binding to the operator would make the *lac* operon constitutive only in the absence of glucose. However, this mutation would not be entirely constitutive because transcription would be inhibited in the presence of glucose. The disadvantage of constitutive expression of the *lac* operon is that the bacterial cell would waste a lot of energy transcribing the genes and translating the mRNA when lactose was not present.

C13. AraC protein binds to the *araI*, $araO_1$, and $araO_2$ operator sites. The binding of the AraC protein to the $araO_1$ and $araO_2$ sites inhibits the transcription of the *araC* gene. Similarly, the AraC proteins bound at $araO_2$ and *araI* repress the *ara* operon. The AraC proteins at $araO_2$ and *araI* can bind to each other by causing a loop in the DNA. This DNA loop prevents RNA polymerase from transcribing the *ara* operon. In the absence of arabinose, the *ara* operon is turned off. When cells are exposed to arabinose, however, arabinose binds to the AraC protein and breaks the interaction between the AraC proteins at the $araO_2$ and *araI* sites and thereby breaks the DNA loop. In addition, a second AraC protein binds at the *araI* site. This AraC dimer at the *araI* operator site activates transcription.

C14. A. Without $araO_2$ the repression of the *ara* operon could not occur. The operon would be constitutively expressed at high levels because AraC protein could still activate transcription of the *ara* operon by binding to *araI.* The presence of arabinose would have no effect. Note: The binding of arabinose to AraC is not needed to form an AraC dimer at *araI.* The dimer is able to form because the loop has been broken. This point may be figured out if you notice that an AraC dimer is bound to $araO_1$ in the presence and absence of arabinose (see Figure 14.12).

B. Without $araO_1$, the AraC protein would be overexpressed. It would probably require more arabinose to alleviate repression. In addition, activation might be higher because there would be more AraC protein available.

C. Without *araI,* transcription of the *ara* operon cannot be activated. You might get a very low level of constitutive transcription.

D. Without $araO_2$ the repression of the *ara* operon could not occur. However, without *araI,* transcription of the *ara* operon cannot be activated. You might get a very low level of constitutive transcription.

C15. Attenuation means that transcription is ended before it has reached the end of an operon. Since it causes an end to transcription, it is a form of transcriptional regulation even though the translation of the *trpL* region plays a key role in the attenuation mechanism.

C16. A. Attenuation will not occur because loop 2–3 will form.

B. Attenuation will occur because 2–3 cannot form, so 3–4 will form.

C. Attenuation will not occur because 3–4 cannot form.

D. Attenuation will not occur because 3–4 cannot form.

C17. A defective tryptophanyl-tRNA synthetase would make attenuation less likely. This is because the bacterial cell would have a lower amount of charged $tRNA^{trp}$. Therefore, it would be more likely for the ribosome to stall at the tryptophan codons found within the *trpL* gene, even if the concentration of tryptophan amino acids in the cell was high. When the ribosome stalls at these tryptophan codons, this prevents attenuation.

C18. The addition of Gs and Cs into the U-rich sequence would prevent attenuation. The U-rich sequence promotes the dissociation of the mRNA from the DNA, when the terminator stem-loop forms. This causes RNA polymerase to dissociate from the DNA and thereby causes transcriptional termination. The UGGUUGUC sequence would probably not dissociate because of the Gs and Cs. Remember that GC base pairs have three hydrogen bonds and are more stable than AU base pairs, which only have two hydrogen bonds.

C19. If you look very carefully at the RNA sequence in Figure 14.14, you will notice that a UAA codon is found just past region 2. Therefore, in this mutant strain, the UGA stop codon at the end of region 1 could be read by the mutant $tRNA^{gly}$ and then the ribosome would stop at the UAA codon that is found just past region 2. If the ribosome paused here, it would probably cover up a portion of region 3, and therefore the terminator 3–4 stem-loop would not form. According to this scenario, attenuation could not occur. However, we should also keep in mind the issue of timing. The ribosome would have to be really close to RNA polymerase to prevent attenuation in this nonsense suppressor strain. It is possible that the 3–4 stem-loop might form before the ribosome reaches the UAA stop codon that is just past region 2. Therefore, attenuation might occur anyway because the 3–4 stem-loop might form before the ribosome reaches the UAA stop codon.

C20. It takes a lot of cellular energy to translate mRNA into a protein. A cell wastes less energy if it prevents the initiation of translation rather than a later stage such as elongation or termination.

C21. Antisense RNA is RNA that is complementary to a functional RNA such as mRNA. The binding of antisense RNA to mRNA inhibits translation.

C22. One mechanism is that histidine could act as corepressor that shuts down the transcription of the histidine synthetase gene. A second mechanism would be that histidine could act as an inhibitor via feedback inhibition. A third possibility is that histidine inhibits the ability of the mRNA encoding histidine synthetase to be translated. Perhaps it induces a gene that encodes an antisense RNA. If the amount of histidine synthetase protein was identical in the presence and absence of extracellular histidine, a feedback inhibition mechanism is favored, because this affects only the activity of the histidine synthetase enzyme, not the amount of the enzyme. The other two mechanisms would diminish the amount of this protein.

C23. *lac* operon: The binding of allolactose causes a conformational change in the repressor protein and removes it from the operator site.

ara operon: The binding of arabinose to AraC breaks the looping interaction and leads to the activation of the *ara* operon.

trp operon: The binding of tryptophan to the *trp* repressor causes it to bind to the operator site and inhibits transcription.

C24. The two proteins are similar in that both bind to a segment of DNA and repress transcription. They are different in three ways. (1) They recognize different effector molecules (i.e., the *lac* repressor recognizes allolactose and the *trp* repressor recognizes tryptophan. (2) Allolactose causes the *lac* repressor to release from the operator, while tryptophan causes the *trp* repressor to bind to its operator. (3) The sequences of the operator sites that these two proteins recognize are different from each other. Otherwise, the *lac* repressor could bind to the *trp* operator and the *trp* repressor could bind to the *lac* operator.

C25. A. Antisense RNA or a translational repressor would shut down protein synthesis the fastest. A transcriptional repressor would also shut down the synthesis of mRNA, so it would eventually shut down protein synthesis once all of the preexisting mRNA had been degraded. Feedback inhibition would have no effect on protein synthesis.

B. Only a transcriptional repressor protein would shut down the synthesis of mRNA.

C. Feedback inhibition is the fastest way to shut down the function of a protein. Antisense RNA and transcriptional repressors eventually prevent protein function once all of the preexisting mRNA and protein have been degraded.

C26. In the lytic cycle, the virus directs the bacterial cell to make more virus particles until eventually the cell lyses and releases them. In the lysogenic cycle, the viral genome is incorporated into the host cell's genome as a prophage. It remains there in a dormant state until some stimulus causes it to erupt into the lytic cycle.

C27. A. The lysogenic cycle would occur because cro protein is necessary to initiate the lytic cycle.

B. The lytic cycle would occur because *cI* encodes the λ repressor, which prevents the lytic cycle.

C. The lytic cycle would occur because cII protein is necessary to initiate the lysogenic cycle.

D. Both cycles could try to initiate but the lysogenic cycle would fail because it would be unable to integrate into the host chromosome.

E. Neither cycle could occur.

C28. The O_R region contains three operator sites and two promoters. P_{RM} and P_R transcribe in opposite directions. The λ repressor will first bind to O_{R1} and then O_{R2}. The binding of the λ repressor to O_{R1} and O_{R2} inhibits transcription from P_R and thereby switches off the lytic cycle. Early in the lysogenic cycle, the λ repressor protein concentration may become so high that it will occupy O_{R3}. Later, when the λ repressor concentration begins to drop, it will first be removed from O_{R3}. This allows transcription from P_{RM} and maintains the lysogenic cycle. By comparison, the cro protein has its highest affinity for O_{R3}, and so it binds there first. This blocks transcription from P_{RM} and thereby switches off the lysogenic cycle. The cro protein has a similar affinity for O_{R2} and O_{R1}, and so it may occupy either of these sites next. It will bind to both O_{R2} and O_{R1}. This turns down the expression from P_R, which is not needed in the later stages of the lytic cycle.

C29. P_{RE} is activated by the cII-cIII complex. However, later in the lysogenic cycle the amount of the cII-cIII complex falls. This would prevent further synthesis of the λ repressor. However, the λ repressor can activate its own transcription from P_{RM}. This will maintain the lysogenic cycle.

C30. It would first increase the amount of cro protein so that the lytic cycle would be favored.

C31. A cell that has a λ prophage is making a significant amount of the λ repressor. If another phage infects the cell, the λ repressor inhibits transcription from P_R and P_L and thereby inhibits the early steps that are required for either the lytic or lysogenic cycles.

C32. Neither cycle could be followed. As shown in Figure 14.19, N protein is needed to make a longer transcript from P_L for the lysogenic cycle and also to make a longer transcript from P_R for the lytic cycle.

C33. A genetic switch is a DNA sequence that governs a choice between two genetic pathways. The segment of DNA usually contains several operator sites that are recognized by two or more transcriptional regulatory proteins. A genetic switch is similar to a simple operator in that the binding of proteins to the DNA is the underlying event that controls transcription. The main difference between a genetic switch and simple operator is the degree of complexity. A simple operator usually controls a single gene or a single operon, while a genetic switch often has multiple operator sites and controls the choice between two (or more) promoters.

C34. If the F^- strain is lysogenic for phage λ, the λ repressor is already being made in that cell. If the F^- strain receives genetic material from an *Hfr* strain, you would not expect it to have an effect on the lysogenic cycle, which is already established in the F^- cell. However, if the *Hfr* strain is lysogenic for λ and the F^- strain is not, the *Hfr* strain could transfer the integrated l DNA (i.e., the prophage) to the F^- strain. The cytoplasm of the F^- strain would not contain any λ repressor. Therefore, this λ DNA could choose between the lytic and lysogenic cycle. If it follows the lytic cycle, the F^- recipient bacterium will lyse.

Experimental Questions

E1. A. *ß*-ONPG was used to measure the level of expression of the *lac* operon. More specifically, the cleavage of *ß*-ONPG to produce a yellow compound requires the action of *ß*-galactosidase. So the assay is indirectly measuring the amount of *ß*-galactosidase protein. The reason why there was no yellow color in one of the tubes was because the repressor was preventing the expression of the *lac* operon, and this prevents the expression of *ß*-galactosidase enzyme. Some other methods to measure the expression level of the *lac* operon could include the following:

1. Conduct a Northern blot to measure the amount of mRNA that is produced from the *lac* operon.
2. Conduct a Western blot using antibodies against one of the three proteins encoded by the *lac* operon.
3. Measure the uptake of radiolabeled-lactose into the cells. This would measure the amount of the lactose permease that is expressed from the *lac* operon.

B. The merozygote has two copies of the *lacZ* gene so it makes twice as much *ß*-galactosidase.

E2. In samples loaded in lanes 1 and 4, we expect the repressor to bind to the operator because there is no lactose present. In the sample loaded into lane 4, the CAP protein could still bind cAMP because there is no glucose. However, there really is no difference between lanes 1 and 4, so it does not look like the CAP can activate transcription when the *lac* repressor is bound. If we compare samples loaded into lanes 2 and 3, the *lac* repressor would not be bound in either case, and the CAP would not be bound in the sample loaded into lane 3. There is less transcription in lane 3 compared to lane 2, but since there is some transcription seen in lane 3, we can conclude that the removal of the CAP (because cAMP levels are low) is not entirely effective at preventing transcription. Overall, the results indicate that the binding of the *lac* repressor is much more effective at preventing transcription of the *lac* operon compared to the removal of the catabolite activator protein.

E3. In the normal strain, the *lac* operon would be fully induced, so there would be maximal expression of the mRNA (lane 1). The same thing would happen in lane 2, since the *lac* repressor has been inactivated. In lane 3, we would not see any mRNA because the *lac* repressor does not bind to allolactose. Therefore, the *lac* repressor would remain bound to the *lac* operator even in the presence of lactose. In lane 4, the CAP would not activate transcription although the repressor would not prevent transcription. You would probably see a little bit of transcription, but not as much as the maximal level seen in lane 1.

In media without lactose, you would see a band only in lane 2. The only strain that would be induced would be the one that lacked a functional *lac* repressor.

E4. A. Yes, if you do not sonicate, then *ß*-galactosidase will not be released from the cell, and not much yellow color will be observed. (Note: You may observe a little yellow color because some of *ß*-ONPG may be taken into the cell.)

B. No, you should still get yellow color in the first two tubes even if you forgot to add lactose because the unmated strain does not have a functional *lac* repressor.

C. Yes, if you forgot to add *ß*-ONPG you could not get yellow color because the cleavage of *ß*-ONPG by *ß*-galactosidase is what produces the yellow color.

E5. The data indicate that two operators are needed because you need O_1 plus either O_2 or O_3 to get a high level of repression. If all three operators were needed, the deletion of any one of them should have prevented high levels of repression. This was not observed.

E6. You could mate a strain that has an F′ factor carrying a normal *lac* operon and a normal *lacI* gene to this mutant strain. Since the mutation is in the operator site, you would still continue to get expression of *ß*-galactosidase, even in the absence of lactose.

E7. It appears to be a mutation in the operator site so that a repressor protein cannot bind there and prevent *lac* operon expression in the absence of lactose.

E8. In this case, things are more complex because AraC acts as a repressor and an activator protein. If AraC were missing due to mutation, there would not be repression or activation of the *ara* operon in the presence or absence of arabinose. It would be expressed constitutively at low levels. The introduction of a normal *araC* gene into the bacterium on an F′ factor would restore normal regulation (i.e., a *trans*-effect).

E9. The results suggest that there is a mutation in the AraC protein so it cannot bind arabinose although it still binds to the operator sites correctly.

E10. Antisense RNA prevents the translation of the mRNA to which it is complementary. If the polypeptide that is encoded by the mRNA has an important function during early embryonic stages, the embryo may not develop properly. In this case, one may observe gross developmental abnormalities. For example, if the polypeptide is important in the formation of anterior structures, one may see that the head does not develop properly when the antisense RNA is injected into the fertilized oocyte. In contrast, if the polypeptide does not play an important role during the early stages of development, the embryo should develop properly.

Questions for Student Discussion/Collaboration

1. An advantage of transcriptional regulation is that it is very efficient. The cells do not waste their energy expressing genes they do not need. On the negative side, it is rather slow. It takes several minutes to turn a gene on and many minutes to degrade preexisting proteins. By comparison, posttranslational control is very fast, since a protein's activity can be abruptly turned on or off. Translational control is faster than transcriptional control but slower than posttranslational control.

2. A DNA loop may inhibit transcription by preventing RNA polymerase from recognizing the promoter. Or it may inhibit transcription by preventing the formation of the open complex. The bend may expose the base sequence that the sigma factor of RNA polymerase is recognizing. The bend may "open up" the major groove so that this base sequence is more accessible to the binding by sigma factor and RNA polymerase.

3. The loss of the λ repressor would allow the transcription from P_R and P_L. Transcription from P_L would eventually lead to the expression of the *xis* gene and excision of the prophage. Under these conditions, the phage would then follow the steps of the lytic cycle.

Chapter 15: Gene Regulation in Eukaryotes

Student Learning Objectives

Upon completion of this chapter you should be able to:

1. Understand the types of regulatory transcription factors and their physical structures that make them able to regulate gene expression.
2. Understand the relationship between chromatin structure and gene expression.
3. Understand how RNA modifications can influence gene expression.

15.1 Regulatory Transcription Factors

Overview

As would be expected, the regulation of eukaryotic genes has a much greater level of complexity than prokaryotic regulation. Yet in many ways the processes are similar, such as the use of proteins to block or enhance transcription. This section of the chapter examines the role of eukaryotic transcription factors. Some of these were previously introduced (Chapter 12). This section goes in detail on how these transcription factors are themselves regulated.

One important aspect of this section is the comparison of steroid hormones and CREB proteins. The differences between steroid and non-steroid hormones is discussed in many introductory and cell biology classes. The discussion here covers the interesting differences in these systems from a genetic perspective.

Key Terms

Activator
Bidirectional
cAMP response element
cAMP response element-binding protein
Control elements
Cyclic-adenosine monophosphate
Domains
Down regulation
Enhancer
General transcription factors
Glucocorticoid receptors
Heteodimer
Homodimer
Mediator
Motif
Orientation dependent
Regulatory elements
Regulatory transcription factors
Repressors
Response elements
Silencers
Steroid receptor
TFIID
Transcription factor
Up regulate

Focal Points

- Overview of regulatory transcription factors (Figure 15.2)
- The differences in the action of steroid proteins and CREB protein (Figure 15.6; 15.7)

Exercises and Problems

For questions 1 to 8, complete the statement regarding regulatory transcription factors.

1. An _____ transcription factor binds to DNA sequences called enhancers.
2. An area of a transcription factor that has a specific function is called a ______.
3. Helix-turn-helix and zinc finger are examples of _____ in transcription factors.
4. A repressor transcription factor binds to DNA sequences called _______.
5. ____ transcription factors influence the binding of the RNA polymerase to the core promoter.
6. An increase in transcription by a transcription factor is called _____ regulation.
7. Silencers and repressors are associated with the ____ regulation of a gene.
8. The combination of two different transcription factors creates a ________.

For questions 9 to 13, match each of the following with its correct description.

_____ 9. Effector molecule

_____ 10. TFIID

_____ 11. Protein-protein interactions

_____ 12. Mediator

_____ 13. Phosphorylation

a. The formation of a homodimer is an example of this.
b. Acts to covalently “turn-on” the transcription factor.
c. This molecule interacts between the transcription factors and the RNA polymerase.
d. These may bind to a transcription factor and regulate its activity.
e. A general transcription factor that attracts the RNA polymerase to the core promoter.

For each of the following, indicate whether the statement applies to a steroid hormone or CREB protein.

_____ 14. The molecule physically enters the cytosol of the cell.

_____ 15. Activates a glucocorticoid receptor that binds to glucocorticoid response elements.

_____ 16. The molecule binds to a receptor on the surface of the cell.

_____ 17. Changes the concentration of cAMP in the cell.

_____ 18. Interacts with heat shock proteins.

_____ 19. Activates a second messenger system

_____ 20. Interacts with cAMP response elements.

15.2 Changes in Chromatin Structure

Overview

There are four mechanisms by which alterations in chromatin structure may influence the regulation of eukaryotic genes. A brief description of each is provided in Table 15.1. Two of these (gene amplification and gene rearrangement) are very rare, and will be discussed in later chapters. Basically, alterations that produce chromatin in an open structure will increase the level of transcription, while changes that result in a more closed configuration will reduce the level of gene transcription.

Key Terms

ATP-dependent chromatin remodeling
Closed conformation
CpG islands
de novo methylation
DNA methylase
DNA methylation
Gene amplification
Gene rearrangement
Histone acetyltransferases
Housekeeping genes
Locus control region
Maintenance methylation
Methyl-CpG-binding proteins
Open conformation
Tissue-specific genes

Focal Points

- Regulation by chromatin changes (Table 15.1)
- Levels of DNA methylation (Figure 15.15)

Exercises and Problems

For each of the following, indicate what form of change is being described by the statement.

a. DNA methylation
b. changes in chromatin structure

_____ 1. Can be detected using the DNase I enzyme.

_____ 2. In general, this chemically silences gene expression.

_____ 3. Studies of lampbrush chromosomes have been used to show that this changes with transcription of the genes.

_____ 4. This is a heritable condition and explains genomic imprinting.

_____ 5. This does not have an influence on housekeeping genes.

_____ 6. Studies of thalassemias have indicated that this involves a change in nucleosome location.

_____ 7. May involve the use of histone acetyltransferases

_____ 8. Involves CpG islands near promoter regions.

_____ 9. This is controlled by locus control regions.

_____ 10. Utilizes a group of proteins belonging to the SWI/SNF family.

15.3 Regulation of RNA Processing and Translation

Overview

Chapter 12 introduced the concept that transcription produces a pre-mRNA that is then modified by a variety of mechanisms. Many of these mechanisms serve to regulate gene expression at the mRNA level. This may occur in one of two methods. First, the physical characteristics of the mRNA may be changed by alternative splicing and RNA editing. Second, the concentration of the mRNA may be altered. Generally, the lower the level of mRNA, the lower the amount of the gene being expressed in the cell.

Before proceeding, read Table 15.2 to provide an overview of these processes. Each of the examples in the text is linked to a specific example in a eukaryotic organism. Not every eukaryotic species uses all of these mechanisms, but almost all species use at least one of these mechanisms.

Key Terms

3'-untranslated region
Alternative exons
Alternative splicing
AU-rich element
Constitutive exons
Dicer
Exon skipping
Guide RNA
Iron regulatory protein
Iron response element
PolyA-binding protein
RNA induced silencing complex
RNA interference
Splicing factors
SR proteins

Focal Points

- Overview of gene regulation by RNA processing (Table 15.2)

Exercises and Problems

For questions 1 to 13, determine which of the following mechanisms of RNA processing regulation is associated with the statement.

a. alternative splicing
b. mRNA concentration
c. RNA editing
d. RNA interference

_____ 1. When polyA binding protein can no longer bind to the mRNA, it is degraded by endo- and exonucleases.

_____ 2. The mRNA from this process will always contain a set of constitutive exons.

_____ 3. The action of dicer and RISC act on double-stranded RNA, resulting in the inactivation of specific mRNA molecules.

_____ 4. Uses guide RNA.

_____ 5. The length of the polyA tail determines the longevity of the mRNA in the cell.

_____ 6. This may serve as a host defense mechanism against viruses and transposable elements.

_____ 7. Genes introduced into the genome may produce antisense RNA.

_____ 8. SR proteins choice specific sites for cutting the pre-mRNA.

_____ 9. Can also be called exon skipping.

_____ 10. May change the base content of a pre-mRNA following transcription.

_____ 11. Uses an endonuclease to remove part of the pre-mRNA and replace it with specific bases.

_____ 12. Produces families of proteins with similar functions.

_____ 13. Internal regions called 3'-UTR and ARE dictate that the mRNA be short-lived.

For questions 14 to 19, choose from the following translational regulatory mechanisms.

a. RNA binding proteins
b. phosphorylation
c. both a and b

_____ 14. Specifically targets the eIF2 and eIF4F proteins.

_____ 15. Depending on the target, may serve to either increase or decrease translation.

_____ 16. Example is the IRP protein.

_____ 17. Interactions of regulatory elements influence the stability of the mRNA molecule.

_____ 18. Frequently targets the initiation factors of translation.

_____ 19. Conditions such as heat shock and viral infections may activate this pathway.

Chapter Quiz

1. Which of the following is not a regulatory region of a eukaryotic gene?
   a. operator
   b. enhancer
   c. silencers
   d. core promoter
   e. all of the above are present in eukaryotic cells

2. Which of the following regulatory mechanisms involves the use of a second messenger system?
   a. glucosteroid hormones
   b. ATP-dependent chromatin remodeling
   c. CREB protein
   d. formation of a homodimer

3. Zinc finger and helix-turn-helix are examples of ____________.
a. response elements
b. phosphorylation sites
c. domains
d. motifs

4. CpG islands are associated with what type of regulation?
a. RNA editing
b. DNA methylation
c. alternative splicing
d. RNA interference

5. The formation of double-stranded RNA is involved in what mechanism?
a. RNA editing
b. RNA binding proteins
c. phosphorylation
d. RNA interference

6. This uses the product of a single gene to produce a family of similar proteins.
a. phosphorylation
b. DNA methylation
c. chromatin packaging
d. alternative splicing

7. This form of regulation is heritable.
a. ATP-dependent chromatin remodeling
b. phosphorylation
c. DNA methylation
d. DNase I sensitivity

8. Iron regulation in mammalian cells is an example of _______ regulation.
a. RNA editing
b. RNA binding proteins
c. phosphorylation
d. RNA interference

9. In addition to regulation, this provides a host defense against viruses and transposable elements.
a. RNA editing
b. chromatin remodeling
c. phosphorylation
d. RNA interference

10. The binding of polyA-binding proteins in this system increases mRNA stability.
a. mRNA concentration
b. DNA methylation
c. alternative splicing
d. RNA interference

Answer Key for Study Guide Questions

This answer key provides the answers to the exercises and chapter quiz for this chapter. Answers in parentheses ( ) represent possible alternate answers to a problem, while answers marked with an asterisk (*) indicate that the response to the question may vary.

15.1

1. activator
2. domain
3. motifs
4. silencers
5. general
6. up
7. down
8. heterodimer
9. d
10. e
11. a
12. c
13. b
14. steroid
15. steroid
16. CREB
17. CREB
18. steroid
19. CREB
20. CREB

15.2

1. b
2. a
3. b
4. a
5. a
6. b
7. b
8. a
9. b
10. b

15.3

1. b
2. a
3. d
4. c
5. b
6. d
7. d
8. a
9. a
10. c
11. c
12. a
13. b
14. b
15. c
16. a
17. a
18. c
19. b

Quiz

1. a
2. c
3. d
4. b
5. d
6. d
7. c
8. b
9. d
10. a

Answers to Conceptual and Experimental Questions

Conceptual Questions

C1. The common points of control are as follows:

1. DNA-chromatin structure. This includes gene amplification—increase in copy number; gene rearrangement—as in immunoglobulin genes; DNA methylation—attachment of methyl groups, which inhibits transcription; locus control regions—sites that control chromatin conformation.
2. Transcription. This includes transcription factors/response elements—interactions that can activate or inhibit transcription.
3. RNA level. This includes RNA processing—regulation of splicing via SR proteins; RNA stability—regulation of RNA half-life; RNA translation—regulation of the ability of RNAs to be translated.
4. Protein level. This includes feedback inhibition—small molecules that modulate enzyme activity; posttranslational modification—covalent changes to protein structure that affect protein activity.

C2. Response elements are relatively short genetic sequences that are recognized by regulatory transcription factors. After the regulatory transcription factor(s) has bound to the response element, it will affect the rate of transcription, either activating it or repressing it, depending on the action of the regulatory protein. Response elements are typically located in the upstream region near the promoter, but they can be located almost anywhere (i.e., upstream and downstream) and even quite far from the promoter.

C3. Transcription factor modulation refers to different ways that the function of transcription factors can be regulated. The three general ways are the binding of an effector molecule, protein-protein interactions, and posttranslational modifications.

C4. Transactivation occurs when a regulatory transcription factor binds to a response element and activates transcription. Transactivators may interact with TFIID and/or mediator to promote the assembly of RNA polymerase and general transcription factors at the promoter region. They also could alter the structure of chromatin so that RNA polymerase and transcription factors are able to gain access to the promoter. Transinhibition occurs when a regulatory transcription factor inhibits transcription. Transinhibitors or repressors also may interact with TFIID and/or mediator to inhibit RNA polymerase.

C5. A. True.

B. False.

C. True.

D. False, it causes up regulation.

C6. A. DNA binding.

B. DNA binding.

C. Protein dimerization.

C7. Glucocorticoid receptor: binding of an effector molecule and protein-protein interactions

CREB protein: covalent modification and protein-protein interactions

C8. For the glucocorticoid receptor to bind to a GRE, the cell must be exposed to the hormone and it must enter the cell. It binds to the receptor, which releases HSP90. After HSP90 is released, the receptor dimerizes and then enters the nucleus. Once inside the nucleus, the dimer will recognize a pair of GREs and bind to them, thereby leading to the activation of specific genes that have two GREs next to them.

C9. It could be in the DNA-binding domain so that the receptor would not recognize GREs.

It could be in the HSP90 domain so that HSP90 would not be released when the hormone binds.

It could be in the dimerization domain, so that the receptor would not dimerize.

It could be in the nuclear localization domain so that the receptor would not travel into the nucleus.

It could be in the transactivation domain so that the receptor would not activate transcription, even though it could bind to GREs.

C10. Phosphorylation of the CREB protein causes it to act as a transactivator. The unphosphorylated CREB protein can still bind to CREs, but it does not stimulate transcription.

C11. A. No effect.

B. No effect.

C. It would be inhibited.

D. No effect.

C12. A. Eventually, the glucocorticoid hormone will be degraded by the cell. The glucocorticoid receptor binds the hormone with a certain affinity. The binding is a reversible process. Once the concentration of the hormone falls below the affinity of the hormone for the receptor, the receptor will no longer have the glucocorticoid hormone bound to it. When the hormone is released, the glucocorticoid receptor will change its conformation, and it will no longer bind to the DNA.

B. An enzyme known as a phosphatase will eventually cleave the phosphate groups from the CREB protein. When the phosphates are removed, the CREB protein will stop transactivating transcription.

C13. Possibility 2 is the correct one. Since we already know that the E protein and the Id protein form heterodimers with myogenic bHLH, we expect all three proteins to have a leucine zipper. Leucine zippers promote dimer formation. We also need to explain why the Id protein inhibits transcription, while the E protein enhances transcription. As seen in possibility 2, the Id protein does not have a DNA-binding domain. Therefore, if it forms a heterodimer with myogenic bHLH, the heterodimer probably will not bind to the DNA very well. In contrast, when the E protein forms a heterodimer with myogenic bHLH, there will be two DNA-binding domains, which would promote good binding to the DNA. Note: A description of these proteins is found in Chapter 23.

C14. The enhancer found in A would work, but the ones found in B and C would not. The sequence that is recognized by the transcriptional activator is 5′–GTAG–3′ in one strand and 3′–CATC–5′ in the opposite strand. This is the same arrangement found in A. In B and C, however, the arrangement is 5′–GATG–3′ and 3′–CATC–5′. In the arrangement found in B and C, the two middle bases (i.e., A and T) are not in the correct order.

C15. There are four types of bases (A, T, G, and C) and this CRE sequence contains 8 bp, so according to random chance, it should occur every 4^8 bp, which equals every 65,536 bp. If we divide 3 billion by 65,536, this sequence is expected to occur approximately 45,776 times. This is much greater than a few dozen. There are several reasons why the CREB protein does not activate over 45,000 genes.

1. To create a functional CRE, there needs to be two of these sequences close together, because the CREB protein functions as a homodimer.
2. CREs might not be near a gene.
3. The conformation of chromatin containing a CRE might not be accessible to binding by the CREB protein.

C16.

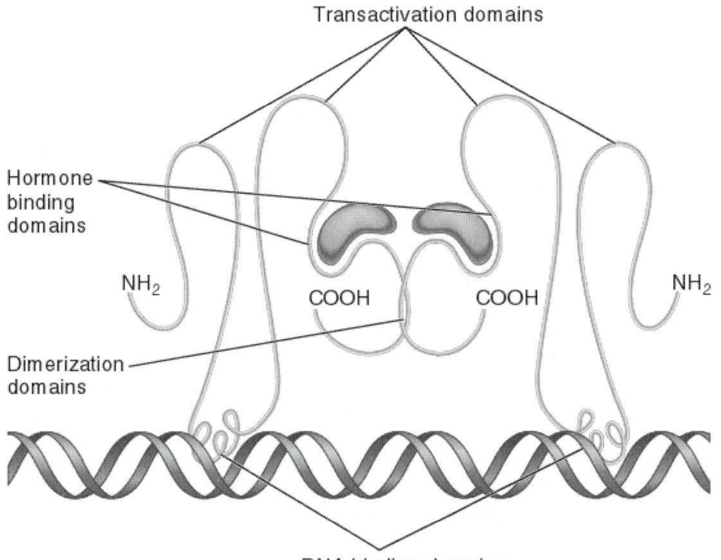

This is one hypothetical drawing of the glucocorticoid receptor. As discussed in your textbook, it forms a homodimer. The dimer shown here has rotational symmetry. If you flip the left side around, it has the same shape as the right side. The hormone-binding, DNA-binding, dimerization, and transactivation domains are labeled. The glucocorticoid hormone is shown in *orange.*

C17. The mutation could cause a defect in the following:

1. adrenaline receptor
2. G protein
3. adenylate cyclase
4. protein kinase A
5. CREB protein
6. CREs of the tyrosine hydroxylase gene

If other genes were properly regulated by the CREB protein, we would conclude that the mutation is probably within the tyrosine hydroxylase gene itself. Perhaps a CRE has been mutated and no longer recognizes the CREB protein.

C18. The 30 nm fiber is the predominant form of chromatin during interphase. The chromatin must be converted from a closed conformation to an open conformation for transcription to take place. This "opening" process involves less packing of the chromatin and may involve changes in the locations of histone proteins. Transcriptional activators recruit histone acetyltransferase and ATP-dependent remodeling enzymes to the region, which leads to a conversion to the open conformation.

C19. The attraction between DNA and histones occurs because the histones are positively charged and the DNA is negatively charged. The covalent attachment of acetyl groups decreases the amount of positive charge on the histone proteins and thereby may decrease the binding of the DNA. In addition, histone acetylation may attract proteins to the region that loosen chromatin compaction.

C20. The binding of the DNA to the core histones could

1. prevent transcriptional activators from recognizing enhancer elements that are required for transcriptional activation.
2. prevent RNA polymerase from binding to the core promoter.
3. prevent RNA polymerase from forming an open complex in which the DNA strands separate.

C21. The function of a locus control region is to control the chromatin conformation in a region containing one or more genes. The control region will govern whether or not the genes in this region are accessible to transcription factors and RNA polymerase.

C22. The translocation breakpoint occurred between the β-globin gene and the locus control region. Therefore, the β-globin gene is not expressed because it needs the locus control region to allow the chromatin to be in an accessible conformation.

C23. DNA methylation is the attachment of a methyl group to a base within the DNA. In many eukaryotic species, this occurs on cytosine at a CG sequence. After *de novo* methylation has occurred, it is passed from mother to daughter cell via a maintenance methylase. Since DNA replication is semiconservative, the newly made DNA contains one strand that is methylated and one that is not. The maintenance methylase recognizes this hemimethylated DNA and methylates the cytosine in the unmethylated DNA strand.

C24. Perhaps the methylase is responsible for methylating and inhibiting a gene that causes a cell to become a muscle cell. The methylase is inactivated by the mutation.

C25. A CpG island is a stretch of 1,000 to 2,000 bp in length that contains several CG sequences. CpG islands are often located near promoters. When the island is methylated, this inhibits transcription. This inhibition may be the result of the inability of the transcriptional activators to recognize the methylated promoter and/or the effects of methyl-CpG-binding proteins, which may promote a closed chromatin conformation.

C26. The function of splicing factors is to influence the selection of splice sites in RNA. In certain cell types, the concentration of particular splicing factors is higher than in other tissues. The high concentration of particular splicing factors, and the regulation of their activities, may promote the selection of particular splice sites and thereby lead to tissue-specific splicing.

C27. As shown in Figure 15.18, the unique feature of the smooth muscle mRNA for α–tropomyosin is that it contains exon 2. Splicing factors that are only found in smooth muscle cells may recognize the splice junction at the 3′ end of intron 1 and the 5′ end of intron 2 and promote splicing at these sites. This would cause exon 2 to be included in the mRNA. Furthermore, since smooth muscle mRNA does not contain exon 3, a splicing suppressor may bind to the 3′ end of intron 2. This would promote exon skipping so that exon 3 would not be contained in the mRNA.

C28. This person would be unable to make ferritin, because the IRP would always be bound to the IRE. The amount of transferrin receptor mRNA would be high, even in the presence of high amounts of iron, because the IRP would always remain bound to the IRE and stabilize the transferrin receptor mRNA. Such a person would not have any problem taking up iron into his/her cells. In fact, this person would take up a lot of iron via the transferrin receptor, even when the iron concentrations were high. Therefore, he/she would not need more iron in the diet. However, excess iron in the diet would be very toxic for two reasons. First, the person cannot make ferritin, which prevents the toxic buildup of iron in the cytosol. Second, when iron levels are high, the person would continue to synthesize the transferrin receptor, which functions in the uptake of iron.

C29. A cell may need to respond rather quickly to a toxic substance in order to avoid cell death. Translational and posttranslational mechanisms are much faster. By comparison, transcriptional activation takes a lot of time. In this case, it is necessary to up regulate the gene, synthesize the mRNA, and then translate the mRNA to make a functional protein.

C30. A disadvantage of mRNAs with a short half-life is that the cells probably waste a lot of energy making them. If a cell needs the protein encoded by a short-lived mRNA, the cell has to keep transcribing the gene that encodes the mRNA because the mRNAs are quickly degraded. An advantage of short-lived mRNAs is that the cell can rapidly turn off protein synthesis. If a cell no longer needs the polypeptide encoded by a short-lived mRNA, it can stop transcribing the gene, and the mRNA will be quickly degraded. This will shut off the synthesis of more proteins rather quickly. With most long-lived mRNAs, it will take much longer to shut off protein synthesis after transcription has been terminated.

C31. Conditions such as viral infection, nutrient deprivation, heat shock, and toxic heavy metals lead to the phosphorylation of eIF2*a*. Under these conditions, a cell is in a state where it should not divide because this would not be good for the survival of the multicellular organism of which it is a part.

C32. If mRNA stability is low, this means that it is degraded more rapidly. Therefore, low stability results in a low mRNA concentration. The length of the polyA tail is one factor that affects stability. A longer tail makes mRNA more stable. Certain mRNAs have sequences that affect their half-lives. For example, AREs are found in many short-lived mRNAs. The AREs are recognized by cellular proteins that cause the mRNAs to be rapidly degraded.

C33. The binding of IRP to the IRE inhibits the translation of ferritin mRNA and enhances the stability of the transferrin receptor mRNA. The increase in the stability of transferrin receptor mRNA increases the concentration of this mRNA and ultimately leads to more transferrin receptor protein. Conditions of low iron promote the binding of IRP to the IRE, leading to a decrease in ferritin protein and an increase in transferrin receptor protein. When the iron concentration is high, iron binds to IRP, causing it to be released from the IRE. This allows the ferritin mRNA to be translated and also causes a decrease in the stability of transferrin receptor mRNA. Under these conditions, more ferritin protein is translated, and less transferrin receptor is made.

Experimental Questions

E1. A DNase-I sensitive site is one that is very easily cleaved by DNase-I compared to most other sites in the chromosomal DNA. In other words, its conformation is very accessible to cleavage. When a gene becomes transcriptionally active, it will become more susceptible to DNase-I digestion because the DNA in this region has been converted from a closed to an open conformation. This will make it easier for DNase-I to gain access to the DNA and cleave it.

E2. S1 nuclease cuts single-stranded DNA but not double-stranded DNA. If a gene is in an open conformation and cut by DNase-I, it would not hybridize to the probe that is complementary to it. In this case, the single-stranded probe would be chewed up by the S1 nuclease. In contrast, if the gene is in a closed conformation and resistant to DNase-I, the probe will hybridize to the gene and will not be digested with S1 nuclease. Therefore, the ability of S1 nuclease to digest the probe tells you whether or not the gene is being cleaved by DNase-I.

The precipitation step is needed to separate the DNA fragments from free nucleotides. If the radioactive probe is bound to a complementary DNA strand, it will be protected from degradation by S1 nuclease; it will be found in the pellet. This occurs when the globin gene is in a closed conformation. In contrast, if the radioactive probe does not bind to a complementary DNA strand, it will be digested into free nucleotides; the free nucleotides will remain in the supernatant. This occurs when the globin gene is in an open conformation.

E3. Certain individuals had defects in globin gene expression even though the structural gene itself was intact. This led researchers to suspect that other regions nearby were important in controlling the chromosomal conformation in a way that would make the globin genes accessible to RNA polymerase and transcription factors.

E4. These results indicate that the fibroblasts have a maintenance methylase because they can replicate and methylate DNA if it has already been methylated. However, the cells do not express a *de novo* methylase that recognizes this DNA segment, because the DNA in the daughter cells remains unmethylated if the donor DNA was unmethylated.

E5. If the DNA band is 3,800 bp, it means that the site is unmethylated because it has been cut by *Not*I (and *Not*I cannot cut methylated DNA). If the DNA fragment is 5,300 bp, the DNA is not cut by *Not*I, so we assume that it is methylated. Now let's begin our interpretation of these data with lane 4. When the gene is isolated from root tissue, the DNA is not methylated because it runs at 3,800 bp. This suggests that gene *T* is expressed in root tissue. In the other samples, the DNA runs at 5,300 bp, indicating that the DNA is methylated. The pattern of methylation seen here is consistent with the known function of gene *T*. We would expect it to be expressed in root cells since it functions in the uptake of phosphate from the soil. It would be silenced in the other parts of the plant via methylation.

E6. Based on these results, there are enhancers that are located in regions A, D, and E. When these enhancers are deleted, the level of transcription is decreased. There also appears to be a silencer in region B, because a deletion of this region increases the rate of transcription. There do not seem to be any response elements in region C, or at least not any that function in muscle cells.

E7. A. Based on the transformation of kidney cells, a silencer is in region B. Based on the transformation of pancreatic cells, an enhancer is in region A.

B. The pancreatic cells must not express the repressor protein that binds to the silencer in region B.

C. The kidney cells must express a repressor protein that binds to region B and represses transcription. Kidney cells express the downstream gene only if the silencer is removed. As mentioned in part B, this repressor is not expressed in pancreatic cells.

E8. The results indicate that protein X binds to the DNA fragment and retards its mobility (lanes 3 and 4). However, the hormone is not required for DNA binding. Because we already know that the hormone is needed for transactivation, it must play some other role. Perhaps, the hormone activates a signaling pathway that leads to the phosphorylation of the transcription factor, and phosphorylation is necessary for transactivation. This situation would be similar to the CREB protein, which is activated by phosphorylation. The CREB protein can bind to the DNA whether or not it is phosphorylated.

E9. The gel shown in (a) is correct for ferritin mRNA regulation. The presence and absence of iron does not affect the amount of mRNA, it affects the ability of the mRNA to be translated. The gel shown in (b) is correct for transferrin receptor mRNA. In the absence of iron, the IRP binds to the IRE in the mRNA and stabilizes it, so the mRNA level is high. In the presence of iron, the iron binds to IRP and IRP is released from the IRE, and the transferrin receptor mRNA is degraded.

Questions for Student Discussion/Collaboration

1. DNA methylation can regulate gene expression in a tissue-specific way because *de novo* methylation may only occur next to particular genes in particular tissues. This would tend to keep them permanently turned off during later stages of development. Genes may be methylated in the germ line to prevent their expression. If a gene was methylated in the germ line, it would have to be demethylated during gamete formation or at an early stage of embryonic development in the offspring so that it could be expressed in particular tissues in the offspring.
2. Probably the most efficient method would be to systematically make deletions of progressively smaller sizes. For example, you could begin by deleting 20,000 bp on either side of the gene and see if that affects transcription. If you found that only the deletion on the 5′ end of the gene had an effect, you could then start making deletions from the 5′ end, perhaps in 10,000 bp or 5,000 bp increments until you localized response elements. You would then make smaller deletions in the putative region until it was down to a hundred or a few dozen nucleotides. At this point, you might conduct site-directed mutagenesis, as described in Chapter 18, as a way to specifically identify the response element sequence.
3. Transcription factors and regulatory splicing factors are similar in that they recognize a particular nucleotide sequence. They can act in a tissue-specific or developmentally specific manner. Likewise, they are similar in that they influence the activity of some other multisubunit complex. Regulatory transcription factors affect the activity of RNA polymerase and general transcription factors; splicing factors affect the splice sites of the spliceosome. Regulatory transcription factors and splicing factors have multiple domains that play different functional roles; they have modular structures. For example, regulatory transcription factors have domains for nucleotide binding and may have domains for dimerization, transactivation, and so forth. Regulatory transcription factors and splicing factors are different in that the former recognizes DNA, the latter RNA. With regard to differences, the role of regulatory transcription factors is to influence the rate of transcription, whereas the role of splicing factors is to affect the specificity of splice site recognition.

Chapter 16: Gene Mutation and DNA Repair

Student Learning Objectives

Upon completion of this chapter you should be able to:

1. Recognize the different classes of mutations and their effects upon the organism.
2. Understand the concept of mutation rate.
3. Understand the mechanisms of DNA repair and the types of mutations that they recognize.

16.1 Consequences of Mutation

Overview

The term mutation typically is associated with a negative change in the genetic material. However, detrimental or lethal mutations are actually but one class of mutations. A mutation represents any change to the genetic material and can occur at the genome, chromosome or gene levels. The purpose of this first section of the chapter is to establish the basic types of mutations and the terminology associated with the study of mutations.

This chapter contains a significant amount of descriptive terminology. It is important that you have a good understanding of this material before proceeding into the remainder of the chapter. Before you finish reading a section on a certain type of mutation, take a good look at the associated figure. As you do so, ask yourself if you understand how this mutation effects the gene or sequence of DNA, and what the possible effects may be on the phenotype. In many cases an example of a disease or condition is provided. It is a good idea to know these as well.

Key Terms

Anticipation
Base substitution
Beneficial mutation
Chromosome mutation
Conditional mutants
Deleterious mutation
Down promoter mutation
Forward mutation
Frameshift mutations
Genetic mosaic
Genome mutations
Germ line
Germ-line mutation
Intergenic suppressor
Intragenic suppressor
Lethal mutation
Missense mutations
Mutation
Neutral mutation
Nonsense mutations
Point mutation
Polarity
Position effect
Reverse mutation
Reversion
Silent mutations
Single-gene mutations
Somatic cells
Somatic mutation
Suppressor mutations
Suppressors
Transition
Transversion
Trinucleotide expansion repeats
Up promoter mutation
Variant
Wild type

Focal Points

- Effects of point mutations (Table 16.1)
- Somatic versus germ-line mutations (Figure 16.4)

Exercises and Problems

For questions 1 to 10, match the form of mutation with its correct definition.

_____ 1. Frameshift mutations

_____ 2. Transition

_____ 3. Silent mutation

_____ 4. Missense mutation

_____ 5. Nonsense mutations

_____ 6. Point mutation

_____ 7. Transversion

_____ 8. Genome mutation

_____ 9. Chromosome mutation

_____ 10. Neutral mutation

a. A change in a single base pair in the DNA.
b. A mutation that does not change the amino acid sequence of the polypeptide.
c. The addition or deletion of one or two nucleotides.
d. A change that alters a single amino acid in the polypeptide.
e. A physical change in the structure of the chromosome.
f. Change a normal codon to a termination codon.
g. A point mutation involving a change from a purine to a purine.
h. A type of missense mutation that has no effect on protein function.
i. A change in the number of chromosomes.
j. A point mutation involving a change from a purine to a pyrimidine.

In questions 11 to 20, choose the correct type of mutation that matches the description provided.

_____ 11. Increases the chances of survival or the reproductive success of the organism.

_____ 12. A second mutation that restores the wild-type condition.

_____ 13. A suppressor that occurs within the same gene as the first mutation.

_____ 14. Causes death.

_____ 15. Decreases the chances of survival, but is not necessarily lethal.

_____ 16. Temperature-sensitive mutations.

_____ 17. Converts a variant back to the wild-type.

_____ 18. A suppressor that occurs in a different gene than the first mutation.

_____ 19. Converts the wild-type to a new variant.

_____ 20. The most common genotype in the population.

a. intragenic suppressor
b. conditional mutants
c. suppressor mutations
d. reversion
e. beneficial mutation
f. deleterious mutation
g. intergenic suppressor
h. lethal mutation
i. wild type
j. forward mutation

For questions 21 to 23, indicate whether the statement is true (T) or false (F). If false, correct the statement.

_____ 21. A mutation in a regulatory region of a gene that causes a decrease in the rate of transcription is called an up mutation.

_____ 22. The term adaptation refers to the fact that TNREs tend to get progressively longer in future generations.

_____ 23. The movement of a regulatory region due to a chromosomal mutation is called a position effect.

For questions 24 to 27, indicate whether the statement is associated with a germ-line mutation (G) or somatic cell mutation (S).

_____ 24. May produce a genetic mosaic.

_____ 25. Is heritable.

_____ 26. Effects all of the cells of the organism.

_____ 27. This form of mutation would not be carried by the gametes.

16.2 Occurrence and Causes of Mutation

Overview

The second section of the chapter explores the debate over whether mutations are directed events, or represent random changes in the genetic material. In order to study these competing theories, it is necessary to understand the difference between mutation rate and mutation frequency. Mutation rate is the likelihood that a gene will be changed by a new mutation, while frequency represents the abundance of the mutant gene in a given population.

After determining that mutations are random (Figure 16.7), the chapter moves into a discussion of the forms of spontaneous mutations. This is followed by a discussion of induced mutations and the mutagens that cause them. You should be familiar with the major mutagen classes in Table 16.5 and their effects on the DNA, before leaving this section. The Ames test (Figure 16.16) remains a commonly used mechanism of determining whether a compound is actually a mutagen.

Key Terms

2-aminopurine
5-bromouracil
Acridine dyes
Ames test
Apurinic site
Chemical mutagens
Deamination
Depurination
Directed mutation hypothesis
Ethyl methanesulfonate
Fluctuation test
Induced mutations
Mutagens
Mutation frequency
Mutation rate
Nitrogen mustards
Nitrous acid
Physical mutagens
Proflavin
Random mutation theory
Spontaneous mutations
Tautomeric shift
Thymine dimers

Focal Points

- Causes of mutations (Table 16.4)
- Tautomeric shifts (Figure 16.10)
- Classes of mutagens (Table 16.5)

Exercises and Problems

Complete the following statements regarding mutations and mutation rates.

1. A(n) ____ mutation is one that is caused by an environmental agent.
2. A mutagen may be either a chemical compound or a ________.
3. The _____ test of Luria and Delbrück distinguished between the opposing theories of physiological adaptation and spontaneous mutation.
4. In a fluctuation test, the rate of mutation is predicted to be _____ as physiological adaptation is occurring.
5. Lamarck's theory of use and disuse favored the ________ mutation theory (hypothesis).
6. In the Lederberg experiment, a process called _______ was used to distinguish between the directed mutation hypothesis and random mutation theory.
7. The probability that a new mutation will alter a given gene is called the ________.
8. The average mutation rate is _____ to _____ per cell generation.
9. Mutation ______ indicates the prevalence of a specific mutation in the population.
10. A region of a gene that is more likely to mutate than another region is called a _______.

For questions 11 to 16, indicate the type of spontaneous mutation that the statement is referring to.

a. deamination
b. tautomeric shift
c. depurination

_____ 11. Site where this occurs tend to be hot spots of mutation.

_____ 12. A temporary change in base structure.

_____ 13. This typically occurs at a cytosine in the DNA.

_____ 14. This must occur immediately prior to replication in order to cause a mutation.

_____ 15. The most common form of spontaneous mutation.

_____ 16. Occurs as a result of a spontaneous reaction with water, which releases the base from the sugar.

For questions 17 to 22, match each of the following mutagens with its correct description.

_____ 17. Nitrous acid

_____ 18. 5-bromouracil

_____ 19. Ethyl methanosulfonate

_____ 20. Acridine dyes

_____ 21. X-rays

_____ 22. UV light

a. A nucleotide base analog that enters into the DNA during replication.
b. A nonionizing radiation that causes thymine dimers.
c. Distort the double-helix structure of DNA by integrating into the helix.
d. An alkylating agent.
e. The first mutagen to be discovered. A form of ionizing radiation.
f. Changes cytosine to uracil by adding a keto functional group.

16.3 DNA Repair

Overview

The final section of this chapter examines the various mechanisms by which cells may repair damage to the DNA caused by spontaneous or induced mutations. This section builds upon the information provided in section 16.2. A summary of the major mechanisms is given in Table 16.6.

You will notice below that the focal points for this chapter include every diagram. That is because it is almost impossible to understand DNA repair by reading the text. As you review this material you should frequently refer back to the diagrams. As you do so, remember that these graphics actually represent three-dimensional structures in the cell and what you are observing are three-dimensional interactions.

Key Terms

AP-endonuclease
Base excision repair
Base mismatch
DNA-N-glycosylases
Methyl-directed mismatch repair
Nick translation
Nucleotide excision repair
Recombinational repair
SOS response
Translesion synthesis
Transcription-repair coupling factor

Focal Points

- DNA repair systems (Table 16.6)
- DNA repair mechanisms (Figures 16.17 to 16.23)

Exercises and Problems

For questions 1 to 10, choose the repair mechanism that is best associated with the statement. Some statements may have more than one answer.

a. nucleotide excision repair
b. methyl-directed mismatch repair
c. translesion synthesis
d. recombinational repair
e. base excision repair

_____ 1. Uses an enzyme called the DNA-N-glycosylases.

_____ 2. Utilizes as group of proteins called Mut.

_____ 3. The disease xeroderma pigmentosum is caused by a defect in this system.

_____ 4. This is associated with the SOS response in bacteria.

_____ 5. Removes a short segment of DNA around the damaged bases.

_____ 6. Removes a single abnormal base from the DNA.

_____ 7. This may be used to repair an unreplicated gap in the DNA.

_____ 8. Identifies the parental strand by methylation patterns, then uses it as a template for repairing the damage to the unmethylated strand.

_____ 9. Abnormal nucleotides are recognized by AP-endonuclease.

_____ 10. Uses the Uvr proteins in the response.

Chapter Quiz

1. Which of the following types of mutations changes a normal codon to a stop codon?
a. silent mutation
b. frameshift mutation
c. nonsense mutation
d. missense mutation
e. none of the above

2. A ____ mutation returns a variant to the wild type condition.
a. forward
b. reverse
c. lateral
d. vertical
e. none of the above

3. A mutation that influences the phenotype only under specific environmental conditions is called a ________ mutation
a. conditional
b. silent
c. transitional
d. missense

4. Position effect is associated with which of the following?
   a. a change in chromosome number
   b. a point mutation
   c. a change in chromosome structure
   d. anticipation
   e. physiological adaptation

5. A deamination event is an example of a ______ mutation.
   a. induced
   b. physiological
   c. conditional
   d. spontaneous

6. The first agent demonstrated to induce mutations was _________.
   a. UV light
   b. acridine dyes
   c. X rays
   d. uranium

7. Which of the following tests is used to determine if a chemical is a mutagen?
   a. fluctuation test
   b. replica plating
   c. tautomeric test
   d. Ames test

8. Which of the following compounds is an example of a base analog?
   a. acridine dyes
   b. ethyl methanosulfonate
   c. 5-bromouracil
   d. nitrous acid

9. Gaps in the DNA are best repaired by which of the following repair mechanisms?
   a. base excision repair
   b. nucleotide excision repair
   c. SOS response
   d. recombinational repair

10. DNA-N-glycosylases and AP-endocnucleases are used in which of the following systems?
   a. base excision repair
   b. nucleotide excision repair
   c. recombinational repair
   d. thymine dimer repair

Answer Key for Study Guide Questions

This answer key provides the answers to the exercises and chapter quiz for this chapter. Answers in parentheses ( ) represent possible alternate answers to a problem, while answers marked with an asterisk (*) indicate that the response to the question may vary.

16.1

1. c
2. g
3. b
4. d
5. f
6. a
7. j
8. i
9. e
10. h
11. e
12. c
13. a
14. h
15. f
16. b
17. d
18. g
19. j
20. I
21. T
22. F, anticipation
23. T
24. S
25. G
26. G
27. S

16.2

1. induced
2. physical agent
3. fluctuation
4. constant
5. directed
6. replica plating
7. mutation rate
8. 10^{-5}, 10^{-9}
9. frequency
10. hot spot
11. a
12. b
13. a
14. b
15. c
16. c
17. f
18. a
19. d
20. c
21. e
22. b

16.3

1. e
2. b
3. a
4. c
5. a
6. e
7. d
8. b
9. e
10. a

Quiz

1. c
2. b
3. a
4. c
5. d
6. c
7. d
8. c
9. d
10. a

Answers to Conceptual and Experimental Questions

Conceptual Questions

C1. A. G→A, which is a transition.

B. T→G, which is a transversion.

C. A single-nucleotide deletion.

D. C→G, which is a transversion.

C2. A gene mutation is a relatively small mutation that is localized to a particular gene. A chromosome mutation is a large enough change in the genetic material so that it can be seen with the light microscope. This would affect several genes. Genome mutations are changes in chromosome number.

C3. It is a gene mutation, a point mutation, a base substitution, a transition mutation, a forward mutation, a deleterious mutation, a mutant allele, a nonsense mutation, a conditional mutation, and a temperature-sensitive lethal mutation.

C4. A suppressor mutation suppresses the phenotypic effects of some other mutation. Intragenic suppressors are within the same gene as the first mutation. Intergenic suppressors are in some other gene.

C5. A. It would probably inhibit protein function, particularly if it was not near the end of the coding sequence.

B. It may or may not affect protein function depending on the nature of the amino acid substitution and whether the substitution is in a critical region of the protein.

C. It would increase the amount of functional protein.

D. It may affect protein function if the alteration in splicing changes an exon in the mRNA that results in a protein with a perturbed structure.

C6. The X rays did not produce a mutation because a mutation is a heritable change in the genetic material. In this case, the X rays have killed the cell, so changes in DNA structure cannot be passed from cell to cell or from parent to offspring.

C7. A. Not appropriate, because the second mutation is at a different codon.

B. Appropriate.

C. Not appropriate, because the second mutation is in the same gene as the first mutation.

D. Appropriate.

C8. An efficient nonsense suppressor would probably inhibit cell growth because all of the genes that have their stop codons in the correct location would make proteins that would be too long. This would waste cellular energy, and in some cases, the elongated protein may not function properly.

C9. A. Silent, because the same amino acid (glycine) is encoded by GGA and GGT.

B. Missense, because a different amino acid is encoded by CGA compared to GGA.

C. Missense, because a different amino acid is encoded by GTT compared to GAT.

D. Frameshift, because an extra base is inserted into the sequence.

C10. Here are two possible examples:

The consensus sequences for many bacterial promoters are –35: 5′–TTGACA–3′ and –10: 5′–TATAAT–3′. Most mutations that alter the consensus sequence would be expected to decrease the rate of transcription. For example, a mutation that changed the –35 region to 5′–GAGACA–3′ would decrease transcription.

The sequence 5′–TGACGTCA–3′ is recognized by the CREB transcription factor. If this sequence was changed to 5′–TGAGGTCA–3′ the CREB protein would not recognize it very well, and the adjacent gene would not be regulated properly in the presence of cAMP.

C11. One possibility is that a translocation may move a gene next to a heterochromatic region of another chromosome and thereby diminish its expression; or it could be moved next to a euchromatic region and increase its expression. Another possibility is that the translocation breakpoint may move the gene next to a new promoter or regulatory sequences that may now influence the gene's expression.

C12. Random mutations are more likely to be harmful. The genes within each species have evolved to work properly. They have functional promoters, coding sequences, terminators, etc., that allow the genes to be expressed. Mutations are more likely to disrupt these sequences. For example, mutations within the coding sequence may produce early stop codons, frameshift mutations, and missense mutations that result in a nonfunctional polypeptide. On rare occasions, however, mutations are beneficial; they may produce a gene that is expressed better than the original gene or produce a polypeptide that functions better. As discussed in Chapter 25, beneficial mutations may be acted upon by natural selection over the course of many generations and eventually become the prevalent allele in the population.

C13. A. No, the position (i.e., chromosomal location) of a gene has not been altered.

B. Yes, the expression of a gene has been altered because it has been moved to a new chromosomal location.

C. Yes, the expression of a gene has been altered because it has been moved to a new chromosomal location.

C14. Yes, a person with cancer is a genetic mosaic. The cancerous tissue contains gene mutations that are not found in noncancerous cells of the body.

C15. If a mutation within the germ line is passed to an offspring, all the cells of the offspring's body will carry the mutation. A somatic mutation affects only the somatic cell in which it originated and all of the daughter cells that the somatic cell produced. If a somatic mutation occurs early during embryonic development, it may affect a fairly large region of the organism. Since germ-line mutations affect the entire organism, they are potentially more harmful (or beneficial), but this is not always the case. Somatic mutations can cause quite harmful effects such as cancer. Germ-line mutations may be passed to offspring.

C16. The drawing should show the attachment of a methyl or ethyl group to a base within the DNA. The presence of the alkyl group disrupts the proper base pairing between the alkylated base and the normal base in the opposite DNA strand.

C17. A thymine dimer can interfere with DNA replication because DNA polymerase cannot slide past the dimer and add bases to the newly growing strand. Alkylating mutagens such as nitrous acid will cause DNA replication to make mistakes in the base pairing. For example, an alkylated cytosine will base pair with adenine during DNA replication, thereby creating a mutation in the newly made strand. A third example is bromouracil, which is a thymine analogue. It may pair with guanine during DNA replication.

C18. A. Nitrous acid causes A—>G and C—>T mutations, which are transition mutations.

B. 5-bromouracil causes G—>A mutations, which are transitions.

C. Proflavin causes small additions or deletions, which may result in frameshift mutations.

C19. During TNRE, a trinucleotide repeat sequence gets longer. If someone was mildly affected with a TNRE disorder, he/she would be concerned that an expansion of the repeat might occur during gamete formation, yielding offspring more severely affected with the disorder.

C20. A spontaneous mutation originates within a living cell. It may be due to spontaneous changes in nucleotide structure, errors in DNA replication, or products of normal metabolism that may alter the structure of DNA. The causes of induced mutations originate from outside the cell. They may be physical agents, such as UV light or X rays, or chemicals that act as mutagens. Both spontaneous and induced mutations may cause a harmful phenotype such as a cancer. In many cases, induced mutations are avoidable if the individual can prevent exposure to the environmental agent that acts as a mutagen.

C21. Spontaneous mutations are random events in the sense that there is no outside force that is directing the mutation to a particular gene or a particular site within a gene. However, the structure of chromatin may cause certain regions of the DNA to be more susceptible to random mutations. For example, DNA in an open conformation may be more accessible to mutagens and more likely to incur mutations. Likewise, hot spots for mutation can occur within a single gene.

C22. Nitrous acid can change a cytosine to uracil. Excision repair systems could remove the defect and replace it with the correct base.

C23. Excision repair systems could fix this damage. Also, recombinational repair could fix the damage.

C24. A. True

B. False, the TNRE is not within the promoter, it is within the coding sequence.

C. True, CAG is a codon for glutamine.

D. False, CCG is a codon for proline.

C25. A spontaneous mutation is a mutation that happens as a result of events that occur within a living cell. A spontaneous mutation is not caused by an environmental agent. An induced mutation is caused directly or indirectly by an environmental agent. An individual can avoid an induced mutation if he/she avoids the harmful environmental agent.

C26. The mutation rate is the number of new mutations per gene per generation. The mutation frequency is the number of copies of a mutant gene within a population, divided by the total number of copies (mutant and nonmutant) of that gene. The mutation frequency may be much higher than the mutation rate if new mutations accumulate within a population over the course of many generations.

C27. The mutation frequency is the total number of mutant alleles divided by the total number of alleles in the population. If there are 1,422,000 babies, there are 2,844,000 copies of this gene (because each baby has 2 copies). The mutation frequency is 31/2,844,000, which equals 1.09×10^{-5}. The mutation rate is the number of new mutations per generation. There are 13 babies who did not have a parent with achondroplasia. Thirteen is the number of new mutations. If we calculate the mutation rate as the number of new mutations in a given gene per generation, then we should divide 13 by 2,844,000. In this case, the mutation rate would be 4.6×10^{-6}.

C28.

A. TTGGHTGUTGG

HHUUTHUGHUU

↓ First round of replication

TTGGHTGUTGG CCAAACACCAA

AACCCACAACC HHUUTHUGHUU

↓ Second round of replication

B. TTGGHTGUTGG TTGGGTGTTGG CCAAACACCAA CCAAACACCAA

AACCCACAACC AACCCACAACC GGTTTGTGGTT HHUUTHUGHUU

C29. The effects of mutations are cumulative. If one mutation occurs in a cell, this mutation will be passed to the daughter cells. If a mutation occurs in the daughter cell, now there will be two mutations. These two mutations will be passed to the next generation of daughter cells, and so forth. The accumulation of many mutations eventually kills the cells. That is why mutagens are more effective at killing dividing cells compared to nondividing cells. It is because the number of mutations accumulates to a lethal level.

There are two main side effects to this treatment. First, some normal (noncancerous) cells of the body, particularly skin cells and intestinal cells, are actively dividing. These cells are also killed by chemotherapy and radiation therapy. Secondly, it is possible that the therapy may produce mutations that will cause noncancerous cells to become cancerous. For these reasons, there is some maximal dose of chemotherapy or radiation therapy that is recommended.

C30. The answer is 5-bromouracil. If this chemical is incorporated into DNA, it can change an AT base pair into a CG base pair. A lysine codon, AAG, could be changed into a glutamic acid codon, CAG, with this chemical. By comparison, UV light is expected to produce thymine dimers, which would not lead to a mutation creating a glutamic acid codon, and proflavin is expected to cause frameshift mutations.

C31. A. Yes.

B. No, the albino trait affects the entire individual.

C. No, the early apple-producing trait affects the entire tree.

D. Yes.

C32. A. If UvrA was missing, the repair system would not be able to identify a damaged DNA segment.

B. If UvrC was missing, the repair system could identify the damaged segment, but it would be unable to make cuts in the damaged DNA strand.

C. If UvrD was missing, the repair system could identify the damaged segment and make cuts in the damaged strand, but it could not unwind the damaged and undamaged strands and thereby remove the damaged segment.

D. If DNA polymerase was missing, the repair system could identify the damaged segment, make cuts in the damaged strand, and unwind the damaged and undamaged strands to remove the damaged segment, but it could not synthesize a complementary DNA strand using the undamaged strands as a template.

C33. Methyl-directed mismatch repair is aimed at eliminating mismatches that may have occurred during DNA replication. In this case, the wrong base is in the newly made strand. The binding of MutH, which occurs on a hemimethylated sequence, provides a sensing mechanism to distinguish between the unmethylated and methylated strands. In other words, MutH binds to the hemimethylated DNA in a way that allows the mismatch repair system to know which strand is methylated and which is not.

C34. Recombinational repair occurs when the DNA is being replicated. If DNA damage occurred in exactly the same location in both replicated helices, recombinational repair could not work because it relies on the use of a strand in one replicated helix to act as a template to repair the complementary strand of the other replicated helix.

C35. The DNA-N-glycosylase would first make a cut between the base and the sugar. This would release the thymine base from the nucleotide to which it was directly attached, but the thymine would still be connected to the adjacent (3′) thymine. AP endonuclease would then make a nick in this DNA strand and the thymine dimer would be removed and repaired by nick translation. Finally, the strand would be sealed by DNA ligase.

C36. In *E. coli,* the TRCF recognizes when RNA polymerase is stalled on the DNA. This stalling may be due to DNA damage such as a thymine dimer. The TRCF removes RNA polymerase and recruits the excision DNA repair system to the region, thereby promoting the repair of the template strand of DNA. It is beneficial to preferentially repair transcribed DNA because it is functionally important. It is a DNA region that encodes a gene.

C37. A. Nucleotide excision repair.

B. Mismatch repair.

C. Recombinational repair and nucleotide excision repair.

C38. The underlying genetic defect that causes xeroderma pigmentosum is a defect in one of the genes that encode a polypeptide involved with nucleotide excision repair. These individuals are defective in repairing DNA abnormalities such as thymine dimers, abnormal bases, etc. Therefore, they are very sensitive to environmental agents such as UV light. Since they are defective at repair, UV light is more likely to cause mutations in these people, compared to unaffected individuals. For this reason, people with XP develop pigmentation abnormalities and premalignant lesions and have a high predisposition to skin cancer.

C39. The mismatch repair system, which detects mistakes made by DNA polymerase, would not function properly. As described in Figure 16.21, the MutH protein recognizes the sequence GATC when it is hemimethylated. If an *E. coli* strain was missing the dam methylase, this sequence would not be methylated, and MutH could not bind there. Therefore, mistakes, involving base pair mismatches caused by DNA polymerase, would not be repaired. This would increase the spontaneous mutation rate in this strain.

C40. Both types of repair systems recognize an abnormality in the DNA and excise the abnormal strand. The normal strand is then used as a template to synthesize a complementary strand of DNA. The systems differ in the types of abnormalities they detect. The mismatch repair system detects base pair mismatches, while the excision repair system recognizes thymine dimers, chemically modified bases, missing bases, and certain types of cross-links. The mismatch repair system operates immediately after DNA replication, allowing it to distinguish between the daughter strand (which contains the wrong base) and the parental strand. The excision repair system can operate at any time in the cell cycle.

Experimental Questions

E1. If the physiological adaptation theory had been correct, mutations should have occurred after the cells were plated on the media containing T1 bacteriophages. Since the same numbers of bacteria were streaked on each plate, we would have expected to see roughly the same number of resistant colonies on all of the plates. The number of resistant colonies would not have depended on the timing of the mutation. In contrast, what was actually observed was quite different. If a random mutation occurred early in the growth of a bacterial population (within a single tube), there were a large number of T1-resistant colonies on the plate. Whereas, if a random mutation occurred late, or not at all, there were few or no colonies on the plate.

E2. When cells from a master plate were replica plated onto two plates containing selective media with the T1 phage, T1-resistant colonies were observed at the same locations on both plates. These results indicate that the mutations occurred randomly while on the master plate (in the absence of T1) rather than occurring as a result of exposure to T1. In other words, mutations are random events, and selective conditions may promote the survival of mutant strains that occur randomly.

To show that antibiotic resistance is due to random mutation, one could follow the same basic strategy except the secondary plates would contain the antibiotic instead of T1 phage. If the antibiotic resistance arose as a result of random mutation on the master plate, one would expect the antibiotic-resistant colonies to appear at the same locations on two different secondary plates.

E3. The paracentric inversion does not actually prevent crossing over. However, if crossing happens to occur, this produces an acentric X chromosome and a dicentric X chromosome that will eventually break. Offspring inheriting these abnormalities will probably not survive, so we do not observe them in the crosses. What the *ClB* chromosome actually does is to prevent us from observing living offspring that are the product of a crossover between the X chromosomes in the *ClB* mothers.

E4. Perhaps the X rays also produce mutations that make the *ClB* daughters infertile. Many different types of mutations could occur in the irradiated males and be passed to the *ClB* daughters. Some of these mutations could prevent the *ClB* daughter from being fertile. These mutations could interfere with oogenesis, etc. Such *ClB* daughters would be unable to have any offspring.

E5. Müller's experiment is not measuring the mutation rate within a single gene. There are many genes along the X chromosome, which may mutate to produce a lethal phenotype. Müller's experiment is a measure of the rate at which X-linked mutations in many different genes can occur to produce a recessive lethal phenotype.

E6. You would conclude that chemical A is not a mutagen. The percentage of *ClB* daughters (whose fathers had been exposed to chemical A) that did not produce sons was similar to the control (compare 3 out of 2,108 with 2 out of 1,402). In contrast, chemical B appears to be a mutagen. The percentage of *ClB* daughters (whose fathers had been exposed to chemical B) that did not produce sons was much higher than the control (compare 3 out of 2,108 with 77 out of 4,203).

E7. Haploid cells are more sensitive to mutation because they have only a single copy of each gene. Therefore, recessive mutations that inhibit cell growth are easily detected. In diploid cells, it would take a mutation in both copies of the same gene to detect its phenotypic effects.

E8. You would expose the bacteria to the physical agent. You could also expose the bacteria to the rat liver extract, but it is probably not necessary for two reasons. First, a physical mutagen is not something that a person would eat. Therefore, the actions of digestion via the liver are probably irrelevant, if you are concerned that the agent might be a mutagen. Second, the rat liver extract would not be expected to alter the properties of a physical mutagen.

E9. Absence of mutagen: $17/10{,}000{,}000 = 17 \times 10^{-7} = 1.7 \times 10^{-6}$

Presence of mutagen: $2{,}017/10{,}000{,}000 = 2.0 \times 10^{-4}$

The mutagen increases the rate of mutation more than 100-fold.

E10. The results suggest that the strain is defective in excision repair. If we compare the normal and mutant strains that have been incubated for 2 hours at 37°C, much of the radioactivity in the normal strain has been transferred to the soluble fraction because it has been excised. In the mutant strain, however, less of the radioactivity has been transferred to the soluble fraction, suggesting that it is not as efficient at removing thymine dimers.

Questions for Student Discussion/Collaboration

1. A strain that has a defect in any of the DNA repair systems discussed in Chapter 16 would be a mutator strain because it would be defective at repairing mutations. Also, a strain that has a defect in the proofreading function of DNA polymerase would be a mutator strain. These types of strains are known, and they tend to have the highest rates of spontaneous mutations. Other defects in the DNA replication enzymes may also increase the mutation rate. Defects in the enzymes that promote recombination can have high rates of mutation. In addition, strains that tend to produce mutagenic metabolic products could be mutator strains. Finally, strains that carry transposable elements can have high rates of mutation.

2. The worst time to be exposed to mutagens would be at very early stages of embryonic development. An early embryo is most sensitive to mutation because it will affect a large region of the body. Adults must also worry about mutagens for several reasons. Mutations in somatic cells can cause cancer, a topic to be discussed in Chapter 22. Also, adults should be careful to avoid mutagens that may affect the ovaries or testes since these mutations could be passed along to offspring.

3. It is a matter of opinion. Some ideas might be the following:

   1. Testing of mutagens would enable us to know what the mutagens are and thereby avoid them. On the other hand, one might argue that there are so many now that it is difficult to avoid them anyway.

   2. Investigating molecular effects may help us find a cure for diseases such as cancer or help us to prevent mutations. On the other hand, it may not.

   3. Similarly, investigating DNA repair mechanisms may lead to ways of preventing mutations. On the other hand, it may not.

   4. Other places: educating the public about mutagens; tighter regulations of substances that contain mutagens; alternative methods of agriculture that may diminish the level of mutagens in food; and many others.

Chapter 17: Recombination and Transposition at the Molecular Level

Student Learning Objectives

Upon completion of this chapter you should be able to:

1. Understand the molecular mechanisms of homologous recombination.
2. Know the principles of the Holliday model of recombination.
3. Understand the importance of site-specific recombination and the molecular mechanisms that are involved in this process.
4. Understand the experimental procedures by which transposable elements were first discovered.
5. Understand the mechanisms by which transposable elements move in the genome, and the consequences of this movement.
6. Understand the use of transposons as molecular tools.

17.1 Sister Chromatid Exchange and Homologous Recombination

Overview

Recombination is a powerful factor in genetics. You already know that recombination produces new combinations of alleles and that recombination can be used as a mechanism of repairing gaps within the DNA. This chapter will explore the mechanisms of genetic recombination from the common method of homologous recombination, to the less common site-specific recombination.

Before proceeding into the chapter you need to review the basic principles of crossing over. An overview is provided in Figure 17.1. Note that we can detect crossing over in non-sister chromatids of homologous chromosomes. This is called sister chromatid exchange (SCE). The Holliday model explains the molecular basis of SCE. Although there have been some modifications to this model since it was first proposed, the basic principles still apply.

A second concept of this chapter is that of gene conversion. In gene conversion, the sequence of one gene is altered so that it is the same as a second gene. This is not the same as a mutation, since no new alleles are being formed. Two mechanisms of gene conversion are presented. You should be familiar with the basic principles of each.

Key Terms

5-bromodeoxyuridine
Branch migration
DNA gap repair synthesis
Double-stranded break model
Gene conversion
Genetic recombination
Heteroduplexes
Holliday junction
Holliday model
Homologous recombination
Resolution
Sister chromatid exchange
Site-specific recombination
Transposition
Transposons

Focal Points

- Review of crossing over (Figure 17.1)
- Holliday model (Figure 17.4)
- Mechanisms of gene conversion (Figures 17.6 and 17.7)

Exercises and Problems

For questions 1 to 4, complete the statement.

1. Harlequin chromosomes are generated by treating chromosomes with _______ and then applying different dyes.
2. The form of crossing over most often described in the process of meiosis is _______.
3. In the process of _______, the allele of one chromosome is altered so that it is the same as the allele on the homologous chromosome.
4. Gene conversion can be explained by the ___________.

For questions 5 to 11, place the steps of the Holliday model in their correct order. The first event is the answer for question 5, the second for question 6, etc.

_____ 5.

_____ 6.

_____ 7.

_____ 8.

_____ 9.

_____ 10.

_____ 11.

a. A Holliday junction is formed.
b. A double-stranded break occurs in the two homologous chromatids.
c. Branch migration occurs.
d. Isomerization may occur.
e. Two homologous chromosomes align.
f. The broken strands invade the opposite helices and base pair with the complementary strands.
g. Resolution of the chromosomes.

For questions 12 to 15, indicate which form of gene conversion the statement applies to.

a. gap repair synthesis
b. mismatch DNA repair
c. both forms

_____ 12. One allele is removed by digestion of the DNA due to a double-stranded break.

_____ 13. Involves the use of DNA repair mechanisms.

_____ 14. Occurs as a result of the Holliday model.

_____ 15. This occurs as the result of a heteroduplex forming during the Holliday model.

17.2 Site-specific Recombination

Overview

The next section of the chapter examines a non-homologous form of recombination called site-specific recombination. In this process, dissimilar DNA sequences recombine to form new genetic combinations. This chapter presents two different forms of this process. The first is the activity of viral integration, and the second is the generation of antibody diversity in mammals. This second is of extreme importance to us since it allows it provides relatively small number of genes the ability to generate a large number of antibody proteins.

Key Terms

Antibodies
Antigens
Attachment sites
Immunoglobulins
Lysogenic
Nonhomologous DNA end-joining proteins
Phophage
RAG1
RAG2
Recombination signal sequence
V(D)J recombination

Focal Points

- Viral integration (Figure 17.8)
- Immunoglobin diversity (Figure 17.9)

Exercises and Problems

For each of the following, match the term to its correct definition.

_____ 1. Prophage

_____ 2. RAG1 and RAG2

_____ 3. Attachment sites

_____ 4. V(D)J recombination

_____ 5. Recombination signal sequence

_____ 6. Antibodies

_____ 7. NHEJ proteins

_____ 8. Integrase

a. The process that forms the tremendous diversity of mammalian antibodies.
b. During antibody site-specific recombination, these rejoin the DNA fragments.
c. Indicates the location of site-specific recombination at the end of every V domain.
d. Generate the double-stranded breaks for immunoglobin recombination.
e. DNA sequences that are recognized by the virus for integration.
f. The enzyme used for viral site-specific recombination.
g. A virus that has integrated into the host genome.
h. Tetrameric proteins containing two heavy and two light chains

17.3 Transposition

Overview

The last section of this chapter introduces one of the more interesting forms of recombination, the action of transposable elements. Transposable elements, also called transposons or mobile elements, have been known since the work of McClintock in the 1950s. However, it has only been within a past two decades that the importance of transposons in the generation of genetic diversity has been recognized.

It is important that you take a good look at the experimental system of McClintock (Experiment 17B). You should appreciate that here work was done before the start of molecular biology and biotechnology. Her work was not recognized by the genetic community for almost 30 years, mostly because the methods did not exist to verify her conclusions!

Following an examination of McClintock's work, the text explores the various mechanisms by which transposons may move within the genome. Figure 17.12 outlines these three for you to review and Figure 17.13 gives the common structure, or organization, of transposable elements. You should study these carefully. The chapter concludes with a brief section regarding the importance of transposons in modern genetics.

Key Terms

Autonomous
Complete
Composite transposons
Direct repeats
Ds
Exon shuffling
Hybrid dysgenesis
Incomplete
Insertion sequences
Integrase
Inverted repeats
LINEs
Long terminal repeats
Mutable locus
Mutable site
Nonautonomous
Replicative transposition
Retroelements
Retroposons
Retrotransposons
Reverse transcriptase
Selfish DNA hypothesis
Simple transposition
SINEs
Transposable elements
Transposase
Transposon tagging
Viral-like retroelements

Focal Points

- McClintock's experiments (Experiment 17B)
- Mechanisms of transposition (Figure 17.12)
- Organization of transposable elements (Figure 17.13)

Exercises and Problems

For questions 1 to 10, choose the type of transposable element that is indicated by the statement. Some questions may use more than one answer.

a. replicative transposition
b. simple transposition
c. retrotransposition

_____ 1. The viral like group of these contain long terminal repeats.

_____ 2. "Cut and paste" transposition.

_____ 3. Uses the resolvase enzyme.

_____ 4. Includes a reverse transcriptase enzyme.

_____ 5. Uses an RNA intermediate.

_____ 6. "Copy and paste" transposition.

_____ 7. Contains direct repeats.

_____ 8. Simple transposable elements found in bacteria move by this mechanism.

_____ 9. Contains a transposase gene.

_____ 10. Produces multiple copies of the transposon in the genome.

For questions 11 to 16, match the term with its correct definition.

_____ 11. Selfish DNA hypothesis

_____ 12. Composite transposon

_____ 13. Autonomous transposon

_____ 14. Viral-like retroelements

_____ 15. LINEs and SINES

_____ 16. Transposon tagging

a. Related to the retroelements.
b. May contain genes that are not related to transposon activity.
c. An evolutionary explanation for the persistence of TEs in the genome.
d. A transposon that contains all of the genetic information to transposition to occur.
e. A mechanism used in molecular biology to clone genes using TEs.
f. Repetitive sequences in the DNA that are probably the result of tranposon activity.

Chapter Quiz

1. Viral integration into the genome is an example of _______.
    a. site specific recombination
    b. sister chromatid exchange
    c. gene conversion
    d. transposition
    e. homologous crossing-over

2. The Holliday model is used to explain _____.

a. retrotransposition
b. DNA repair
c. molecular basis of recombination
d. the formation of harlequin chromosomes
e. none of the above

3. McClintock's research in corn demonstrated the existence of ______.

a. retroviruses
b. transposable elements
c. crossing over
d. mismatch repair

4. Harlequin chromosomes have been used to demonstrate which of the following?

a. recombination
b. existence of SINEs and LINEs
c. transposon tagging
d. retrotransposition
e. none of the above

5. V(D)J recombination is an example of what?

a. sister chromatid exchange
b. site specific recombination
c. translocations
d. transposition
e. viral integration

6. Retroelements may contain all of the following, except _____.

a. reverse transcriptase
b. direct repeats
c. long terminal repeats
d. integrase
e. resolvase

7. What is the final stage of the Holliday model?

a. formation of a Holliday junction
b. formation of a double-stranded break
c. gene conversion
d. resolution
e. anticipation

8. 5-bromodeoxyuridine is used in which of the following?

a. cloning
b. transposon tagging
c. identification of TEs
d. creation of harlequin chromosomes
e. none of the above

9. Recombination signal sequences are used in ______.
   a. simple transposition
   b. retrotransposition
   c. transposon tagging
   d. sister chromatid exchange
   e. site-specific recombination

10. Attachment sites are necessary for which of the following?
   a. integration of a virus into the host genome
   b. attachment of 5-bromodeoxyuridine to the chromosome
   c. sister chromatid exchange
   d. retrotransposition
   e. transposon tagging

Answer Key for Study Guide Questions

This answer key provides the answers to the exercises and chapter quiz for this chapter. Answers in parentheses ( ) represent possible alternate answers to a problem, while answers marked with an asterisk (*) indicate that the response to the question may vary.

17.1	1. 5-bromodeoxyuridine	6. b	11. a
	2. sister chromatid exchange	7. f	12. a
	3. gene conversion	8. a	13. c
	4. Holliday model	9. c	14. c
	5. e	10. d	15. b

17.2	1. a	4. a	7. b
	2. d	5. c	8. f
	3. d	6. h	

17.3	1. c	7. a, b, c	13. d
	2. b	8. b	14. a
	3. a	9. a, b	15. f
	4. c	10. a, c	16. e
	5. c	11. c	
	6. a	12. b	

Quiz

1. a	5. b	9. e
2. c	6. e	10. a
3. b	7. d	
4. a	8. d	

Answers to Conceptual and Experimental Questions

Conceptual Questions

C1. At the molecular level, sister chromatid exchange and homologous recombination are very similar. Identical (sister chromatid exchange) or similar (homologous recombination) segments of DNA line up and then cross over. Due to the molecular similarities of the two processes, one would expect that the same types of proteins would catalyze both events. At the genetic level, the events are different, however. Sister chromatid exchange does not result in the recombination of alleles because the chromatids are genetically identical. Homologous recombination usually results in a new combination of alleles after a crossover has taken place.

C2. Branch migration will not create a heteroduplex during sister chromatid exchange because the sister chromatids are genetically identical. There should not be any mismatches between the complementary strands. Gene conversion cannot take place because the sister chromatids carry alleles that are already identical to each other.

C3. Breakage would have to occur at the arrows labeled *2* and *4*. This would connect the *A* allele with the *B* allele. The *a* allele in the other homologue would become connected with the *b* allele. In other words, one chromosome would be *AB* and the homologue would be *ab*.

C4. The steps are described in Figure 17.6.

A. The ends of the broken strands would not be recognized and degraded.

B. RecA protein would not recognize the single-stranded ends, and strand invasion of the homologous double helix would not occur.

C. Holliday junctions would not form.

D. Branch migration would not occur without these proteins. And resolution of the intertwined helices would not occur.

C5. The two molecular mechanisms that can explain the phenomenon of gene conversion are mismatch DNA repair and gap repair synthesis. Both mechanisms could occur in the double-stranded break model.

C6. Usually, the overall net effect is not to create any new mutations in particular genes. However, homologous recombination does rearrange the combinations of alleles along particular chromosomes. This can be viewed as a mutation, since the sequence of a chromosome has been altered in a heritable fashion.

C7. A recombinant chromosome is one that has been derived from a crossover and contains a combination of alleles that is different from the parental chromosomes. A recombinant chromosome is a hybrid of the parental chromosomes.

C8. It depends on which way the breaks occur in the DNA strands during the resolution phase. If the two breaks occur in the crossed DNA strands, nonrecombinant chromosomes result. If the two breaks occur in the uncrossed strands, the result is a pair of recombinant chromosomes.

C9. Gene conversion occurs when a pair of different alleles is converted to a pair of identical alleles. For example, a pair of *Bb* alleles could be converted to *BB* or *bb*.

C10. Holliday model—proposes two breaks, one in each chromatid, and then both strands exchange a single strand of DNA. Double-stranded break model—proposes two breaks, both in the same chromatid. As in the Holliday model, single-strand migration occurs between both homologues. The double-stranded break model also proposes strand degradation and gap repair synthesis. Finally, the double-stranded break model generates two Holliday junctions, not just one.

C11. Gene conversion is likely to take place near the breakpoint. According to the double-stranded break model, a gap may be created by the digestion of one DNA strand in the double helix. Gap repair synthesis may result in gene conversion. A second way that gene conversion can occur is by mismatch repair. A heteroduplex may be created after DNA strand migration. This heteroduplex may be repaired in such a way as to cause gene conversion.

C12. The RecA protein binds to single-stranded DNA and forms a filament. The formation of the filament promotes the sliding of the filament along another DNA region until it recognizes homologous sequences. This recognition process may involve the formation of triplex DNA. After recognition has occurred, RecA protein mediates the movement of the invading DNA strand and the displacement of the original strand.

C13. No, it could involve mismatch repair, but it could also involve gap repair synthesis. The double-stranded break model involves a migration of DNA strands and the digestion of a gap. Therefore, in this gap region, only one chromatid is providing the DNA strands. As seen in Figure 17.6, the top chromosome is using the top DNA strand from the bottom chromosome in the gap region. The bottom chromosome is using the bottom DNA strand from the bottom chromosome. After DNA synthesis, both chromosomes may have the same allele.

C14. RecG and RuvABC bind to Holliday junctions. They do not necessarily recognize a DNA sequence but, instead, recognize a region of crossing over (i.e., a four-stranded structure).

C15. First, gene rearrangement of V, D, and J domains occurs within the light- and heavy-chain genes. Second, within a given B cell, different combinations of light and heavy chains are possible. And third, imprecise fusion may occur between the V, D, and J domains.

C16. The function of the RAG1 and RAG2 proteins is to recognize the recombination signal sequences and make double-stranded cuts. In the case of V/J recombination, a cut is made at the end of one V region and the beginning of one J region. The NHEJ proteins recognize these ends and join them together. This is a form of DNA splicing. This creates different combinations of the V, J, (D), and constant regions, thereby creating a large amount of diversity in the encoded antibodies.

C17. One segment (which includes some variable sequences and perhaps one or more joining sequences) of DNA is removed from the κ light-chain gene. One segment (which may include one or more joining regions and the region between the joining regions and constant region) is removed during pre-mRNA splicing.

C18. Integrase is an enzyme that recognizes the attachment site sequences within the l DNA and the *E. coli* chromosome. It brings the l DNA close to the chromosome and then makes staggered cuts in the attachment sites. The strands are exchanged, and then integrase catalyzes the covalent attachment of the strands to each other. In this way, the l DNA is inserted at a precise location within the *E. coli* chromosome.

C19. The ends of a short region would be flanked by direct repeats. This is a universal characteristic of all transposable elements. In addition, many elements contain inverted repeats or LTRs that are involved in the transposition process. In transposons, one might also look for the presence of a transposase gene, although this is not an absolute requirement since the transposase gene is missing in incomplete elements. In retroviral elements, one might also locate a reverse transcriptase gene, but this may not be present.

C20. In Figure 17.15, the TE has transposed prior to DNA replication. The transposon is single stranded at both locations. DNA replication (i.e., gap repair synthesis) copies the single-stranded elements into double-stranded elements. Therefore, the end result is two double-stranded TEs at two distinct locations.

C21. Direct repeats occur because transposase or integrase produces staggered cuts in the two strands of chromosomal DNA. The transposable element is then inserted into this site, which temporarily leaves two gaps. The gaps are filled in by DNA polymerase. Since this gap filling is due to complementarity of the base sequences, the two gaps end up with the exact same sequence. This is how the two direct repeats are formed.

C22. Retroelements have the greatest potential for proliferation because the element is transcribed into RNA as an intermediate. Many copies of this RNA could be transcribed and then copied into DNA by reverse transcriptase. Theoretically, many copies of the element could be inserted into the genome in a single generation.

C23. Transposable elements are mutagens since they alter (disrupt) the sequences of chromosomes and genes within chromosomes. They do this by inserting themselves into genes.

C24. Keep in mind that each type of transposase recognizes only the inverted repeats of a particular type of transposable element. The mosquitoes must express a transposase that recognizes the Z elements. This explains why the Z elements are mobile. This Z element transposase must not recognize the inverted repeats of the X elements; the inverted repeats of the X elements and Z elements must have different sequences. This same group of mosquitoes must not express a transposase that recognizes the X elements. This explains why the X elements are very stable.

C25. A. Viral-like retroelements and nonviral-like retroelements.

B. Insertion sequences, composite transposons, and replicative transposons.

C. All five types have direct repeats.

D. Insertion sequences, composite transposons, and replicative transposons.

C26. A transposon is excised as a segment of DNA, and then it transposes to a new location via transposase. A retroelement is transcribed as RNA, and then reverse transcriptase makes a double-stranded copy of DNA. Integrase then inserts this DNA copy into the chromosome. All transposable elements have direct repeats. Transposons that move as DNA have inverted repeats, and they may encode transposase (as well as other genes). Retroelements do not have inverted repeats, but they may or may not have LTRs. Autonomous retroelements encode reverse transcriptase and integrase.

C27. A. As shown in solved problem S2, a crossover between direct repeats will excise the region between the repeats. In this case, it will excise a large chromosomal region, which does not contain a centromere, and will be lost. The remaining portion of the chromosome is shown here. It has one direct repeat that is formed partly from DR-1 and partly from DR-4. It is designated DR-1/4.

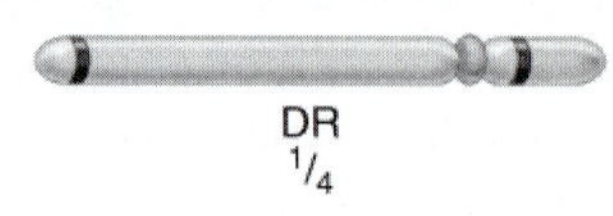

B. A crossover between IR-1 and IR-4 will cause an inversion between the intervening region. This is a similar effect to the crossover within a single transposable element, as described in solved problem S2. The difference is that the crossover causes a large chromosomal region to be inverted.

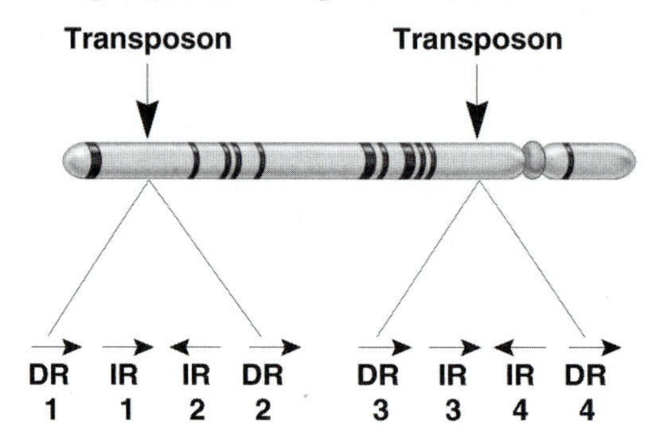

C28. An autonomous element has the genes that are necessary for transposition. For example, a cut-and-paste transposon that was autonomous would also have the transposase gene. A nonautonomous element does not have all the genes that are necessary for transposition. However, if a cell contains an autonomous element and a nonautonomous element of the same type, the nonautonomous element can move. For example, if a *Drosophila* cell contained two P elements, one autonomous and one nonautonomous, the transposase expressed from the autonomous P element could recognize the nonautonomous P element and catalyze its transposition.

C29. At a frequency of about 1 in 10,000, recombination occurs between the inverted repeats within this transposon. As shown in solved problem S2, recombination between inverted repeats causes the sequence within the transposon to be reversed. If this occurred in a strain that expressed the *H2/rH1* operon, the promoter would be flipped in the opposite direction, so that the *H2* gene and the *rH1* gene would not be expressed. The *H2* flagellar protein would not be made, and the *H1* repressor protein would not be made. The *H1* flagellar protein would be made because the expression of the *H1* gene would not be repressed. A strain expressing the *H1* gene could also "switch" at a frequency of about 1 in 10,000 by the same mechanism. If a crossover occurred between the inverted repeats within the transposable element, the promoter would be flipped around again, and the *H2* and *rH1* gene would be expressed again.

C30. The formation of deletions and inversions is illustrated in the answer to conceptual question C27. A deletion occurs when there is recombination between the direct repeats in two different elements. The region between the elements is deleted. An inversion can happen when recombination occurs between inverted repeats between two different elements. This occurs when the inverted repeats are in the opposite orientation. A translocation can result when recombination occurs between transposable elements that are located on different (nonhomologous) chromosomes. In other words, the TEs on different chromosomes align themselves and a crossover occurs. This will produce a translocation, as shown here.

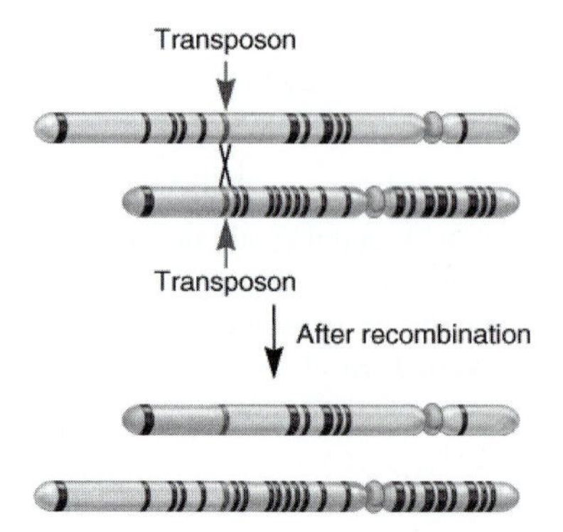

Experimental Questions

E1. Due to semiconservative DNA replication, one of the sister chromatids has both of the DNA strands that contain BrdU while the other sister chromatid only has one of its two DNA strands that contain BrdU.

E2. You would conclude that the substance is a mutagen. Substances that damage DNA tend to increase the level of genetic exchange such as sister chromatid exchange.

E3. You would add it after the second round of DNA replication but before crossing over occurs during mitosis (i.e., add it during the G_2 phase). In this experiment you are detecting SCEs that occur after the second round of DNA replication. If the mutagen can persist in the cells a long time, you might be able to add it earlier.

E4. The drawing here shows the progression through three rounds of BrdU exposure. After one round, all of the chromosomes would be dark. After two rounds, all of the chromosomes would be harlequin. After three rounds, the number of light sister chromatids would be twice as much as the number of light sister chromatids found after two rounds of replication.

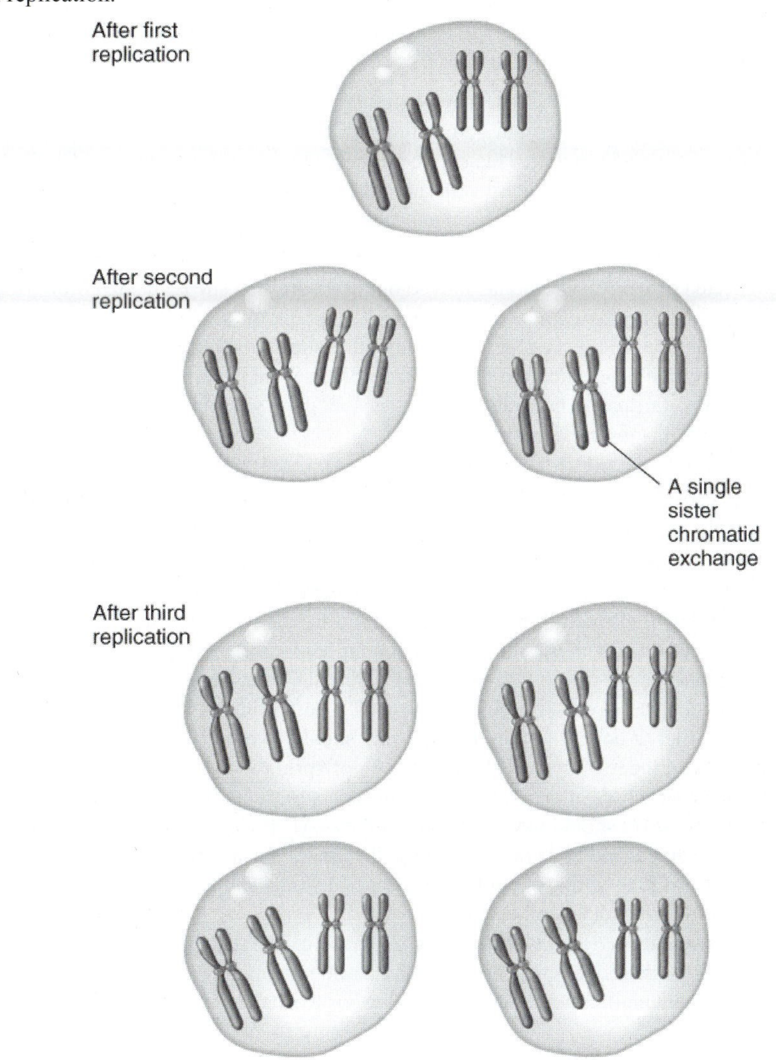

E5. BrdU inhibits Giemsa binding. As described in Chapter 8, the binding of Giemsa to chromosomes produces G bands, which are dark. The chromatids that contain both strands of BrdU are light, indicating that BrdU inhibits Giemsa binding.

E6. When McClintock started with a colorless strain containing *Ds,* she identified 20 cases where *Ds* had moved to a new location to produce red kernels. This identification was possible because the 20 strains had a higher frequency of chromosomal breaks at a specific site, and because of the mutability of particular genes. She also had found a strain where *Ds* had inserted into the red-color-producing gene, so that its transposition out of the gene would produce a red phenotype. Overall, her analysis of the data showed that the sectoring (i.e., mutability) phenotype was consistent with the transposition of *Ds.*

E7. It is due to the excision of the transposable element. This conclusion was based on the reversion of the phenotype (namely colorless to red), which suggests a restoration of gene function, and the nonmutability of the phenotype.

E8. Transposon tagging is an experimental method that is aimed at cloning genes. In this approach, a transposon is introduced into a strain, and the experimenter tries to identify individuals in which the gene of interest has lost its function. In many cases, the loss of function has occurred because the transposon has been inserted into the gene. If so, one can make a library and then use a labeled transposon probe to identify clones in which the transposon has been inserted. This provides a way to clone the gene of interest. Later, one would use the transposon-inserted clone to screen a library from an individual with an active version of the gene. In this way, one could then identify a normal copy (nontransposon-inserted) of the gene.

E9. One could begin with the assumption that the inactivation of a tumor-suppressor gene would cause cancerous cell growth. If so, one could begin with a normal human line and introduce a transposon. The next step would be to identify cells that have become immortal. This may be possible by identifying clumps of cells that have lost contact inhibition. One could then grow these cells and make a genomic DNA library from them. The library would be screened using the transposon as a labeled probe.

E10. A transposon creates a mutable site because the excision of a transposon causes chromosomal breakage if the two ends are not reconnected properly. After it has moved out of its original site (causing chromosomal breakage), it may be inserted into a new site somewhere else. You could experimentally determine this by examining a strain that has incurred a chromosomal breakage at the first mutable site. You could microscopically examine many cells that had such a broken chromosome and see if a new mutable site had been formed. This new mutable site would be the site into which the transposon had moved. On occasion, there would be chromosomal breakage at this new mutable site, which could be observed microscopically.

E11. First, you would inject the plasmid carrying the P element and *rosy*$^+$ gene into embryos that were homozygous for the recessive rosy allele (i.e., *rosy rosy*). Allow these embryos to mature into adult flies, and then mate them to *rosy rosy* flies. Analyze the color of the F_1 offspring. If the eyes are red, at least one copy of the P element has hopped into the chromosome of these flies. Possibly, the P element has hopped into a gene that is involved with wing development. However, at this point, such a fly would have one normal copy of the gene and one copy that has been inactivated by the P element. If the inactivation of the gene were inherited as a dominant phenotype, these flies would have deformed wings. Alternatively (and more likely), however, the inactivation of one copy of most genes would be inherited as a recessive trait. If this were the case, the wings of the red-eyed flies would still appear normal. To identify these recessive genes, take each F_1 offspring with red eyes (and normal wings) and mate them to *rosy rosy* flies. About 1/4 of the F_2 offspring should have red eyes (assuming that one P element has hopped into the genome of the red-eyed flies). Take the F_2 offspring with red eyes (from a single cross) and mate them to each other. In other words, mate F_2 brothers and sisters to each other. From these crosses, 1/4 of the F_3 flies should have rosy eyes, 1/2 should have one copy of the *rosy*$^+$ P element and red eyes, and 1/4 should be homozygous for the P element and have red eyes. If the P element has inserted into a gene that is required for wing development, 1/4 of the offspring should have deformed wings and red eyes. In this way, you could identify genes that are required for wing development.

Note: This experiment is a lot of work! You would have to identify thousands of flies with red eyes and set up thousands of crosses to identify a few genes that play a role in wing development. Nevertheless, this approach has been successful in the identification of genes that play a role in many aspects of *Drosophila* development.

Questions for Student Discussion/Collaboration

1. Overall, these three processes are very similar to each other at the molecular level. All three involve the specific recognition of DNA sequences, the cutting of DNA strands, and the ligation of DNA strands. Also, all three processes are catalyzed by enzymes that recognize specific DNA sequences and cut and ligate the DNA at the appropriate locations. As for differences, homologous recombination is unique in that homologous DNA sequences synapse with each other. At the molecular level, homologous recombination is much more complex, since it also involves strand migration, strand degradation, and resolution of Holliday structures. Homologous recombination and transposition both involve DNA gap repair synthesis. Transposition is unique in that it often involves a replication of the transposable element and the movement of the TE to many places in the genome.

2. Harmful consequences: The level of genetic diversity would be decreased, because linked combinations of alleles would not be able to recombine. You would not be able to produce antibody diversity in the same way. Gene duplication could not occur, so the evolution of new genes would be greatly inhibited.

   Beneficial consequences: You would not get (as many) translocations, inversions, and the accumulation of selfish DNA.

3. It seems that the selfish DNA theory could be correct. From the point of view of natural selection, one would think that the accumulation of worthless DNA would be a disadvantage. However, if the transposition of DNA occurs rather frequently, and the accumulation of this DNA is not significantly detrimental, it may occur anyway and result in the accumulation of DNA that does not provide any advantage to the organism.

Chapter 18: Recombinant DNA Technology

Student Learning Objectives

Upon the completion of this chapter you should be able to:

1. Understand the experimental procedures used in the study of recombinant DNA and DNA cloning.
2. Understand the process of gene cloning and the concept of a vector.
3. Understand the concept of a library, and how libraries are screened for gene products.
4. Understand the processes of DNA sequencing and the process of in vitro site-directed mutagenesis.

18.1 Gene Cloning

Overview

Although recombinant DNA is present in any cell that undergoes crossing-over, site-directed recombination, or has transposon activity, the ability to duplicate this outside the cell has only been possible over the past thirty years. However, in that time scientists have developed a variety of techniques and procedures to manipulate DNA and unlock the secrets of gene action.

The first step for most scientists is to produce large numbers of copies of the gene of interest. This process is called cloning. While the term cloning typically is now associated with organismal cloning, such as sheep and humans, at the molecular level it simply means copying. In order to understand the cloning process, study the diagrams (specifically Figure 18.2) in the text.

This section then introduces a number of different procedures that researchers can use to manipulate the DNA and study its function. For each of these experimental procedures, be familiar with the general process and the key terms that are associated with the technique. As you read scientific papers and articles on genetics, you will begin to notice how widespread these techniques are in the study of genetics.

Key Terms

Annealing
BACs
Competent cells
Complementary DNA
Cosmids
DNA ligase
Gene cloning
Gene therapy
Genomics
Hybrid vector
Palindromic
Polymerase chain reaction
Recircularized vector
Recombinant DNA molecules
Recombinant DNA technology
Restriction endonucleases
Restriction enzymes
Restriction fragment length polymorphism
Restriction mapping
Reverse transcriptase
Reverse transcriptase PCR
Selectable marker
Subcloning
Taq polymerase
Template DNA
Thermocycler
Transfection
Transformation
Transgenic
Vector
YACs

Focal Points

- Action of restriction enzymes (Figure 18.1)
- Steps in gene cloning (Figure 18.2)
- cDNA synthesis (Figure 18.4)
- Restriction mapping (Figure 18.5)
- Polymerase chain reaction (Figure 18.6)

Exercises and Problems

Questions 1 to 11 refer to the process of cloning. For each, select the correct definition for the term.

_____ 1. Plasmids

_____ 2. Restriction enzymes

_____ 3. Selectable marker

_____ 4. Vector

_____ 5. Recircularized vector

_____ 6. Viral vector

_____ 7. Competent cells

_____ 8. Transformation

_____ 9. Transfection

_____ 10. Hybrid vector

_____ 11. Kanamycin

a. A carrier for DNA in a cloning experiment.
b. A vector that does not contain a piece of chromosomal DNA.
c. Enzymes that recognize palindromic sequences in the DNA and cut them using endonuclease activity.
d. Small circular pieces of extranuclear DNA.
e. The process by which a bacteria receives the cloned DNA from the environment.
f. The first cloned gene.
g. The process by which a virus introduces the cloned DNA into the cell.
h. Cells that are capable of taking up DNA from the environment.
i. A vector that contains a piece of chromosomal DNA.
j. The use of a virus to carry the cloned DNA into a host cell.
k. A gene contained within a vector that allows a researcher to choose bacterial colonies that contain only hybrid vectors.

For questions 12 to 16, place the following steps of a cloning experiment in order. The letter of the first step of the experiment is the answer for question 12, the second step is the answer for question 13, etc.

_____ 12.

_____ 13.

_____ 14.

_____ 15.

_____ 16.

a. Mix the cut plasmid and chromosomal DNA together.
b. Cut the plasmid and chromosomal DNA using the same restriction endonuclease.
c. Treat the mixture with DNA ligase.
d. Screen the colonies for those that indicate hybrid vectors.
e. Plate cells on a media containing the substrate for the selectable marker.

For questions 17 to 23, match the term with its correct description.

_____ 17. Subcloning

_____ 18. Reverse transcriptase

_____ 19. *Taq* polymerase

_____ 20. Polymerase chain reaction

_____ 21. cDNA

_____ 22. Restriction fragment length polymorphism

_____ 23. Thermocycler

a. A machine in which a PCR reaction is conducted.
b. Allows a rapid amplification of small segments of DNA.
c. Variations in the location of restriction enzymes sites.
d. A thermostable enzyme that makes PCR possible.
e. An enzyme that uses RNA as a template to make DNA
f. DNA that lacks introns and is derived from expressed RNA.
g. This follows a cloning experiment and serves to reduce the size of the cloned piece.

18.2 Detection of Genes and Gene Products

Overview

The second section of this chapter examines the creation and screening of a DNA library. Many times researchers do not know what gene product they are looking for precisely, and therefore must screen the entire collection of the organism's DNA for a gene of interest. The screening of chromosomal or cDNA libraries for gene(s) of interest is called Southern blotting.

In addition to DNA, researchers may screen for a specific RNA or protein in a sample. These experiments use techniques that are similar to the Southern blots, but are called Northern blots and Western blots, respectively. As you proceed through this section, notice the similarities between the procedures, and the minor differences in the types of probes that are used to screen the samples.

Key Terms

Antibody
Antigens
Band shift assay
cDNA library
Colony hybridization
DNA footprinting
DNA library
Epitopes
Gel retardation assay
Gene detection
Gene family
Genomic library
High stringency
Low stringency
Northern blotting
Southern blotting
Western blotting

Focal Points

- Differences in Southern, Northern, and Western blotting (pgs. 507-510)

Exercises and Problems

For each of the following questions, choose the correct definition for the term.

_____ 1. Stringency
_____ 2. Colony hybridization
_____ 3. Genomic library
_____ 4. cDNA library
_____ 5. DNA footprinting
_____ 6. Gel retardation assay

a. Identifies DNA that is interacting with proteins by a change in its rate of movement in an electrical field.
b. The conditions at which a probe will hybridize with its target molecule.
c. Uses a probe to screen a library for a gene of interest.
d. A collection of hybrid vectors that use chromosomal DNA as the source.
e. Identifies region of the DNA that interact with specific proteins.
f. A collection of hybrid vectors that are derived from expressed RNA in a cell.

For each of the following, select the type of hybridization experiment that is indicated by the statement. Some questions may have more than one answer.

a. Western blotting
b. Southern blotting
c. Northern blotting

_____ 7. Identifies a specific gene from a collection of chromosomal DNA fragments.
_____ 8. Identifies if a specific protein is present in the sample.
_____ 9. Uses antibodies as the probe.
_____ 10. Uses a gel to separate the molecules by size or molecular mass.
_____ 11. This process may also be done by RT-PCR.
_____ 12. May be used to identify gene families.
_____ 13. Identifies a specific RNA from a collection of expressed RNAs.
_____ 14. Frequently uses a radioactive label on the probe.
_____ 15. Typically uses a fluorescent secondary antibody to indicate the presence of the primary antibody.

18.3 Analysis and Alteration of DNA Sequences

Overview

The final section of this chapter examines the process of DNA sequencing. Although DNA sequencing is largely an automated process now, this was not always the case. However, technological advances, mostly as a result of the Human Genome Project, have made DNA sequencing easy and inexpensive. This section provides an overview of the process, as well as an examination of site-directed mutagenesis, which allows researchers to change the sequence of cloned DNA to observe the effects on the phenotype of the organism.

Key Terms

Automated sequencing
Chain termination
Dideoxy sequencing
Dideoxyribonucleotides
DNA sequencing
Sequencing ladder
Site-directed mutagenesis

Focal Points

- Site-directed mutagenesis (Figure 8.17)

Exercises and Problems

Match each of the following terms to its correct statement.

_____ 1. Chain termination

_____ 2. Dideoxy sequencing

_____ 3. Site-directed mutagenesis

_____ 4. Dideoxyribonucleotides

_____ 5. Automated sequencing

_____ 6. Sequencing ladder

a. A synthetic nucleotide used in sequencing reactions.
b. A dideoxy sequencing reaction that used fluorescent labels.
c. The change in a specific base of a cloned DNA segment.
d. The process by which a dideoxyribonucleotide stops DNA replication.
e. The pattern of bands produced by traditional dideoxy sequencing reactions.
f The most common method of DNA sequencing.

Chapter Quiz

1. Plasmids, BACs and YACs represent what?
    a. vectors
    b. selectable markers
    c. the first cloned genes
    d. forms of cDNA

2. Which of the following can rapidly make copies of specific regions of chromosomal DNA?
   a. southern blotting
   b. restriction digests
   c. cloning
   d. PCR

3. Which of the following is able to identify specific areas of DNA and cut them using an endonuclease activity?
   a. PCR
   b. Northern blotting
   c. dideoxy sequencing
   d. restriction enzymes

4. ________ blotting can detect the presence of a specific protein from a cell.
   a. Southern
   b. Northern
   c. Western
   d. all of the above

5. _______ blotting can be used to screen for gene families.
   a. Southern
   b. Northern
   c. Western
   d. all of the above

6. A collection of hybrid vectors is called a ________.
   a. Southern blot
   b. clone
   c. library
   d. antigen

7. cDNA is made using an enzyme called ________.
   a. dideoxyribonucleotide
   b. integrase
   c. reverse transcriptase
   d. restriction endonuclease

8. R factors are an example of _______.
   a. host cells
   b. viral vectors
   c. restriction enzymes
   d. plasmids
   e. none of the above

9. Which of the following is not a step in a cloning experiment?
   a. Plasmid and chromosomal DNA is cut using a restriction enzyme.
   b. The plasmid and chromosomal DNA is mixed.
   c. *Taq* polymerase is added to combine the DNA fragments.
   d. The DNA is mixed with bacterial cells and plated.
   e. All of the above are correct and are part of a cloning experiment.

10. A thermocycler is used for _____.

    a. dideoxy sequencing
    b. DNA footprinting
    c. PCR reactions
    d. colony hybridization

Answer Key for Study Guide Questions

This answer key provides the answers to the exercises and chapter quiz for this chapter. Answers in parentheses ( ) represent possible alternate answers to a problem, while answers marked with an asterisk (*) indicate that the response to the question may vary.

18.1

1. d
2. c
3. k
4. a
5. b
6. j
7. h
8. e
9. g
10. i
11. f
12. b
13. a
14. c
15. e
16. d
17. a
18. e
19. d
20. b
21. f
22. c
23. a

18.2

1. b
2. c
3. d
4. f
5. e
6. a
7. b
8. a
9. a
10. a b c
11. c
12. b
13. c
14. b c
15. a

18.3

1. d
2. f
3. c
4. a
5. b
6. e

Quiz

1. a
2. d
3. d
4. c
5. a
6. c
7. c
8. d
9. c
10. c

Answers to Conceptual and Experimental Questions

Conceptual Questions

C1. First, cloned genes can be used for DNA sequencing. This has allowed researchers to understand genetics at the molecular level. Second, cloned DNA can be mutated using site-directed mutagenesis to see how specific mutations alter the structure and function of DNA. Third, cloning is useful in biotechnology. This topic is discussed more thoroughly in Chapter 20. And many others.

C2. A restriction enzyme recognizes a DNA sequence and then cleaves a (covalent) phosphoester bond in each of two DNA strands.

C3. Here is an example:

GGGCCCATATATATGGGCCC
CCCGGGTATATATACCCGGG

C4. The term *cDNA* refers to DNA that is derived from mRNA. Compared to genomic DNA, it would lack introns.

C5. A dideoxynucleotide is missing the 3′ —OH group. When the 5′ end of a dideoxynucleotide is added to a growing strand of DNA, another phosphoester bond cannot be formed at the 3′ position. Therefore, the dideoxynucleotide terminates any further addition of nucleotides to the growing strand of DNA.

Experimental Questions

E1. Sticky ends, which are complementary in their DNA sequence, will promote the binding of DNA fragments to each other. This binding is due to hydrogen bonding.

E2. Remember that AT base pairs form two hydrogen bonds while GC base pairs form three hydrogen bonds. The order (from stickiest to least sticky) would be:

*Bam*HI = *Pst* I = *Sac* I > *Eco*RI > *Cla* I.

E3. All vectors have the ability to replicate when introduced into a living cell. This ability is due to a DNA sequence known as an origin of replication. Modern vectors also contain convenient restriction sites for the insertion of DNA fragments. These vectors also contain selectable markers, which are genes that confer some selectable advantage for the host cell that carries them. The most common selectable markers are antibiotic-resistance genes, which confer resistance to antibiotics that would normally inhibit the growth of the host cell.

E4. In conventional gene cloning, many copies are made because the vector replicates to a high copy number within the cell, and the cells divide to produce many more cells. In PCR, the replication of the DNA to produce many copies is facilitated by primers, nucleotides, and *Taq* polymerase.

E5. First, the chromosomal DNA that contains the source of the gene that you want to clone must be obtained from a cell (tissue) sample. A vector must also be obtained. The vector and chromosomal DNA are digested with a restriction enzyme. They are mixed together to allow the sticky ends of the DNA fragments to bind to each other, hopefully to create a hybrid vector. DNA ligase is then added to promote covalent bonds. The DNA is then transformed or transfected into a living cell. The vector will replicate and the cells will divide to produce a colony of cells that contain "cloned" DNA pieces. To identify colonies that contain the gene you wish to clone, you must use a probe that will specifically identify a colony containing the correct hybrid vector. A probe may be a DNA probe that is complementary to the gene you want to clone; or it could be an antibody that recognizes the protein that is encoded by the gene.

E6. A hybrid vector is a vector that has a piece of "foreign" DNA inserted into it. The foreign DNA came from somewhere else, like the chromosomal DNA of some organism. To construct a hybrid vector, the vector and source of foreign DNA are digested with the same restriction enzyme. The complementary ends of the fragments are allowed to hydrogen bond to each other (i.e., sticky ends are allowed to bind), and then DNA ligase is added to create covalent phosphoester bonds. If all goes well, a piece of the foreign DNA will become ligated to the vector, thereby creating a hybrid vector.

As described in Figure 18.2, the insertion of foreign DNA can be detected using X-Gal. As seen here, the insertion of the foreign DNA causes the inactivation of the *lacZ* gene. The *lacZ* gene encodes the enzyme *ß*-galactosidase, which is necessary to convert X-gal to a blue compound. If the *lacZ* gene is inactivated by the insertion of foreign DNA, the bacterial colonies will be white. If the vector has simply recircularized, and the *lacZ* gene remains intact, the colonies will be blue.

E7. Probably not. It was necessary for the researchers to obtain kanamycin-resistant colonies, and this would occur only if the entire kan^R gene were present. If the kan^R gene had been cut in half, it would have been difficult to insert that whole gene into pSC101. It is possible, if the two pieces of pSC101 DNA (carrying the kan^R gene) were inserted into the same pSC101 plasmid. However, the chance of this happening is much less than inserting a single piece of DNA, carrying the entire kan^R gene, into pSC101.

E8. If the *Eco*RI fragment containing the kan^R gene also had an origin of replication, it is possible that this fragment could circularize and become a very small plasmid. The electrophoresis results would be consistent with the idea that the same bacterial cells could contain two different plasmids: pSC101 and a small plasmid corresponding to the segment of DNA (now circularized) that carried the kan^R gene. However, the results shown in step 9 rule out this possibility. The density gradient centrifugation showed a single peak, corresponding to a plasmid that had an intermediate size between pSC101 and pSC102. In contrast, if our alternative explanation had been correct (i.e., that bacterial cells contain two plasmids), there would be two peaks from the density gradient centrifugation. One peak would correspond to pSC101 and the other peak would indicate a very small plasmid (i.e., smaller than pSC101 and pSC102).

E9. The map is shown here.

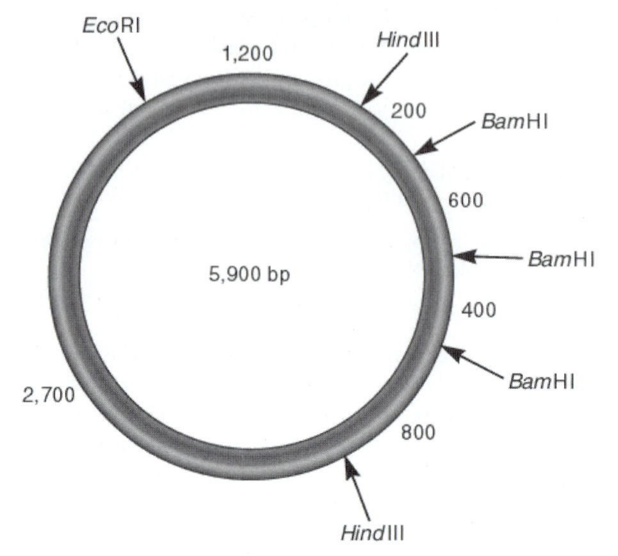

E10. 3×2^{27}, which equals 4.0×10^8, or about 400 million copies.

E11. A thermostable form of DNA polymerase (e.g., *Taq* polymerase) is used in PCR because each PCR cycle involves a heating step to denature the DNA. This heating step would inactivate most forms of DNA polymerase. However, *Taq* polymerase is thermostable and can remain functional after many cycles of heating and cooling. It is not necessary to use a thermostable form of DNA polymerase in the techniques of dideoxy DNA sequence or site-directed mutagenesis. In these methods, DNA polymerase can be added after the annealing step, and the sample can be incubated at a temperature that does not inactivate most forms of DNA polymerase.

E12. Initially, the mRNA would be mixed with reverse transcriptase and nucleotides to create a complementary strand of DNA. Reverse transcriptase also needs a primer. This could be a primer that is known to be complementary to the *ß*-globin mRNA. Alternatively, mature mRNAs have a polyA tail, so one could add a primer that consists of many Ts. This is called a poly-dT primer. After the complementary DNA strand has been made, the sample would then be mixed with primers, *Taq* polymerase, and nucleotides, etc., and subjected to the standard PCR protocol. Note: the PCR reaction would have two kinds of primers. One primer would be complementary to the 5′ end of the mRNA and would be unique to the *ß*-globin sequence. The other primer would be complementary to the 3′ end. This second primer could be a poly-dT primer or it could be a unique primer that would bind slightly upstream from the polyA-tail region.

E13. One interpretation would be that the gene is part of a gene family. In this case, the family would contain four homologous members. At high stringency, the probe binds only to the gene that is its closest match, but at low stringency it recognizes the three other homologous genes.

E14. A DNA library is a collection of hybrid vectors that contain different pieces of DNA from a source of chromosomal DNA. Because it is a diverse collection of many different DNA pieces, the name *library* seems appropriate.

E15. It would be necessary to use cDNA so that the gene would not carry any introns. Bacterial cells do not contain spliceosomes (which are described in Chapter 15). To express a eukaryotic protein in bacteria, a researcher would clone cDNA into bacteria since the cDNA does not contain introns.

E16. Hybridization occurs due to the hydrogen bonding of complementary sequences. Due to the chemical properties of DNA and RNA strands, they form double-stranded regions when the base sequences are complementary. In a Southern and Northern experiment, the probe is labeled.

E17. The purpose of gel electrophoresis is to separate the many DNA fragments, RNA molecules, or proteins that were obtained from the sample you want to probe. This separation is based on molecular mass and allows you to identify the molecular mass of the DNA fragment, RNA molecule, or protein that is being recognized by the probe.

E18. The purpose of a Northern blot experiment is to determine if a gene is transcribed into RNA using a piece of cloned DNA as a probe. It can tell you if a gene is transcribed in a particular cell or at a particular stage of development. It can also tell you if a pre-mRNA is alternatively spliced.

E19. The Northern blot is shown here. The female mouse expresses the same total amount of this mRNA compared to the male. In the heterozygous female, there would be 50% of the 900 bp band and 50% of the 825 bp band.

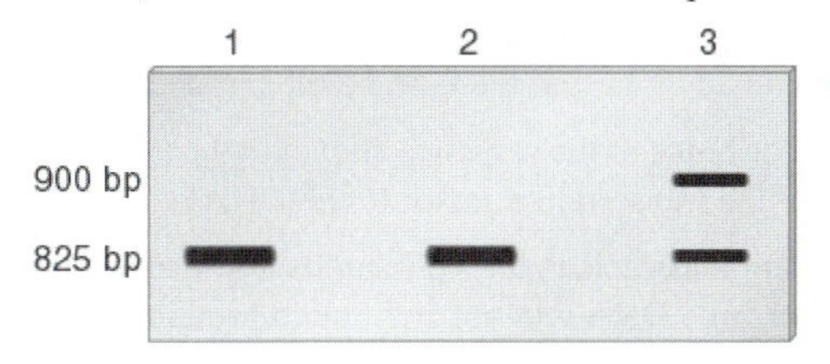

E20. It appears that this mRNA is alternatively spliced to create a high molecular mass and a lower molecular mass product. Nerve cells produce a very large amount of the larger mRNA, whereas spleen cells produce a moderate amount of the smaller mRNA. Both types are produced in small amounts by the muscle cells. It appears that kidney cells do not transcribe this gene.

E21. Restriction enzymes recognize many sequences throughout the chromosomal DNA. If two fragments from different samples have the same molecular mass in a Southern blot, it is likely (though not certain) that the two fragments are found at the same chromosomal site in the genome. In this Southern blot, most of the transposable elements are found at the same sites within the genomes of these different yeast strains. However, a couple of bands are different among the three strains. These results indicate that the *Ty* element may occasionally transpose to a new location or that chromosomal changes (point mutations, chromosomal rearrangements, deletions, etc.) may have slightly changed the genomes among these three strains of yeast in a way that changes the distances between the restriction sites.

E22. 1. β (detected at the highest stringency)

2. δ

3. γ_A and ε

4. α_1

5. *Mb* (detected only at the lowest stringency)

Note: At the lowest stringency, all of the globin genes would be detected. At the highest stringency, only the β-globin gene would be detected.

E23. Lane 1 shows that β-globin is made in normal red blood cells. In red blood cells from a thalassemia patient (lane 2), however, very little is made. Perhaps this person is homozygous for a down promoter mutation, which diminishes the transcription of the gene. As shown in lanes 3 and 4, β-globin is not made in muscle cells.

E24. The Western blot is shown here. The sample in lane 2 came from a plant that was homozygous for a mutation that prevented the expression of this polypeptide. Therefore, no protein was observed in this lane. The sample in lane 4 came from a plant that is homozygous for a mutation that introduces an early stop codon into the coding sequence. As seen in lane 4, the polypeptide is shorter than normal (13.3 kDa). The sample in lane 3 was from a heterozygote that expresses about 50% of each type of polypeptide. Finally, the sample in lane 5 came from a plant that is homozygous for a mutation that changed one amino acid to another amino acid. This type of mutation, termed a missense mutation, may not be detectable on gel. However, a single amino acid substitution could affect polypeptide function.

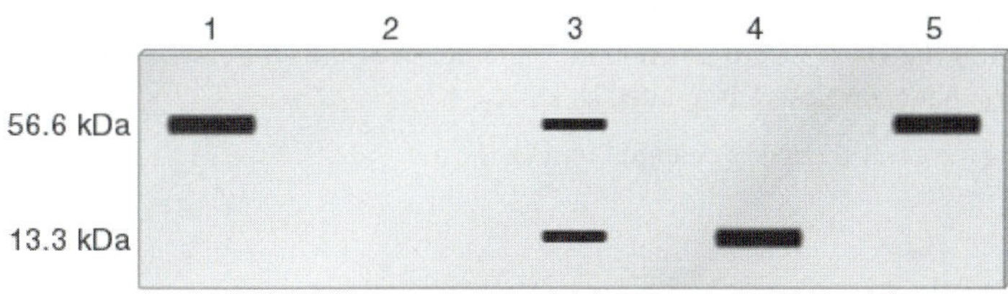

E25. Western blotting.

E26. The products of structural genes are proteins with a particular amino acid sequence. Antibodies can specifically recognize proteins due to their amino acid sequence. Therefore, an antibody can detect whether or not a cell is making a particular type of protein.

E27. You would first make a DNA library using chromosomal DNA from pig cells. You would then radiolabel the human β-globin gene and use it as a probe in a colony hybridization experiment; each colony would contain a different cloned piece of the pig genome. You would identify "hot" colonies that hybridize to the human β-globin probe. You would then go back to the master plate and pick these colonies and grow them in a test tube. You would then isolate the plasmid DNA from the colonies and subject the DNA to DNA sequencing. By comparing the DNA sequences of the human β-globin gene and the putative clones, you could determine if the putative clones were homologous to the human clone and likely to be the pig homologue of the β-globin gene.

E28. In this case, the transcription factor binds to the response element when the hormone is present. Therefore, the hormone promotes the binding of the transcription factor to the DNA and thereby promotes transactivation.

E29. The rationale behind a gel retardation assay is that a segment of DNA with a protein bound to it will migrate more slowly through a gel than will the same DNA without any bound protein. A shift in a DNA band to a higher molecular mass provides a way to identify DNA-binding proteins.

E30. The levels of cAMP affect the phosphorylation of CREB, and this affects whether or not it can transactivate transcription. However, CREB can bind to CREs whether or not it is phosphorylated. Therefore, in a gel retardation assay, we would expect CREB to bind to CREs and retard their mobility using a cell extract from cells that were or were not retreated with adrenalin.

E31. TFIID can bind to this DNA fragment by itself, as seen in lane 2. However, TFIIB and RNA polymerase II cannot bind to the DNA by themselves (lanes 3 and 4). As seen in lane 5, TFIIB can bind, if TFIID is also present, because the mobility shift is higher than TFIID alone (compare lanes 2 and 5). In contrast, RNA polymerase II cannot bind to the DNA when only TFIID is present. The mobility shift in lane 6 is the same as that found in lane 2, indicating that only TFIID is bound. Finally, in lane 7, when all three components are present, the mobility shift is higher than when TFIIB and TFIID are present (compare lanes 5 and 7). These results mean that all three proteins are bound to the DNA. Taken together, the results indicate that TFIID can bind by itself, TFIIB needs TFIID to bind, and RNA polymerase II needs both proteins to bind to the DNA.

E32. The glucocorticoid receptor will bind to GREs if glucocorticoid hormone is also present. The glucocorticoid receptor does not bind to CREs. The CREB protein will bind to CREs (with or without hormone), but it will not bind to GREs. The expected results are shown here. In this drawing, the binding of CREB protein to the 700 bp fragment results in a complex with a higher mass compared to the glucocorticoid receptor binding to the 600 bp fragment.

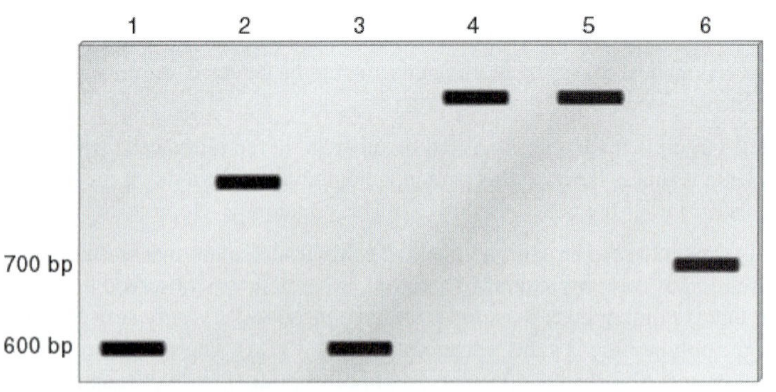

E33. The region of the gel from about 350 bp to 175 bp does not contain any bands. This is the region being covered up; it is about 175 bp long.

E34. The rationale behind a footprinting experiment has to do with accessibility. If a protein is bound to the DNA, it will cover up the part of the DNA where it is bound. This region of the DNA will be inaccessible to the actions of chemicals or enzymes that cleave the DNA, such as DNase I.

E35. The A closest to the bottom of the gel is changed to a G.

E36. A. AGGTCGGTTGCCATCGCAATAATTTCTGCCTGAACCCAATA

B. Automated sequencing has several advantages. First, the reactions are done in a single tube as opposed to four tubes. Second, the detector can "read" the sequence and provide the researcher with a printout of the sequence. This is much easier than looking at an X-ray film and writing the sequence out by hand. It also avoids human error. Finally, automated sequencing does not require the use of radioisotopes, which are more expensive and require more laboratory precautions, compared to fluorescently labeled compounds.

E37. 5′–CCCCCGATC**G**ACATCATTA–3′. The mutagenic base is underlined.

E38. There are lots of different strategies one could follow. For example, you could mutate every other base and see what happens. It would be best to make very nonconservative mutations such as a purine for a pyrimidine or a pyrimidine for a purine. If the mutation prevents protein binding in a gel retardation assay, then the mutation is probably within the response element. If the mutation has no effect on protein binding, it probably is outside the response element.

E39. You would conclude that it might be important. The only amino acid substitution that gave a substantial amount of functional activity was an aspartate. Glutamate and aspartate have very similar amino acid side chains (see Chapter 13); they both contain a carboxyl (COOH) group. Based on these results, you would suspect that a carboxyl group at this location in the protein might be important for its function.

Questions for Student Discussion/Collaboration

1. 1. To know what type of cell a gene is expressed in.
   2. To know what stage of development it is expressed in.
   3. To know if alternative splicing is occurring.
   4. To know if environmental factors affect the level of gene expression.
   5. To know if mutations affect the level of gene expression.
2. 1. Does a particular amino acid within a protein sequence play a critical role in the protein's structure or function?
   2. Does a DNA sequence function as a promoter?
   3. Does a DNA sequence function as a regulatory site?
   4. Does a DNA sequence function as a splicing junction?
   5. Is a sequence important for correct translation?
   6. Is a sequence important for RNA stability?

   And many others.

Chapter 19: Biotechnology

Student Learning Objectives

Upon completion of this chapter you should be able to:

1. Understand the different functions of microorganisms in biotechnology.
2. Understand the different procedures that can be used to genetically manipulate plants and animals.
3. Understand the concept of a transgenic plant.
4. Understand the process by which somatic cells may be use for mammalian cloning.
5. Understand the different types of stem cells.
6. Understand the use of transgenic plants and animals.
7. Understand what is meant by DNA fingerprinting and DNA profiling.
8. Understand how gene therapy is being used to treat human diseases.

19.1 Uses of Microorganisms in Biotechnology

Overview

This chapter presents one of the hottest areas in the study of genetics. Biotechnology is the use of organisms to benefit humans. Almost daily the media report advances or applications of genetic principles to solve the mysteries of disease or develop new products. An understanding of the terminology associated with biotechnology is crucial for anyone who is interested in the study of genetics.

The first section of this chapter examines the use of microorganisms (primarily bacteria) in biotechnology. This mostly focuses on the development of recombinant bacteria, although there are other uses of these organisms. The section describes the development of the first recombinant bacteria, and the current efforts to develop bacteria that can assist in bioremediation and biotransformation.

Key Terms

Biodegredation
Biological control
Bioremediation
Biotechnology
Biotransformation
DNA fingerprinting
Gene therapy
Mammalian cloning
Stem cell research
Transgenic

Focal Points

- Uses of microorganisms in biotechnology (Table 19.1)

Exercises and Problems

For the following questions, match the term to its correct definition.

_____ 1. Biological control

_____ 2. Insulin

_____ 3. Somatostatin

_____ 4. Bioremediation

_____ 5. Biotransformation

_____ 6. Biodegradation

a. The use of gene products from *Agrobacterium radioba*cter to prevent crop disease.
b. The first protein to be produced by recombinant bacteria.
c. The use of a bacteria to alter the chemical structure of a pollutant in the environment.
d. The development of this allowed for a more practical treatment for diabetes.
e. Then general term for the use of recombinant microorganisms to reduce pollutants in the environment.
f. The breakdown of a toxic material into a nontoxic compound.

19.2 New Methods for Genetically Manipulating Plants and Animals

Overview

In the second section of the chapter we examine some of the procedures that can be used in biotechnology. It is important to note that this is a rapidly evolving field, with new advances occurring at a rapid pace. The procedures described in this section provide a survey of some of the processes that are used by geneticists to study the relationship between a gene and the phenotype of the organism.

A few areas of this section are worth special mention. The first is the production of transgenic plants and animals. The term transgenic indicates that the organisms contains genetic material that is not native to the species. In other words, it contains introduced gene(s). Transgenics is a rapidly evolving field of research in both animals and plants. The second major area of focus is that of cloning. The creation, and death, of Dolly the sheep has created an intense amount of interest in the process of cloning, especially in ways that it might be applied to humans.

Finally, this section examines the terminology associated with stem cells. As undifferentiated cells, stem cells have an enormous potential for the study for both genetics and medicine. This section provides and overview of the use of stem cells in genetic research.

Key Terms

Biolistic gene transfer
Chimera
Electroporation
Embryonic carcinoma cells
Embryonic germ cells
Embryonic stem cells
Gene addition
Gene knockout
Gene replacement
Microinjection
Multipotent
Organismal cloning
Pluripotent
Stem cells
T DNA
Ti plasmid
Totipotent
Transgene
Unipotent

Focal Points

- Differences between gene replacement and gene addition (Figure 19.5)
- Process of organismal cloning (Figure 19.9)
- Origins of stem cells (Figure 19.11)

Exercises and Problems

For questions 1 to 9, match the definitions of terms associated with the process of gene replacement and transgenics with the appropriate term.

_____ 1. The presence of both normal and clone genes in an organism.

_____ 2. A mechanism that is used to produce transgenic plants.

_____ 3. The use of microscopic needles to inject DNA into a cell.

_____ 4. Allows the study of loss of function for a gene.

_____ 5. A gene from one species that is introduced into a second species.

_____ 6. The use of microprojectiles to introduce transgene DNA into a cell.

_____ 7. An organism that contains cells from two different individuals.

_____ 8. The use of an electrical current to inject DNA into a cell.

_____ 9. A type of cell that can regenerate an entire new organism.

a. gene knockout
b. chimera
c. gene addition
d. Ti plasmid
e. biolistic gene transfer
f. transgene
g. microinjection
h. electroporation
i. totipotent

19.3 Applications for Transgenic Plants and Animals

Overview

The third section of the chapter introduces the concept of transgenic organisms, or those that contain genes from different species. The creation, and use, of transgenic organisms has been controversial since it began, yet it remains a major area of research for both medical and agricultural genetics.

Key Terms

Gene knockin

Gene redundancy

Molecular pharming

Focal Points

- Process of molecular pharming (Figure 19.13)

Exercises and Problems

For the following questions, match the term with the most appropriate definition.

_____ 1. Molecular pharming

_____ 2. Gene knockin

_____ 3. Gene knockout

_____ 4. Gene redundancy

a. Allows researchers to study loss-of-function mutations in organisms.
b. An example is the attachment of a protein in a milk-specific promoter so that it can be expressed in mammals.
c. Following gene knockout, a second gene compensates for the loss of function of the first gene.
d. The procedure that attempts to replace a defective gene with an altered copy.

For each of the following, complete the statement by providing the needed word or phrase.

5. The FlavrSavr tomato contains _____ RNA that is complementary for RNA from the enzyme polygalacturonase.
6. _____ from transgenic animals has become an important source of human proteins with a therapeutic value.
7. Transgenic research in plants has sought to produce strains that are resistant to _____, ______, and ______.

19.4 DNA Fingerprinting

Overview

While humans are remarkably similar genetically, there is enough variation in our genome to be able to distinguish individuals based upon patterns in their DNA. This is the basis of DNA fingerprinting, also known as DNA profiling. Forensic scientists, law enforcement agencies and anthropologists (to name a few) are all intensely interested in the genetic identification of humans. Over the past several years the technology to do this reliably has evolved considerably. In this section of the chapter you will examine some of the ways that humans can be identified by these processes, and the types of sequence data that is the most useful in DNA fingerprinting.

Key Terms

DNA profiling
Microsattelites
Minisatellites
Short tandem repeats
Variable number of tandem repeats

Focal Points

- VNTRs (Figure 19.17)
- DNA fingerprinting (Figures 19.18 and 19.19)

Exercises and Problems

For the questions below, determine which of the following answers apply to the statement.

a. STRs
b. VNTRs
c. both a and b

_____ 1. Microsatellites

_____ 2. Studied using restriction enzymes and radioactive probes.

_____ 3. Short variable sequences of 100 to 450 bp

_____ 4. These are variable in length between humans.

_____ 5. Minisatellites

_____ 6. Studied using PCR reactions and DNA sequencing.

_____ 7. May be used in paternity testing.

19.5 Human Gene Therapy

Overview

The final section of this chapter examines what could be one of the most important developments in genetics in the next few years. Over the past several decades, biotechnology has progressed to the point where it is becoming possible to correct the causes of certain diseases using a process called gene therapy. While there are still obstacles to its widespread use, this section of the chapter examines the basic processes of gene therapy and some of the early successes in its use to treat diseases.

Key Terms

Liposomes

Focal Points

- Process of gene therapy (Figure 19.20)

Exercises and Problems

For each of the following, match the term or phrase with its correct statement.

_____ 1. Liposomes

_____ 2. Cystic fibrosis

_____ 3. Adenosine deaminase

_____ 4. Nonviral approach

_____ 5. Viral approach

a. A lipid molecule that contains DNA for use in gene therapy.
b. The gene of interest is introduced into the cell by endocytosis of a liposome.
c. The first clinical application of gene therapy to reverse a disease.
d. Treatment has involved the delivery of the gene therapy via an aerosol spray.
e. The gene of interest is integrated into a retrovirus.

Chapter Quiz

1. In this procedure for making a transgenic plant, DNA is moved inside of the target cell using an electrical current.
    a. biolistic gene transfer
    b. liposome gene therapy
    c. microinjection
    d. electroporation

2. The first drug produced by recombinant bacteria was _____.
    a. somatostatin
    b. insulin
    c. adenosine deaminanse
    d. glucagon

3. Which of the following is not a pluripotent cell?
    a. embryonic stem cell
    b. embryonic germ cell
    c. embryonic carcinoma cell
    d. adult stem cell
    e. all of the above are plutipotent

4. Which of the following terms specifically applies to the use of microorganisms to breakdown a toxic pollutant into a less complex metabolite?
    a. biotransformation
    b. bioremediation
    c. biodegradation
    d. biotherapy

5. Which of the following is commonly used to study loss-of-function mutations?
  a. gene knockout
  b. gene knockin
  c. biolistic gene transfer
  d. gene redundancy
  e. gene therapy

6. Liposomes are commonly used for which of the following?
  a. organismal cloning
  b. creating transgenic plants
  c. a delivery molecule for gene therapy
  d. producing recombinant bacteria

7. Which of the following best describes molecular pharming?
  a. introduction of new genes into a CF patient.
  b. production of a new cucumber with an improved shelf life
  c. production of drugs by bacteria
  d. use of mammals as biological factories for producing human proteins
  e. DNA fingerprinting

8. STRs and VNTRs are used in which of the following?
  a. molecular pharming
  b. DNA fingerprinting
  c. creation of transgenic organisms
  d. organismal cloning

9. The use of living organisms or their products to alleviate plant disease or damage is called ______.
  a. molecular pharming
  b. biotechnology
  c. biological control
  d. nonviral gene therapy

10. The first disease to successfully be treated using gene therapy was _____.
  a. diabetes
  b. cancer
  c. cystic fibrosis
  d. severe combined immunodeficiency disease
  e. muscular dystrophy

Answer Key for Study Guide Questions

This answer key provides the answers to the exercises and chapter quiz for this chapter. Answers in parentheses ( ) represent possible alternate answers to a problem, while answers marked with an asterisk (*) indicate that the response to the question may vary.

19.1	1.a	3. b	5. c
	2. d	4. e	5. f

19.2	1. c 2. d 3. g	4. a 5. f 6. e	7. b 8. h 9. I
19.3	1. b 2. d 3. a	4. c 5. antisense 6. milk	7. insects, disease, herbicides
19.4	1. a 2. b 3. a	4. c 5. b	6. a 7. c
19.5	1. a 2. d	3. c 4. b	5. e

Quiz

1. d 2. a 3. d 4. c	5. a 6. c 7. d	8. b 9. c 10. d

Answers to Conceptual and Experimental Questions

Conceptual Questions

C1. A recombinant microorganism is one that contains DNA that has been manipulated in vitro and then reintroduced back into the organism. Recombinant microorganisms have been used to synthesize human gene products (e.g., insulin), as biological control agents (e.g., *Ice*$^{-}$ bacteria), and in bioremediation (e.g., oil-eating bacteria).

C2. *A. radiobacter* synthesizes an antibiotic that kills *A. tumefaciens.* The genes, which are necessary for antibiotic biosynthesis and resistance, are plasmid encoded and can be transferred during interspecies matings. If *A. tumefaciens* received this plasmid during conjugation, it would be resistant to killing. Therefore, the conjugation-deficient strain prevents the occurrence of *A. tumefaciens*–resistant strains.

C3. Bioremediation is the use of microorganisms to eliminate pollutants in the environment. Biotransformation is the conversion of a toxic compound into a nontoxic one. Biodegradation occurs when a toxic compound is broken down into a smaller, less toxic one.

C4. A biological control agent is an organism that prevents the harmful effects of some other agent in the environment. Examples include *Bacillus thuringiensis,* a bacterium that synthesizes compounds that act as toxins to kill insects, *Ice*$^{-}$ bacteria that inhibit the proliferation of *Ice*$^{+}$ bacteria, and the use of *Agrobacterium radiobacter* to prevent crown gall disease caused by *Agrobacterium tumefaciens.*

C5. These medicines are difficult and expensive to purify from human sources. The advantage of genetically engineered organisms is that they can produce a large amount of these medical agents, at a fraction of the cost. There are some disadvantages. For example, a medical agent may require posttranslational modifications that do not occur in microorganisms. Public perception of genetic engineering may also be a problem, although it has not been a big problem in this particular area of genetic engineering.

C6. A mouse model is a strain of mice that carries a mutation in a mouse gene that is analogous to a mutation in a human gene that causes disease. For example, after the mutation causing cystic fibrosis was identified, the analogous gene was mutated in the mouse. Mice with mutations in this gene have symptoms similar to the human symptoms (though not identical). These mice can be used to study the disease and to test potential therapeutic agents.

C7. A transgenic organism is one that has recombinant DNA incorporated into its genome. The FlavrSavr tomato is an example of transgenic plant. Plants resistant to glyphosate are another example. The large mouse shown in Figure 19.4 contains a transgene from a human. Sheep that express human hormones in their milk are also transgenics.

C8. The T DNA gets transferred to the plant cell; it then is incorporated into the plant cell's genome.

C9. Gene addition occurs when a gene is introduced into a cell and it integrates into the chromosomal DNA by nonhomologous recombination. The gene is simply added to the genome. Gene replacement occurs by homologous recombination. In gene replacement, the gene that is introduced into a cell is swapped with a homologous gene that is already present in the genome.

A. This is gene replacement. The normal mouse gene is replaced with a mutant gene.

B. This is gene addition. The T DNA that carries a gene of interest can integrate at various sites throughout a plant genome.

C10. A. With regard to maternal effect genes, the phenotype would depend on the animal that donated the oocyte. It is the cytoplasm of the oocyte that accumulates the gene products of maternal effect genes.

B. The extranuclear traits depend on the mitochondrial genome. Mitochondria are found in the oocyte and in the somatic cell. So, theoretically, both cells could contribute extranuclear traits. In reality, however, researchers have found that the mitochondria in Dolly were from the animal that donated the egg. It is not clear why she had no mitochondria from the mammary cell.

C. The cloned animal would be genetically identical to the animal that donated the nucleus with regard to traits that are determined by nuclear genes, which are expressed during the lifetime of the organism. The cloned animal would/could differ from the animal that donated the nucleus with regard to traits that are determined by maternal effect genes and extracytoplasmic genes. Such an animal is not a true clone, but it is likely that it would greatly resemble the animal that donated the nucleus, because the vast majority of genes are found in the cell nucleus.

C11. See Table 19.5.

C12. Some people are concerned with the release of genetically engineered microorganisms into the environment. The fear is that such organisms may continue to proliferate and it may not be possible to "stop them." A second fear is the use of genetically engineered organisms in the food we eat. Some people are worried that genetically engineered organisms may pose an unknown health risk. A third issue is ethics. Some people feel that it is morally wrong to tamper with the genetics of organisms. This opinion may also apply to genetic techniques such as cloning, stem cell research, and gene therapy.

C13. Gene therapy, as described in your textbook, is conducted on somatic cells, not on germ-line cells. Therefore, the genes that are introduced into a person's body via gene therapy cannot be passed to his/her offspring.

Experimental Questions

E1. The human gene that encodes the hormone is manipulated in vitro and then transformed into a bacterium. In many cases, the coding sequence of the human hormone gene is fused with a bacterial gene to prevent the rapid degradation of the human hormone. The bacteria then express the fusion protein and the hormone is separated by cyanogen bromide cleavage.

E2. The plasmid with the wrong orientation would not work because the coding sequence would be in the wrong direction relative to the promoter sequence. Therefore, the region containing the somatostatin sequence would not be transcribed into RNA.

E3. One possibility is to clone the toxin-producing genes from *B. thuringiensis* and introduce them into *P. syringae.* This bacterial strain would have the advantage of not needing repeated applications. However, it would be a recombinant strain and might be viewed in a negative light by people who are hesitant to use recombinant organisms in the field. By comparison, *B. thuringiensis* is a naturally occurring species.

E4. To construct the coding sequence for somatostatin, the researchers synthesized eight oligonucleotides, labeled A–H. When these oligonucleotides were mixed together, they would hydrogen bond to each other due to their complementary sequences. The addition of ligase would covalently link the DNA backbones. The *left* side of the coding sequence had an *Eco*RI site and the *right* side had a *Bam*HI site; these sites made it possible to insert this sequence at the end of the *ß*-galactosidase gene. You may also notice that an ATG codon (AUG in the mRNA) precedes the first alanine codon in somatostatin. This AUG codon specifies methionine, which allows the somatostatin to be cleaved from *ß*-galactosidase using cyanogen bromide. This was necessary because somatostatin, by itself, is rapidly degraded in *E. coli,* whereas the fusion protein is not.

E5. Basically, one can follow the strategy described in Figure 19.6. If homologous recombination occurs, only the Neo^R gene is incorporated into the genome. The cells will be neomycin resistant and also resistant to gancyclovir. If gene addition occurs, the cells will be sensitive to gancyclovir. By growing the cells in the presence of neomycin and gancyclovir, one can select for homologous recombinants. The chimeras are identified by the observation that they have a mixture of light and dark fur.

E6. A kanamycin-resistance gene is contained within the T DNA. Exposure to kanamycin selects for the growth of plant cells that have incorporated the T DNA into their genome. The carbenicillin kills the *A. tumefaciens.* The phytohormones promote the regeneration of an entire plant from somatic cells. If kanamycin were left out, it would not be possible to select for the growth of cells that had taken up the T DNA.

E7. 1. *A. tumefaciens:* Genes are cloned into the T DNA of the Ti plasmid. This T DNA from this plasmid is transferred to the plant cell when it is infected by the bacterium.

2. Microinjection: A microscopic needle containing a solution of DNA is used to inject DNA into plant cells.

3. Electroporation: An electrical current is used to introduce DNA into plant cells.

4. Biolistic gene transfer: DNA is coated on microprojectiles that are "shot" into the plant cell.

5. Also, DNA can be introduced into protoplasts by treatment with polyethylene glycol and calcium phosphate. This method is similar to bacterial transformation procedures.

E8. The term *gene knockout* refers to an organism in which the function of a particular gene has been eliminated. For autosomal genes in eukaryotes, a gene knockout is a homozygote for a defect in both copies of the gene. If a gene knockout has no phenotypic effect, perhaps the gene is redundant. In other words, there may be multiple genes within the genome that can carry out the same function. Another reason why a gene knockout may not have a phenotypic effect is because of the environment. As an example, let's say a mouse gene is required for the synthesis of a vitamin. If the researchers were providing food that contained the vitamin, the knockout mouse that was lacking this gene would have a normal phenotype; it would survive just fine. Sometimes, researchers have trouble knowing the effects of a gene knockout until they modify the environmental conditions in which the animals are raised.

E9. In Mendel's work, and the work of many classical geneticists, an altered (mutant) phenotype is the initial way to identify a gene. For example, Mendel recognized a gene that affects plant height by the identification of tall and dwarf plants. The transmission of this gene could be followed in genetic crosses, and eventually, the gene could be cloned using molecular techniques. Reverse genetics uses the opposite sequence of steps. The gene is cloned first, and a phenotype for the gene (based on the creation of a gene knockout) is discovered later, by making a transgenic animal with a gene knockout.

E10. Gene replacement occurs by homologous recombination. For homologous recombination to take place, two crossovers must occur, one at each end of the target gene. After homologous recombination, only the Neo^R gene, which is inserted into the target gene, can be incorporated into the chromosomal DNA of the embryonic stem cell. In contrast, nonhomologous recombination can involve two crossovers anywhere in the cloned DNA. Since the *TK* gene and target gene are adjacent to each other, nonhomologous recombination usually transfers both the *TK* gene and the Neo^R gene. If both genes are transferred to a stem cell, it will die because the cells are grown in the presence of gancyclovir. The product of the *TK* gene kills the cell under these conditions. In contrast, cells that have acquired the Neo^R gene due to homologous recombination but not the *TK* gene will survive. Stem cells, which have not taken up any of the cloned DNA, will die because they will be killed by the neomycin. In this way, the presence of gancyclovir and neomycin selects for the growth of stem cells that have acquired the target gene by homologous recombination.

E11. A chimera is an organism that is composed of cells from two different individuals (usually of the same species). Chimeras are made by mixing together embryonic cells from two individuals, and allowing the cells to organize themselves and develop into a single individual.

E12. A. Dolly's chromosomes may seem so old because they were already old when they were in the nucleus that was incorporated into the enucleated egg. They had already become significantly shortened in the mammary cells. This shortening was not repaired by the oocyte.

B. Dolly's age does not matter. Remember that shortening does not occur in germ cells. However, Dolly's eggs are older than they seem by about 6 or 7 years, because Dolly's germ-line cells received their chromosomes from a sheep that was 6 years old, and the cells were grown in culture for a few doublings before a mammary cell was fused with an enucleated egg. Therefore, the calculation would be: 6 or 7 years (the age of the mammary cells that produced Dolly's germ-line cells) plus 8 years (the age of Molly), which equals 14 or 15 years. However, only half of Molly's chromosomes would appear to be 14 or 15 years old. The other half of her chromosomes, which she inherited from her father, would appear to be 8 years old.

C. Chromosome shortening is a bit disturbing, because it suggests that aging has occurred in the somatic cell, and this aging is passed to the cloned organism. If cloning was done over the course of many generations, this may eventually have a major impact on the life span of the cloned organism. It may die much earlier than a noncloned organism. However, chromosome shortening may not always occur. It does not seem to occur in mice, which were cloned for six consecutive generations.

E13. You would conduct a Southern blot to determine the number of gene copies. As described in Chapter 18, you will observe multiple bands in a Southern blot if there are multiple copies of a gene. You need to know this information to predict the outcome of crosses. For example, if there are four integrated copies at different sites in the genome, an offspring could inherit anywhere from zero to four copies of the gene. You would want to understand this so you could predict the phenotypes of the offspring.

You would use a Northern blot or a Western blot to monitor the level of gene expression. A Northern blot will indicate if the gene is transcribed into mRNA, and a Western blot will indicate if the gene is translated into protein.

E14. The term *molecular pharming* refers to the practice of making transgenic animals that will synthesize (human) products in their milk. It can be advantageous when bacterial cells are unable to make a functional protein product from a human gene. For example, some proteins are posttranslationally modified by the attachment of carbohydrate molecules. This type of modification does not occur in bacteria, but it may occur correctly in transgenic animals. Also, dairy cows produce large amounts of milk, which may improve the yield of the human product.

E15. Organismal cloning means the cloning of entire multicellular organisms. In plants, this is easy. Most species of plants can be cloned by asexual cuttings. In animals, cloning occurs naturally, as in identical twins. Identical twins are genetic replicas of each other because they begin from the same fertilized egg. (Note: There could be some somatic mutations that occur in identical twins that would make them slightly different.) Recently, as in the case of Dolly, organismal cloning has become possible by fusing somatic cells with enucleated eggs. The advantage, from an agricultural point of view, is that organismal cloning could allow one to choose the best animal in a herd and make many clones from it. Breeding would no longer be necessary. Also, breeding may be less reliable because the offspring inherit traits from both the mother and father.

E16. You would first need to clone the normal mouse gene. Cloning methods are described in Chapter 18. After the normal gene was cloned, you would then follow the protocol shown in Figure 19.6. The normal gene would be inactivated by the insertion of the Neo^R gene, and the *TK* gene would be cloned next to it. This DNA segment would be introduced into mouse embryonic stem cells and grown in the presence of neomycin and gancyclovir. This selects for homologous recombinants. The surviving embryonic stem cells would be injected into early embryos, which would then develop into chimeras. The chimeric mice would be identified by their patches of light and dark fur. At this point, if all has gone well, a portion of the mouse is heterozygous for the normal gene and the gene that has the Neo^R insert. This mouse would be bred, and then brothers and sisters from the litter would be bred to each other. Southern blots would be conducted to determine if the offspring carried the gene with the Neo^R insert. At first, one would identify heterozygotes that had one copy of the inserted gene. These heterozygotes would be crossed to each other to obtain homozygotes. The homozygotes are gene knockouts because the function of the gene has been "knocked out" by the insertion of the Neo^R gene. Perhaps, these mice would be dwarf and exhibit signs of mental retardation. At this point, the researcher would have a mouse model to study the disease.

E17. Male 2 is the potential father, because he contains the bands that are found in the offspring but are not found in the mother. To calculate the probability, one would have to know the probability of having each of the types of bands that match. In this case, for example, male 2 and the offspring have four bands in common. As a simple calculation, we could eliminate the four bands that the offspring shares with the mother. If the probability of having each paternal band is 1/4, the chances that this person is not the father are $(1/4)^4$.

E18. DNA fingerprinting is a method of identification based on the properties of DNA. VNTR and STR sequences are variable with regard to size in natural populations. This variation can be seen when DNA fragments are subjected to gel electrophoresis. Within a population, any two individuals (except for identical twins) will display a different pattern of DNA fragments, which is called their DNA fingerprint.

E19. PCR is used to amplify DNA if there is only a small amount of it (e.g., a small sample at a crime scene). It is also used to amplify STRs. Southern blotting, using a probe that is complementary to a VNTR, is needed to specifically identify a limited number of bands (20 or so) that are variable within human populations.

E20. A VNTR is a sequence that is repeated several times within a genome and is variable in its length. In natural populations, it is common to find length variation. Therefore, any two individuals (who are not genetically identical) will differ with regard to the sizes of many of their VNTRs. When a gel is run and the VNTRs are seen in a Southern blot, the pattern of VNTR sizes provides a fingerprint of the individual's DNA. This fingerprint is a unique feature of each individual.

E21. This percentage is not too high. Based on their genetic relationship, we expect that a father and daughter must share at least 50% of the same bands in a DNA fingerprint. However, the value can be higher than that because the mother and father may have some bands in common, even though they are not genetically related. For example, at one site in the genome, the father may be heterozygous for a 4,100 bp and 5,200 bp VNTR, and the mother may also be heterozygous in this same region and have 4,100 bp and 4,700 bp VNTRs. The father could pass the 5,200 bp band to his daughter and the mother could pass the 4,100 bp band. The daughter would inherit the 4,100 bp and 5,200 bp bands. This would be a perfect match to both of the father's bands, even though the father transmitted only the 5,200 bp band to his daughter. The 4,100 bp band matches because the father and mother happened to have a VNTR in common. Therefore, the 50% estimate of matching bands in a DNA fingerprint based on genetic relationships is a minimum estimate. The value can be higher than that.

E22. The minimum percentage of matching bands is based on the genetic relationships.

A. 50%

B. 50% (on average, but occasionally it could be less)

C. 25% (on average, but occasionally it could be less)

D. 25% (on average, but occasionally it could be less)

E23. Ex vivo therapy involves the removal of living cells from the body and their modification after they have been removed. The modified cells are then reintroduced back into a person's body. This approach works well for cells such as blood cells that are easily removed and replaced. By comparison, this approach would not work very well for many cell types. For example, lung cells cannot be removed and put back again. In this case, in vivo approaches must be sought.

E24. In cystic fibrosis gene therapy, an aerosol spray, containing the normal *CF* gene in a retrovirus or liposome, is used to get the normal *CF* gene into the lung cells. The epithelial cells on the surface of the lung will take up the gene. However, these surface epithelial cells have a finite life span, so it is necessary to have repeated applications of the aerosol spray. Ideally, scientists hope to devise methods whereby the normal *CF* gene will be able to penetrate more deeply into the lung tissue.

E25. It is the gene product (i.e., the polypeptide) of an oncogene that causes cancerous cell growth. The antisense RNA (from the gene introduced via gene therapy) would bind to the mRNA from an oncogene. This would prevent the translation of the mRNA into polypeptides and thereby prevent cancerous cell growth.

Questions for Student Discussion and Collaboration

1. There are several advantages to gene therapy. In Chapter 19, we have considered how it can be used to correct a specific gene defect. It can also be used to fight other types of diseases, such as infectious diseases and cancer. The disadvantages include a high cost, necessity of repeated applications, difficulty in getting the gene to the correct cell type, and possible side effects. As for priorities, it is difficult to make a list. But the priority list could be based on the prevalence of the disease, the likelihood that gene therapy will be effective at combating the disease, the severity of the disease, and so on.
2. From a genetic viewpoint, the recombinant and nonrecombinant strains are very similar. The main difference is their history. The recombinant strain has been subjected to molecular techniques to eliminate a particular gene. The nonrecombinant strain has the advantage of good public relations. People are less worried about releasing nonrecombinant strains into the environment.
3. This could be a very long list. One could alter the characteristics of food, fiber, disease resistance, and on and on. Hopefully, you will be able to draw on the genetic knowledge you have gathered in other chapters to come up with clever ways of making interesting transgenic organisms.

Chapter 20: Structural Genomics

Student Learning Objectives

Upon completion of this chapter you should be able to:

1. Understand the concept of genomic analysis and the forms that it may take.
2. Understand the procedures for determining the location of a gene on a chromosome.
3. Understand the procedures to physically map a gene on a chromosome.
4. Understand the different forms of cloning.
5. Understand the goals of the Human Genome Project.

Introduction

20.1 Cytogenetic and Linkage Mapping

Overview

In this chapter we will begin to look at the genome of an organism as a whole. The study of the genome is called genomics and involves the cytological and molecular study of the genome. In the first section you are going to be introduced to methods of localizing a gene of interest first on a chromosome, and then at a specific location on the chromosome.

There are three levels of gene mapping. These are summarized in Figure 20.2. Notice that each of the methods uses different terms to describe the location of the gene. Cytogenetic mapping uses physical locations along the chromosome, such as 12p2a (see Chapter 8, Figure 8.1 for review). In linkage mapping, the relative order of the genes are determined based upon recombination frequencies. These are expressed as map units. Finally, the exact nucleotide distance, in base-pairs, is determined using physical mapping. Keep this in mind as you progress through the chapter.

Key Terms

- Amplified restriction fragment length polymorphism
- Cytogenetic mapping
- Cytological mapping
- Fluorescence *in situ* hybridization
- Founder
- Functional genomics
- Genome
- Genomics
- *In situ* hybridization
- Linkage mapping
- Lod score method
- Microsatellites
- Molecular marker
- Monomorphic
- Physically mapped
- Polymorphism
- Proteomics
- Restriction fragment length polymorphism
- RFLP map
- Sequence-tagged site
- Structural genomics

Focal Points

- Comparison of mapping types (Figure 20.2)
- Description of RFLPs and linkage mapping (Figures 20.5 and 20.8)
- Microsatellites (Figure 20.12)

Exercises and Problems

For each of the following, determine the level of genomic analysis to which the statement is most applicable.

a. physical mapping
b. cytological mapping
c. linkage mapping

_____ 1. Determines the precise nucleotide distance between genes.

_____ 2. May use STSs as molecular markers.

_____ 3. Uses FISH.

_____ 4. Studies recombination frequencies between genes or markers.

_____ 5. Determines the location of a gene in relation to a banding pattern of a chromosome.

_____ 6. Examples of results would be 3q5a and 14p1c.

_____ 7. May use RFLPs as a molecular marker.

_____ 8. Sequences such as $(CA)_n$ are used as markers.

_____ 9. The end result may be an RFLP map.

_____ 10. Reported as map units.

For questions 11 to 20, match the term to its correct definition.

_____ 11. Molecular marker

_____ 12. Functional genomics

_____ 13. Fluorescent *in situ* hybridization

_____ 14. Structural genomics

_____ 15. Polymorphism

_____ 16. Lod score method

_____ 17. Proteomics

_____ 18. Monomorphic

_____ 19. Founder

_____ 20. Sequence tagged site

a. Studies the interactions of genes in the genome.
b. A genetic segment that varies among members of the population.
c. Determines the likelihood of linkage between two markers.
d. Uses a probe that reacts to a specific wavelength of light.
e. The original member of a population that possessed a disease.
f. A general term for a segment of DNA that is found at a specific location on a chromosome.
g. Examines the relationship between proteins and characteristics of the organism.
h. A genetic segment that is identical among all members of the population.
i. A sequence of DNA that is amplified by a single set of DNA primers.
j. Maps regions of the chromosome.

20.2 Physical Mapping

Overview

As the name implies, physical mapping involves determining the exact location of a gene on a chromosome. There are a variety of methods that may be used to construct a physical map. The major forms of these are introduced in this section. For each, you should familiarize yourself with the procedure, as well as any limitations, such as the size of the chromosome that can be examined.

The final pages of this chapter introduces the Human Genome Project. The Human Genome Project represents one of the most important undertakings in the study of genetics. Its goal, to map the human genome, was completed in 2003, but the analysis of the material may take decades. As students of genetics, you should be familiar with the goals of this project.

Key Terms

Bacterial artificial chromosome
Chromosome sorting
Chromosome walking
Chromosome-specific library
Contig
Cosmid
Positional cloning
Pulsed-field gel electrophoresis
Subcloning
Yeast artificial chromosome

Focal Points

- Examples of physical mapping (Figures 20.14, 20.15, 20.17 and 20.19)
- A comparison of mapping procedures (Figure 20.18)

Exercises and Problems

For each of the following, match the term to its correct definition.

_____ 1. Cloning of a gene based upon its mapped position along a chromosome.

_____ 2. Uses fluorescent dyes to separate large eukaryotic chromosomes.

_____ 3. A hybrid between a plasmid and phage vector.

_____ 4. A collection of hybrid vectors all derived from the same chromosome.

_____ 5. A contig that contains the largest segments of chromosomal DNA inserts.

a. chromosome sorting
b. cosmids
c. positional cloning
d. artificial chromosomes
e. chromosome-specific library

For questions 6 to 13, choose the procedure type that correlates with the statement.

a. pulsed field gel electrophoresis
b. contigs
c. YACs
d. chromosome walking

_____ 6. Hybrid vectors that contain a continuous region of the chromosome.

_____ 7. Overlapping pieces of a chromosome specific library.

_____ 8. May be used to purify large pieces of DNA.

_____ 9. Structures that are similar to human chromosomes, but possess characteristics that are ideal for cloning.

_____ 10. Can be used to coarsely map the locations of genes.

_____ 11. A method uses in positional cloning.

_____ 12. Used in preparing contigs that span long sections of chromosomes.

_____ 13. Involves a series of subclones to isolate the gene to a specific region.

Chapter Quiz

1. Which of the following allows mapping to the precise nucleotide?
   a. cytological mapping
   b. linkage mapping
   c. physical mapping
   d. FISH

2. Which of the following is not a molecular marker used in linkage mapping?
   a. RFLPs
   b. YACs
   c. STSs
   d. AFLPs
   e. microsatellites

3. Positions such as 2p4c and 17q1c would be determined by which of the following?
   a. FISH
   b. PCR
   c. chromosome sorting
   d. AFLPs

4. Subcloning is most commonly used in which of the following?
   a. FISH
   b. RFLP mapping
   c. chromosome walking
   d. pulsed field gel electrophoresis

5. Which of the following represents a hybrid between a plasmid and phage vector?
   a. BACs
   b. YACs
   c. contigs
   d. subclones
   e. cosmids

6. The likelihood of linkage is determined by the _____.
   a. contig
   b. cosmid
   c. log score
   d. RFLP map
   e. none of the above

7. The study of how the components of a genome interact is associated with which of the following fields of study?
   a. functional genomics
   b. structural genomics
   c. proteomics
   d. PGFE
   e. all of the above

8. A segment is called ______ if 99% of the individuals in a population have identical sequences for the segment.
   a. contiguous
   b. continuous
   c. polymorphic
   d. monomorphic

9. A gene is mapped to within 20 map units of a marker. Which of the following was used to determine this?
   a. FISH
   b. PFGE
   c. linkage mapping
   d. cytological mapping
   e. physical mapping

10. A sequence-tagged site occurs how many times in the genome?
   a. none
   b. once
   c. several
   d. multiple

Answer Key for Study Guide Questions

This answer key provides the answers to the exercises and chapter quiz for this chapter. Answers in parentheses ( ) represent possible alternate answers to a problem, while answers marked with an asterisk (*) indicate that the response to the question may vary.

20.1

1. a
2. c
3. b
4. c
5. b
6. b
7. c
8. c
9. c
10. c
11. f
12. a
13. d
14. j
15. b
16. c
17. g
18. h
19. e
20. i

20.2

1. c
2. a
3. b
4. e
5. d
6. b
7. b
8. a
9. c
10. a
11. d
12. c
13. d

Quiz

1. c
2. b
3. a
4. c
5. e
6. c
7. a
8. d
9. c
10. b

Answers to Conceptual and Experimental Questions

Conceptual Questions

C1. One would conclude that she has a deletion of the gene that the probe recognizes. To clone this gene, one could begin with a marker that is known to be near band p11 and walk in either direction. This walking experiment would be done on the DNA from a normal person and compared to the DNA from the person described in the problem. At some point, the walk would yield a clone that contained a deletion in the abnormal person, but the DNA would be present in a normal person. This DNA fragment in the normal person should also hybridize to the probe.

C2. A. Yes.

B. No, this is only one chromosome in the genome.

C. Yes.

D. Yes.

C3. A. False, they do not have to carry genes.

B. True.

C. False, the marker may not carry a gene that affects phenotype.

D. True.

E. True.

Experimental Questions

E1. A. Cytogenetic mapping

B. Linkage mapping

C. Physical mapping

D. Cytogenetic mapping

E. Linkage mapping

F. Physical mapping

E2. They are complementary to each other.

E3. *In situ* hybridization is a cytological method of mapping. A probe that is complementary to a chromosomal sequence is used to locate the gene microscopically within a mixture of many different chromosomes. Therefore, it can be used to cytologically map the location of a gene sequence. When more than one probe is used, the order of genes along a particular chromosome can be determined.

E4. Because normal cells contain two copies of chromosome 14, one would expect that a probe would bind to complementary DNA sequences on both of these chromosomes. If a probe recognized only one of two chromosomes, this means that one of the copies of chromosome 14 has been lost, or it has suffered a deletion in the region where the probe binds. With regard to cancer, the loss of this genetic material may be related to the uncontrollable cell growth.

E5. The term *fixing* refers to procedures that chemically freeze cells and prevent degradation. After fixation has occurred, the contents within the cells do not change their morphology. In a sense, they are frozen in place. For a FISH experiment, this keeps all the chromosomes within one cell in the vicinity of each other; they cannot float around the slide and get mixed up with chromosomes from other cells. Therefore, when we see a group of chromosomes in a FISH experiment, this group of chromosomes comes from a single cell.

It is necessary to denature the chromosomal DNA so that the probe can bind to it. The probe is a segment of DNA that is complementary to the DNA of interest. The strands of chromosomal DNA must be separated (i.e., denatured) so that the probe can bind to complementary sequences.

E6. After the cells and chromosomes have been fixed to the slide, it is possible to add two or more different probes that recognize different sequences (i.e., different sites) within the genome. Each probe has a different fluorescence emission wavelength, so it can be identified by its color. Usually, a researcher will use computer imagery that recognizes the wavelength of each probe and then assigns that region a bright color. The color seen by the researcher is not the actual color emitted by the probe; it is a secondary color assigned by the computer. This can be done with two or more different probes, as a way to color the regions of the chromosomes that are recognized by the probes. In a sense, the probes, with the aid of a computer, are "painting" the regions of the chromosomes that are recognized by a probe. An example of chromosome painting is shown in Figure 20.4. In this example, human chromosome 5 is painted with six different colors.

E7. If the sample was from a normal individual, two spots (one on each copy of chromosome 21) would be observed. Three spots would be observed if the sample was from a person with Down syndrome because the person has three copies of chromosome 21.

E8. A contig is a collection of clones that contain overlapping segments of DNA that span a particular region of a chromosome. To determine if two clones are overlapping, one could conduct a Southern blotting experiment. In this approach, one of the clones is used as a probe. If it is overlapping with the second clone, it will bind to it in a Southern blot. Therefore, the second clone is run on a gel and the first clone is used as a probe. If the band corresponding to the second clone is labeled, this means that the two clones are overlapping.

E9. A YAC vector can contain extremely large pieces of DNA, so they are used as a first step to align the segments of DNA in a physical mapping study. However, it is difficult to work with them in subcloning and DNA sequencing experiments. Cosmids, by comparison, contain smaller segments of the genome. The locations of cosmids can be determined by hybridizing them to YACs. The cosmids can then be used for subcloning and DNA sequencing.

E10. YAC cloning vectors have the replication properties of a chromosome and the cloning properties of a plasmid. To replicate like a chromosome, the YAC vector contains an origin of replication and centromeric and telomeric sequences. Therefore, in a yeast cell, a YAC can behave as a chromosome. Like a plasmid, YACs also contain selectable markers and convenient cloning sites for the insertion of large segments of DNA. The primary advantage is the ability to clone very large pieces of DNA.

E11. A polymorphism refers to genetic variation at a particular locus within a population. If the polymorphism occurs within gene sequences, this is allelic variation. A polymorphism can also occur within genetic markers such as RFLPs. The molecular basis for an RFLP is that two distinct individuals will have variation in their DNA sequences and some of the variation may affect the relative locations of restriction enzyme sites. Since this occurs relatively frequently between unrelated individuals, many RFLPs can be identified. They can be detected by restriction digestion and agarose gel electrophoresis and then Southern blotting. They are useful in mapping studies because it is relatively easy to find many of them along a chromosome. They can be used in gene cloning as a starting point for a chromosomal walk.

E12. The resistance gene appears to be linked to RFLP 4B.

E13. If the genes were unlinked, we would expect a 1:1:1:1 ratio among the four combinations of offspring. Since there are a total of 272 offspring, there are expected to be 68 in each category according to independent assortment.

$$\chi^2 = \sum \frac{(O-E)^2}{E}$$

$$\chi^2 = \frac{(40-68)^2}{68} + \frac{(98-68)^2}{68} + \frac{(97-68)^2}{68} + \frac{(37-68)^2}{68}$$

$$\chi^2 = 51.2$$

With 3 degrees of freedom, this high chi square value would be expected to occur by chance less than 1% of the time. Therefore, we reject the hypothesis that the RFLPs are independently assorting.

In this example, the recombinant offspring contain the 5,200, 4,500, and 2,100 and 4,500, 2,100, and 3,200 RFLPs.

$$\text{Map distance} = \frac{40+37}{40+98+97+37} \times 100$$

$$\text{Map distance} = 28.3 \text{ mu}$$

E14. For most organisms, it is usually easy to locate many RFLPs throughout the genome. The RFLPs can be used as molecular markers to make a map of the genome. This is done using the strategy described in Figure 20.8. To map a functional gene, one would also follow the same general strategy described in Figure 20.8, except that the two strains would also have an allelic difference in the gene of interest. The experimenter would make crosses, such as dihybrid crosses, and determine the number of parental and recombinant offspring based on the alleles and RFLPs they had inherited. If an allele and RFLP are linked, there will be a much lower percentage (i.e., less than 50%) of the recombinant offspring.

E15. Child 1 and child 3 belong to father 2.

Child 2 and child 4 belong to father 1.

Child 5 could belong to either father.

E16.

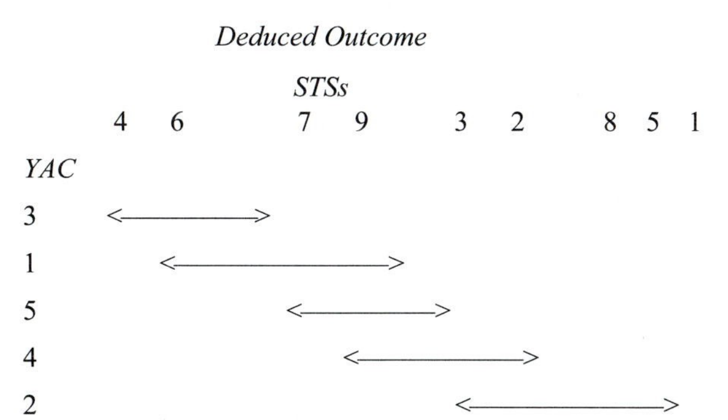

E17. An explanation is that the rate of recombination between homologous chromosomes is different during oogenesis compared to spermatogenesis. Physical mapping measures the actual distance (in bp) between markers. The physical mapping of chromosomes in males and females reveals that they are the same lengths. Therefore, the sizes of chromosomes in males and females are the same. The differences obtained in linkage maps are due to differences in the rates of recombination during oogenesis versus spermatogenesis.

E18. A. One homologue contains the STS-1 that is 289 bp and STS-2 that is 422 bp while the other homologue contains STS-1 that is 211 bp and STS-2 that is 115 bp. This is based on the observation that 28 of the sperm have either the 289 bp and 422 bp bands or the 211 bp and 115 bp bands.

B. There are two recombinant sperm; see lanes 12 and 18. Since there are two recombinant sperm out of a total of thirty:

$$\text{Map distance} = \frac{2}{30} \times 100$$
$$= 6.7 \text{ mu}$$

C. In theory, this method could be used. However, there is not enough DNA in one sperm to carry out an RFLP analysis unless the DNA is amplified by PCR.

E19. They appear to be linked. If they were not linked, we would expect equal amounts of the four types of offspring. However, as seen in the data, there is a much higher proportion of parental combinations (red, small and purple, big) compared to nonparental combinations. The map distance is

$$\text{Map distance} = \frac{111+109}{725+111+109+729} \times 100 = 13.1 \text{ mu}$$

With regard to RFLP inheritance, the following results are expected:

725 red, small flowers with the 4,000 bp and 1,600 bp RFLPs

111 red, big flowers with the 4,000 bp and 7,200 bp RFLPs

109 purple, small flowers with the 3,400 bp and 1,600 bp

729 purple, big flowers with the 3,400 bp and 7,200 bp

E20. One possibility is that the geneticist could try a different restriction enzyme. Perhaps there is sequence variation in the vicinity of the pesticide-resistance gene that affects the digestion pattern of a restriction enzyme other than *Eco*RI. There are hundreds of different restriction enzymes that recognize a myriad of different sequences.

Alternatively, the geneticist could give up on the RFLP approach and try to identify one or more sequence-tagged sites that are in the vicinity of the pesticide-resistance gene. In this case, the geneticist would want to identify STSs that are also microsatellites. As described in Figure 20.12, the transmission of microsatellites can be followed in genetic crosses. Therefore, if the geneticist could identify microsatellites in the vicinity of the pesticide-resistance gene, this would make it possible to predict the outcome of crosses. For example, let's suppose a microsatellite linked to the pesticide-resistance gene existed in three forms: 234 bp, 255 bp, and 311 bp. And let's also suppose that the 234 bp form was linked to the high-resistance allele, the 255 bp form was linked to the moderate-resistance allele, and the 311 bp form was linked to the low-resistance allele. According to this hypothetical example, the geneticist could predict the level of resistance in an alfalfa plant by analyzing the inheritance of these microsatellites.

E21. You would have expected equal amounts of the four patterns of RFLPs. There would have been about 25 of each of the four patterns.

E22. When chromosomal DNA is isolated and digested with a restriction enzyme, this produces thousands of DNA fragments of different sizes. This makes it impossible to see any particular band on a gel, if you simply stained the gel for DNA. Southern blotting allows you to detect one or more RFLPs that are complementary to the radioactive probe that is used.

E23. Besides a selectable marker and an origin that will replicate in *E. coli,* YAC vectors also require two telomere sequences, a centromere sequence, and an ARS sequence. The telomeres are needed to prevent the shortening of the artificial chromosome from the ends. The centromere sequence is needed for the proper segregation of the artificial chromosome during meiosis and mitosis. The ARS sequence is the yeast equivalent of an origin of replication, which is needed so that the YAC DNA can be replicated.

E24. Based on these results, it is likely that the sickle-cell allele originated in an individual with the 13.0 kbp RFLP. This would explain why the Hb^S allele is usually transmitted with the 13.0 kbp RFLP. On occasion, however, a crossover could occur in the region between the β-globin gene and the distal *Hpa*I site.

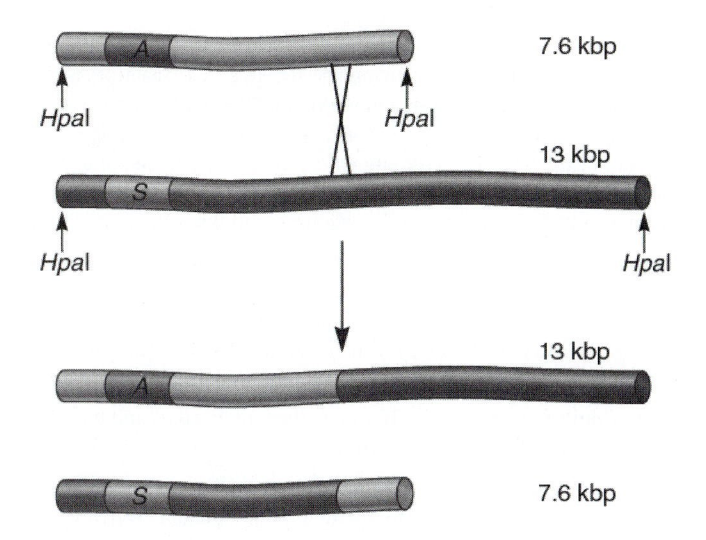

After crossing over, the Hb^S allele is now linked to a 7.6 kbp RFLP.

E25. The 13.0 kbp fragment is not always linked to the Hb^S allele because, on rare occasions, a crossover can occur between the restriction site and the allele. Nevertheless, within the human population, the 13.0 kbp fragment is usually linked to the Hb^S allele. Therefore, if a person is heterozygous for the 13.0 kbp fragment, he/she has a much higher probability of being a heterozygous carrier for the Hb^S allele. This information can be useful in predicting the likelihood of having an affected child.

E26. PFGE is a method of electrophoresis that is used to separate small chromosomes and large DNA fragments. The electrophoresis devices used in PFGE have two sets of electrodes. The two sets of electrodes produce alternating pulses of current, and this facilitates the separation of large DNA fragments.

It is important to handle the sample gently to prevent the breakage of the DNA due to mechanical forces. Cells are first embedded in agarose blocks, and then the blocks are loaded into the wells of the gel. The agarose keeps the sample very stable and prevents shear forces that might mechanically break the DNA. After the blocks are in the gel, the cells within the blocks are lysed, and if desired, restriction enzymes can be added to digest the DNA. For PFGE, a restriction enzyme that cuts very infrequently might be used.

PFGE can be used as a preparative technique to isolate and purify individual chromosomes or large DNA fragments. PFGE can also be used, in conjunction with Southern blotting, as a mapping technique.

E27. The proper order is C, A, D, E, B.

1. Isolate whole chromosomes or large DNA fragments via pulsed-field gel electrophoresis or chromosome sorting.
2. Clone large fragments of DNA to make a YAC library.
3. Subclone YAC fragments to make a cosmid library.
4. Subclone cosmid fragments for DNA sequencing.
5. Determine the DNA sequence of subclones from a cosmid library.

E28. Note: The insert of cosmid B is contained completely within the insert of cosmid C.

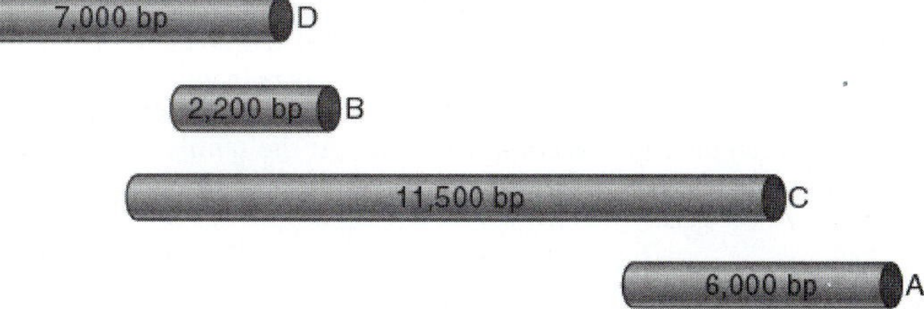

E29. A sequence-tagged site is a segment of DNA, usually quite short (e.g., 100 to 400 bp in length), that serves as a unique site in the genome. STSs are identified using primers in a PCR reaction. STSs serve as molecular markers in genetic mapping studies. Sometimes, the region within an STS may contain a microsatellite. A microsatellite is a short DNA segment that is variable in length, usually due to a short repeating sequence. When a microsatellite is within an STS, the length of the STS will vary among different individuals, or even the same individual may be heterozygous for the STS. This makes the STS polymorphic. Polymorphic STSs can be used in linkage analysis, since their transmission can be followed in family pedigrees and through crosses of experimental organisms.

E30. A. The general strategy is shown in Figure 20.19. The researcher begins at a certain location and then walks toward the gene of interest. You begin with a clone that has a marker that is known to map relatively close to the gene of interest. A piece of DNA at the end of the insert is subcloned and then used in a Southern blot to identify an adjacent clone in a cDNA library. This is the first "step." The end of this clone is subcloned to make the next step. And so on. Eventually, after many steps, you will arrive at your gene of interest.

B. In this example, you would begin at STS-3. If you walked a few steps and happened upon STS-2, you would know that you were walking in the wrong direction.

C. This is a difficult aspect of chromosome walking. Basically, you would walk toward gene *X* using DNA from a normal individual and DNA from an individual with a mutant gene *X*. When you have found a site where the sequences are different between the normal and mutant individual, you may have found gene *X*. You would eventually have to confirm this by analyzing the DNA sequence of this region and determining that it encodes a functional gene.

E31. The first piece of information you would start with is the location of a gene or marker that is known to be close to the gene of interest by previous mapping studies. You would begin with a clone containing this marker (or gene) and follow the procedure of chromosome walking to eventually reach the gene of interest. A contig would make this much easier because you would not have to conduct a series of subcloning experiments to reach your gene. Instead, you could simply analyze the members of the contig.

E32. Two techniques to purify chromosomes are pulsed-field gel electrophoresis and chromosome sorting. They are useful because the isolation of a single chromosome can efficiently lead to the construction of a contig for that particular chromosome. Also, the isolation of a particular chromosome can lead to the mapping of markers or genes on that chromosome.

Questions for Student Discussion/Collaboration

1. After many RFLPs have been identified using molecular techniques, it is possible to determine if they are linked by conducting crosses. Two individuals that are different for many RFLPs are crossed together. The experimenter counts the offspring and categorizes them as being parental or recombinant types. If two RFLPs are linked, there will be a much higher percentage of parental types, as found in the experiment shown in Figure 20.8. RFLPs can be used as markers to map the relative positions of genes. They can also be used as markers to clone a gene by chromosomal walking techniques.

2. A molecular marker is a segment of DNA, not usually encoding a gene, which has a known location within a particular chromosome. It marks the location of a site along a chromosome. RFLPs and STSs are examples. It is easier to use these types of markers because they can be readily identified by molecular techniques such as restriction digestion analysis and PCR. Locating functional genes is usually more difficult because this relies on conventional linkage mapping approaches whereby allelic differences in the gene are mapped by making crosses or following a pedigree. For monomorphic genes, this approach does not work.

3. This is a matter of opinion. Many people would say that the ability to identify many human genetic diseases is the most important outcome. In addition, we will better understand how humans are constructed at the molecular level. As we gain a greater understanding of our genetic makeup, some people are worried that this may lead to greater discrimination. Insurance or health companies could refuse people who are known to carry genetic abnormalities. Similarly, employers could make their employment decisions based on the genetic makeup of potential employees rather than their past accomplishments. At the family level, genetic information may affect how people choose mates, and whether or not they decide to have children.

Chapter 21: Functional Genomics, Proteomics, and Bioinformatics

Student Learning Objectives

Upon completion of this chapter you should be able to:

1. Understand the use of functional genomics and the experimental procedures used by researcher to gather information for functional analysis.
2. Understand the procedures utilized in the study of proteomics and the information that may be obtained from this area of genetics.
3. Understand the relationship of computer science and genetics in the study of bioinformatics.
4. Understand the procedures that researchers in bioinformatics use to analyze genetic information.

Introduction

21.1 Functional Genomics

Overview

The previous chapter introduced the concept of structural genomic analysis, which determined the location of genes in the genome. In this chapter we begin to explore the interrelationships of these genes and their expression in an organism. This is done at a variety levels, including the DNA (genome) and protein (proteome) levels. Much of this work is now associated with assessing the mathematical relationships of genes and proteins. This is called bioinformatics.

In the first section of the chapter we examine the procedures associated with functional genomics. An important advance in this field is the invention of DNA microarrays, which allow researchers to examine the expression of genes under various conditions. The use of microarrays has greatly simplified the study of genomics.

Key Terms

Bioinformatics
cDNA library
Cluster analysis
DNA microarrays
Expressed sequence tag library
Functional genomics
Gene chips
Proteome
Proteomics
Subtractive cDNA library
Subtractive hybridization

Focal Points

- DNA microarrays (Figure 21.2, Experiment 21A)

Exercises and Problems

For questions 1 to 5, match the term to its correct description.

_____ 1. DNA microarrays

_____ 2. Cluster analysis

_____ 3. Expressed sequence tag library (EST)

_____ 4. cDNA

_____ 5. Subtractive hybridization

a. cDNA that can also be used as markers for physical mapping.
b. This is made using RNA as a starting material.
c. Used to study gene expression of thousands of genes simultaneously.
d. Uses two cDNA libraries to examine cDNA that is produced under one environmental condition, but not another.
e. An analysis of microarrays to determine groups of genes that are being expressed together.

21.2 Proteomics

Overview

The proteome includes all of the proteins that can be produced from the genome. Due to factors such as alternative splicing, RNA editing, and posttranslational modifications, the number of proteins far exceed the number of genes in the genome.

This section introduces the techniques that researchers use to classify and identify proteins of interest in the proteome. As was the case with the previous section, you should familiarize yourself the procedures and their applications.

Key Terms

Alternative splicing
Antibody microarrays
Functional protein microarrays
Mass spectrometry
Posttranslational covalent modification
Protein microarrays
RNA editing
Tandem mass spectrometry
Two-dimensional gel electrophoresis

Focal Points

- Procedures to study the proteome (Figures 21.4 and 21.5)

Exercises and Problems

For each of the following, choose the appropriate technique that is best associated with the statement.

a. protein microarrays
b. two-dimensional gel electrophoresis
c. antibody microarrays
d. tandem mass spectrometry
e. mass spectrometry

_____ 1. One step in this procedure uses SDS to neutralize the charges of a protein.

_____ 2. Proteins are isolated by their net charge at a given pH and molecular mass.

_____ 3. This procedure is used to determine the precise amino acid sequence of a protein.

_____ 4. This enables thousands of proteins to be screened simultaneously.

_____ 5. This is used to study protein expression.

_____ 6. This is used to determine the molecular mass of a specific protein.

21.3 Bioinformatics

Overview

The combination of the study of genomics and proteomics and computer science has led to the science of bioinformatics. This powerful field of study attempts to identify the mathematical relationships of proteins and DNA.

This section of the text introduces you to some basic concepts in computer science. Some of these terms might appear elementary at first, but an understanding of them is essential for comprehension of how computer programs can analyze molecular data. The chapter then proceeds to a discussion of some of the major methods that researchers can use to study the desired relationships. You should be aware that these programs are constantly evolving, with improvements occurring at an amazing pace. However, the principles discussed here provide a good foundation for how these programs operate.

Key Terms

Annotated
Basic local alignment search tool
Codon bias
Computer data file
Computer program
Database
Genome databases
Homologous
Homology
Multiple sequence alignment
Open reading frame
Orthologs
Paralogous genes
Paralogs
Pattern recognition
Search by content
Search by signal
Sequence element
Sequence motif
Sequence recognition
Similarity

Focal Points

- Dot matrix similarity (Figure 21.9)

Exercises and Problems

Complete each of the following statements by providing the correct word or phrase.

1. A large number of computer files that are stored in one location is called a(n) __________.
2. ______ databases contain the genetic characteristics of a single species.
3. A genetic sequence that has a particular function is called a sequence element, or ______.
4. The use of certain codons at a higher frequency than others by an organism is called ______.
5. A nucleotide sequence that does not contain any stop codons is called a ________.
6. A random piece of DNA could be searched using the _______ approach to detect promoters, start and stop codons, and splice sites to identify the location of genes.
7. Two genes that share a common ancestor are said to be __________.
8. ________ indicates that the sequences of two genes may be alike.
9. Homologous genes that are found in different species are called _______.
10. Gene families contain ______ genes.

For questions 11 to 15, choose one of the following answers.

a. simple dot matrix analysis
b. multiple sequence alignment
c. BLAST
d. structure prediction

_____ 11. Used in proteomics and the study of RNA.

_____ 12. Is used to detect regions of similarity between two sequences.

_____ 13. Is used to locate homologous sequences within a database using a query sequence.

_____ 14. This has the ability to compare several sequences to detect regions of similarity.

_____ 15. Considered by many to be the most important bioinformatic tool.

Chapter Quiz

1. Which of the following uses a query sequence to search a database for similar sequences?
   a. multiple sequence alignment
   b. dot matrix analysis
   c. BLAST
   d. microarrays
   e. none of the above

2. Two dimensional gel electrophoresis is used in the study of ______.

a. proteins
b. RNA
c. DNA
d. all of the above

3. The specific amino acid sequence of a protein can be determined using _____.

a. antibody microarrays
b. mass spectrometry
c. tandem mass spectrometry
d. pulsed field gel electrophoresis
e. two dimensional gel electrophoresis

4. Which of the following serves to increase the size of the proteome over the size of the genome?

a. posttranslational modification of proteins
b. RNA editing
c. alternative splicing
d. all of the above

5. The term that describes two genes that are derived from the same ancestral gene and are found within a single organism is _____.

a. homologous
b. orthologs
c. paralogs
d. none of the above

6. Expressed sequence tag libraries contain which of the following?

a. proteins
b. cDNA
c. DNA
d. RNA
e. none of the above

7. Subtractive cDNA libraries can provide which of the following information?

a. molecular mass of the final protein
b. sequence similarity
c. genes that are turned on under specific environmental conditions
d. the sequence of amino acids in a protein
e. the presence of introns

8. Protein expression is typically studied using _____.

a. two dimensional gel electrophoresis
b. RFLP analysis
c. DNA microarrays
d. antibody microarrays
e. multiple sequence alignment

9. If you have DNA sequences from several organisms and wish to identify regions of similarity between the sequences, which of the following would you use?

a. dot matrix
b. pulsed field gel electrophoresis
c. tandem mass spectrometry
d. multiple sequence alignment
e. BLAST

10. A genetic sequence with a particular function is called a(n) _______.

a. open reading frame
b. sequence motif
c. codon bias
d. cluster analysis

Answer Key for Study Guide Questions

This answer key provides the answers to the exercises and chapter quiz for this chapter. Answers in parentheses ( ) represent possible alternate answers to a problem, while answers marked with an asterisk (*) indicate that the response to the question may vary.

21.1
1. c
2. e
3. a
4. b
5. d

21.2
1. b
2. b
3. d
4. a
5. c
6. e

21.3
1. database
2. genome
3. sequence motif
4. codon bias
5. open reading frame
6. search by signal
7. homologous
8. similarity
9. orthologs
10. paralogous
11. d
12. a
13. c
14. b
15. c

Quiz

1. c
2. a
3. c
4. d
5. c
6. b
7. c
8. d
9. d
10. b

Answers to Conceptual and Experimental Questions

Conceptual Questions

C1. Structural genomics is the study of genome composition. Researchers attempt to map all of the genes in the genome and ultimately to determine the sequence of all the chromosomes. Functional genomics attempts to understand how genetic sequences function to produce the characteristics of cells and the traits of organisms. Much of functional genomics is aimed at an understanding of gene function. However, it also tries to understand the roles of other genetic sequences such as centromeres and repetitive sequences. Proteomics focuses on the functions of proteins. The ultimate goal is to understand how groups of proteins function as integrated units.

C2. There are two main reasons why the proteome is larger than the genome. The first reason involves the processing of pre-mRNA, a phenomenon that occurs primarily in eukaryotic species. RNA splicing and editing can alter the codon sequence of mRNA and thereby produce alternative forms of proteins that have different amino acid sequences. The second reason for protein diversity is posttranslational modifications. There are many ways that a given protein's structure can be covalently modified by cellular enzymes. These include proteolytic processing, disulfide bond formation, glycosylation, attachment of lipids, phosphorylation, methylation, and acetylation, to name a few.

C3. A database is a collection of many computer files in a single location. These data are usually DNA, RNA, or protein sequences. The data come from the contributions of many research labs. A major objective of a genome database is to organize the genetic information of a single species. A genome database will identify all of the known genes and indicate the map locations within the genome. In addition, a genome database may have other types of information such as a registry of workers and details that are specific to the particular species.

C4. Centromeric sequences, origins of replication, telomeric sequences, repetitive sequences, and enhancers. Other examples are possible.

C5. Sequence recognition involves the recognition of a particular sequence of an already-known function. For example, a program could locate start codons (ATG) within a DNA sequence. By comparison, pattern recognition relies on a pattern of arrangement of symbols but is not restricted to particular sequences.

C6. There are a few interesting trends. Sequences 1 and 2 are similar to each other, as are sequences 3 and 4. There are a few places where amino acid residues are conserved among all five sequences. These amino acids may be particularly important with regard to function.

C7. In genetics, the term *similarity* means that two genetic sequences are similar to each other. Homology means that two genetic sequences have evolved from the same ancestral sequence. Homologous sequences are similar to each other, but not all (short) similar sequences are due to homology.

C8. A. Correct.

B. Correct.

C. This is not correct. These are short genetic sequences that happen to be similar to each other. The *lac* operon and *trp* operon are not derived from the same primordial operon.

D. This is not correct. The two genes are homologous to each other. It is correct to say that their sequences are 60% identical.

C9. A gap is necessary when two homologous sequences are not the same length. Since homologous sequences are derived from the same ancestral gene, two homologous sequences were originally the same length. However, during evolution the sequences can incur deletions and/or additions that make the sequences longer or shorter than the original ancestral gene. If one gene incurs a deletion, a gap will be necessary in this gene's sequence in order to align it with a homologous gene. If an addition occurs in a gene's sequence, a gap will be necessary in a homologous gene sequence in order to align the two sequences.

Experimental Questions

E1. A subtractive cDNA library contains cDNAs that are derived from mRNAs that were produced only under one set of conditions but not under another set of conditions. For example, as described in Figure 21.1, a subtractive cDNA library may only contain cDNAs that were derived from mRNAs that were made when cells were exposed to a hormone. This provides a way to identify genes that are induced by the presence of the hormone.

E2. As described in solved problem S1, one reason for making a subtractive DNA library is to determine which RNAs are produced when environmental conditions change. You want to load a small amount of cDNA on the column that was derived from cells that had been exposed to mercury. You want all of the cDNA made in both the presence and absence of mercury to bind to the column. Remember that the cDNA, which is derived from mRNA that is made in the absence of mercury, is already bound to the column. You want the cDNA that is made in the presence of mercury to bind to this cDNA, if it is complementary. If too much of this cDNA is loaded, all of the cDNAs will have complementary cDNA bound to them, and some of them will not bind to the column, even though they may be complementary to the cDNAs. In other words, if you load too much cDNA (derived from the mercury-exposed cells), you will have saturated the binding sites for cDNAs that are made in both the presence and absence of mercury. You do not want this to happen, because you want only the cDNAs that are derived from mRNAs that are specifically expressed in the presence of mercury to flow through the column. These cDNAs are not complementary to any cDNAs that are attached to the column.

E3. A. A DNA microarray is a small slide that is dotted with many different fragments of DNA. In some microarrays, DNA fragments, which were made synthetically (e.g., by PCR), are individually spotted onto the slide. The DNA fragments are typically 500 to 5,000 bp in length, and a few thousand to tens of thousands are spotted to make a single array. Alternatively, short oligonucleotides can be directly synthesized on the surface of the slide. In this process, the DNA sequence at a given spot is produced by selectively controlling the growth of the oligonucleotide using narrow beams of light. In this case, there can be hundreds of thousands of different spots on a single array.

B. In most cases, fluorescently labeled cDNA is hybridized to the microarray, though labeled genomic DNA or RNA could also be used.

C. After hybridization, the array is washed and placed in a scanning confocal fluorescence microscope that scans each pixel (the smallest element in a visual image). After correction for local background, the final fluorescence intensity for each spot is obtained by averaging across the pixels in each spot. This results in a group of fluorescent spots at defined locations in the microarray.

E4. The cDNA that is labeled with a green dye is derived from mRNA that was obtained from cells at an early time point, when glucose levels were high. The other samples of cDNA were derived from cells collected at later time points, when glucose levels were falling, and when the diauxic shift was occurring. These were labeled with a red dye. The green fluorescence provides a baseline for gene expression when glucose is high. At later time points, if the red:green ratio is high (i.e., greater than one), this means that a gene is induced as glucose levels fall, because there is more red cDNA compared to green cDNA. If the ratio is low (i.e., less than one), this means that a gene is being repressed.

E5. This is a way to analyze DNA microarray data. Using a computer, the data are analyzed to determine if certain groups of genes show the same expression patterns under a given set of conditions. This is illustrated in the data of Figure 21.3. Certain groups of genes form clusters that show very similar patterns of transcription. This is useful because it may identify genes that participate in a common cellular function.

E6. In the first dimension (i.e., in the tube gel), proteins are separated according to the isoelectric point. This is the pH at which a given protein's net charge is zero. In the second dimension (i.e., the slab gel), proteins are coated with SDS and separated according to their molecular mass.

E7. Yes, two-dimensional gel electrophoresis can be used as a purification technique. A spot from a two-dimensional gel can be cut out, and the protein can be eluted from the spot. This purified protein can be subjected to tandem mass spectroscopy to determine peptide sequences within the protein. It should be mentioned, however, that two-dimensional gel electrophoresis would not be used to purify proteins in a functional state. The exposure to SDS in the second dimension would denature proteins and probably inactivate their function.

E8. In tandem mass spectroscopy, the first spectrometer determines the mass of a peptide fragment from a protein of interest. The second spectrometer determines the masses of progressively smaller pieces that are derived from that peptide. Because the masses of each amino acid are known, the molecular masses of these smaller fragments reveal the amino acid sequence of the peptide. With peptide sequence information, it is possible to use the genetic code and produce putative DNA sequences that could encode such a peptide. These sequences, which are degenerate due to the degeneracy of the genetic code, are used as query sequences to search a genomic database. This will (hopefully) locate a match. The genomic sequence can then be analyzed to determine the entire coding sequence for the protein of interest.

E9. The two general types of protein microarrays are antibody microarrays and functional protein arrays. In an antibody microarray, many different antibody molecules are spotted onto the array. Since each antibody recognizes a different peptide sequence, this microarray can be used to monitor protein expression levels. A functional protein microarray involves the spotting of many cellular proteins onto a slide. This type of microarray can be analyzed with regard to substrate specificity, drug binding, and/or protein-protein interactions.

E10. The first strategy is a search by signal approach, which relies on known sequences such as promoters, start and stop codons, and splice sites to help predict whether or not a DNA sequence contains a structural gene. It would try to identify a region that contains a promoter sequence, then a start codon, a coding sequence, and a stop codon. A second strategy is a search by content approach. The goal is to identify sequences whose nucleotide content differs significantly from a random distribution, which is usually due to codon bias. A search by content approach attempts to locate coding regions by identifying regions where the nucleotide content displays a bias. A third method to locate structural genes is to search for long open reading frames within a DNA sequence. An open reading frame is a sequence that does not contain any stop codons.

E11. A motif is a sequence that carries out a particular function. There are promoter motifs, enhancer motifs, and amino acid motifs that play functional roles in proteins. For a long genetic sequence, a computer can scan the sequence and identify motifs with great speed and accuracy. The identification of amino acid motifs helps a researcher to understand the function of a particular protein.

E12. By searching a database, one can identify genetic sequences that are homologous to a newly determined sequence. In most cases, homologous sequences carry out identical or very similar functions. Therefore, if one identifies a homologous member of a database whose function is already understood, this provides an important clue regarding the function of the newly determined sequence.

E13. In a comparative approach, one uses the sequences of many homologous genes. This method assumes that RNAs of similar function and sequence have a similar structure. Computer programs can compare many different 16S rRNA sequences to aid in the prediction of secondary structure.

E14. The basis for secondary structure prediction is that certain amino acids tend to be found more frequently in α-helices or β-sheets. This information is derived from the locations of amino acids within proteins that have already been crystallized. Predictive methods are perhaps 60 to 70% accurate, which is not very good.

E15. The three-dimensional structure of a homologous protein must already be solved before one can attempt to predict the three-dimensional structure of a protein based on its amino acid sequence.

E16. The backtranslate program works by knowing the genetic code. Each amino acid has one or more codons (i.e., three-base sequences) that are specified by the genetic code. This program would produce a sequence file that was a nucleotide base sequence. The backtranslate program would produce a degenerate base sequence because the genetic code is degenerate. For example, lysine can be specified by AAA or AAG. The program would probably store a single file that had degeneracy at particular positions. For example, if the amino acid sequence was lysine–methionine–glycine–glutamine, the program would produce the following sequence:

5′–AA(A/G)ATGGG(T/C/A/G)CA(A/G)

The bases found in parentheses are the possible bases due to the degeneracy of the genetic code.

E17. The advantages of running a computer program are speed and accuracy. Once the program has been made, and a sequence file has been entered into a computer, the program can analyze long genetic sequences with great speed and accuracy.

E18. A. To identify a specific transposable element, a program would use sequence recognition. The sequence of P elements is already known. The program would be supplied with this information and scan a sequence file looking for a match.

B. To identify a stop codon, a program would use sequence recognition. There are three stop codons that are specific three-base sequences. The program would be supplied with these three sequences and scan a sequence file to identify a perfect match.

C. To identify an inversion (of any kind), a program would use pattern recognition. In this case, the program would be looking for a pattern in which the same sequence was running in opposite directions in a comparison of the two sequence files.

D. A search by signal approach uses both sequence recognition and pattern recognition as a means to identify genes. It looks for an organization of sequence elements that would form a functional gene. A search by content approach identifies genes based on patterns, not on specific sequence elements. This approach looks for a pattern in which the nucleotide content is different from a random distribution. The third approach to identify a gene is to scan a genetic sequence for long open read frames. This approach is a combination of sequence recognition and pattern recognition. The program is looking for specific sequence elements (i.e., stop codons) but it is also looking for a pattern in which the stop codons are far apart.

E19. A sequence element is a specific sequence (i.e., a base sequence or an amino acid sequence). Two examples would be a stop codon (i.e., UAA), which is a base sequence element, and an amino acid sequence that is a site for protein glycosylation (i.e., asparagine–any amino acid–serine or threonine), which is an amino acid sequence element or motif. The computer program does not create these sequence elements. The program is given the information about sequence elements, and this information comes from genetic research. Scientists have conducted experiments to identify the sequence of bases that constitute a stop codon and the sequence of amino acids where proteins are glycosylated. Once this information is known from research, it can be incorporated into computer programs, and then the program can analyze new genetic sequences and predict the occurrence of stop codons and glycosylation sites.

E20. A. The amino acids that are most conserved (i.e, the same in all of the family members) are most likely to be important for structure and/or function. This is because a mutation that changed the amino acid might disrupt structure and function, and these kinds of mutations would be selected against. Completely conserved amino acids are found at the following positions: 101, 102, 105, 107, 108, 116, 117, 123 (Note: Asp or Glu are found here, and these two amino acids are very similar), 124, 130, 134, 139, 143, and 147.

B. The amino acids that are least conserved are probably not very important because changes in the amino acid does not seem to inhibit function. (If it did inhibit function, natural selection would eliminate such a mutation.) At one location, position 118, there are five different amino acids.

E21. A. Since most family members contain a histidine, this is likely to be the ancestral codon. The histidine codon mutated into an arginine codon after the gene duplication occurred that produced the ζ-globin gene. This would be after the emergence of primates or within the last 10 or 20 million years.

B. We do not know if the ancestral globin gene had a glycine or proline at codon-121. The mutation probably occurred after the duplication that produced the α–globin family and β-globin family, but before the gene duplications that gave rise to the multiple copies of α– and β-globins on chromosome 16 and chromosome 11, respectively. Therefore, it occurred between 300 million and 200 million years ago.

C. All of the β-globins contain glutamic acid at position 103, and all of the α–globins contain valine, except for θ-globin. We do not know if the ancestral globin gene had a valine or glutamic acid at codon 121. Nevertheless, a mutation, converting one to the other, probably occurred after the duplication that produced the α–globin family and the β-globin family, but before the gene duplications that gave rise to the multiple copies of α– and β-globins on chromosome 16 and chromosome 11, respectively. Therefore, it occurred between 300 million and 200 million years ago. The mutation that produced the alanine codon in the θ-globin gene probably occurred after the gene duplication that produced this gene. This would be after the emergence of mammals (i.e., sometime within the last 200 millions years).

E22. As described in part C of solved problem S2, a serine codon was likely to be the ancestral codon. If we look at the codon table, an AGU or AGC codon for serine could change into an Asn, Thr, or Ile codon by a single-base change. In contrast, UCU, UCC, UCA, and UCG codons, which also code for serine, could not change into Asn or Ile codons by a single-base change. Therefore, the two likely scenarios are shown next. The mutated base is underlined. The mutations would actually occur in the DNA, although the sequences of the RNA codons are shown here.

Ancestral codon

A$\underline{C}$U (Thr) ← AGU (Ser) → A$\underline{A}$U (Asn)

A$\underline{U}$U (Ile)

Ancestral codon

A$\underline{C}$C (Thr) ← AGC (Ser) → A$\underline{A}$C (Asn)

A$\underline{U}$C (Ile)

E23. A. This sequence has two regions that are about twenty amino acids long and very hydrophobic. Therefore, it is probable that this polypeptide has two transmembrane segments.

B.

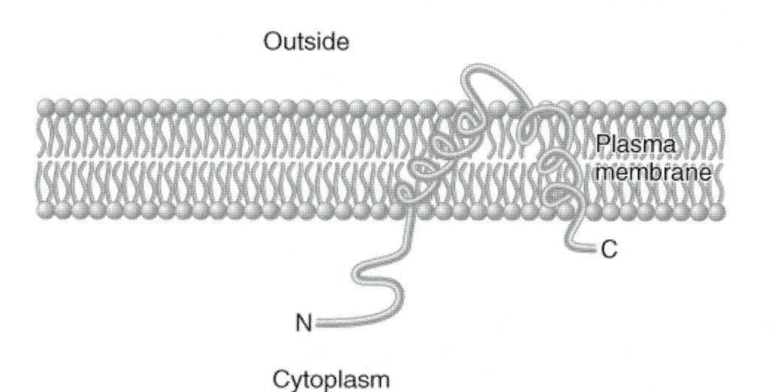

E24. RNA secondary structure is based on the ability of complementary sequences (i.e., sequences that obey the AU/GC rule) to form a double helix. The program employs a pattern recognition approach. It looks for complementary sequences based on the AU/GC rule.

E25. A and C are true. B is false because the programs are only about 60 to 70% accurate.

Questions for Student Discussion/Collaboration

1. Perhaps the key to this question is "organization." The genome database needs to be organized in a way that makes it easy for researchers to access the information. The known genes should be mapped and described in a way that enables someone to easily understand the map locations. A list of alleles for all the known genes might be quite useful to researchers. This could be included in the map, or perhaps in a description of each gene. A directory of researchers might be helpful. A list of pertinent publications that describe the sequencing data is also useful.
2. This is a very difficult question. In 20 years, we may have enough predictive information so that the structure of macromolecules can be predicted from their genetic sequences. If so, it would be better to be a mathematical theoretician with some genetics background. If not, it is probably better to be a biophysicist with some genetics background.
3. We need to answer this with a computer. Students will need access to programs that can translate DNA sequences and search databases. See our textbook website for a few programs.

Chapter 22: Medical Genetics and Cancer

Student Learning Objectives

Upon completion of this chapter you should be able to:

1. Understand the processes by which human genetic diseases are studied.
2. Understand how prions cause human disease.
3. Understand the genetic basis of cancer, including the role of oncogenes and tumor-suppressing genes.

22.1 Genetic Analysis of Human Diseases

Overview

The first section of this chapter introduces some of the terminology and procedures that are associated with the study of generically-related human diseases. It is frequently difficult to distinguish whether a certain human disease is due to environmental factors, including infectious agents, or is due to a genetic cause. The study of human genetics is further complicated by the fact that crosses of humans for the purposes of determining genotypes are unethical, and by the fact that typically humans produce small numbers of offspring and have long generation times.

To examine human genetic analysis, geneticists rely heavily on family information presented in pedigrees as well as a variety of methods for determining the presence of a genetic abnormality in an individual (Table 22.4). This chapter describes how observations of patterns of inheritance over several generations can provide clues as to whether a disease is autosomal recessive, autosomal dominant, or sex-linked. You should familiarize yourselves with the checklists in the text for each of these choices.

The final section of this chapter examines the role of prions, or self-replicating proteins, as a cause of disease in humans (Figure 22.6). While prions are not inherited, susceptibility to them may be, although research in this area is continuing. You should familiarize yourselves with the major prion-related diseases in Table 22.5.

Key Terms

Age of onset
Amniocentesis
Chorionic villus sampling
Concordance
Dizygotic twins
Genetic screening
Genetic testing
Heterogenicity
Monozygotic twins
Prions

Focal Points

- Description of autosomal recessive inheritance (pgs. 607-608), autosomal dominant inheritance (pg. 609), and sex-linked inheritance (pg. 610)
- Molecular mechanism of prion diseases (Figure 22.6).

Exercises and Problems

Match each of the following terms to its correct definition.

_____ 1. Concordance

_____ 2. Dizygotic twins

_____ 3. Heterogenetity

_____ 4. Chorionic villus sampling

_____ 5. Age of onset

_____ 6. Prion

_____ 7. Monozygotic twins.

a. The time during the lifespan when the disease first appears.
b. Also called amniocentesis, this procedure tests for genetic abnormalities in the fetus.
c. The degree to which a disorder can be attributed to inheritance.
d. Individuals who share 50% of their genetic material.
e. A disease may be caused by mutations in more than one gene.
f. An infectious agent composed entirely of proteins.
g. These are effectively clones.

For questions 8 to 10, indicate whether the statement is true (T) or false (F). If false, correct the statement.

_____ 8. The term genetic screening is applied to individual examinations for genetic abnormalities.

_____ 9. Monozygotic twins are genetically identical since they are formed from the same egg and sperm.

_____ 10. The PrP^c prion protein is the abnormal, disease forming form of the protein.

For each of the following, determine the most likely pattern of inheritance described by the statement.

a. sex-linked
b. autosomal dominant
c. autosomal recessive
d. can't be determined

_____ 11. The homozygote is more severely affected with the disorder.

_____ 12. An affected individual, with only one affected parent, is expected to produce 50% affected offspring.

_____ 13. Two affected individuals will always have 100% affected children.

_____ 14. Males are much more likely to exhibit the trait.

_____ 15. The trait occurs in the same frequency in both sexes.

_____ 16. The mothers of infected males often have brothers or fathers who are affected with the same trait.

22.2 Genetic Basis of Cancer

Overview

Simply put, the genetics of cancer is complicated. In most cases, cancer is not an inheritable disease, although some people may be predisposed to some forms more than others. The chapter starts with an introduction to some of the key terms associated with cancer. It then progresses to an important section that describes the actions and interactions of oncogenes and tumor-suppressing genes. An especially important set of interactions is shown in Figure 22.15, which illustrates the regulation of the cell cycle. If these checkpoints aren't obeyed, unregulated cell growth (cancer) may occur. The names of these genes can be somewhat confusing, but the modes of action are important. Focus on the illustrations in the text to understand these interactions.

Key Terms

Acutely transforming viruses
Apoptosis
Benign
Cancer
Carcinogen
Caspaces
Checkpoint proteins
Clonal
Focus
Invasive
Malignant
Metastatic
Oncogene
Proto-oncogene
Provirus
Transformation
Tumor-suppressor genes

Focal Points

- Ras protein cycle (Figure 22.11)
- Cell cycle checkpoints (Figure 22.15)

Exercises and Problems

For each of the following, match the term with its correct definition.

_____ 1. Proto-oncogene
_____ 2. Oncogene
_____ 3. Metastatic
_____ 4. Malignant
_____ 5. Caspaces
_____ 6. Tumor-suppressor genes
_____ 7. Provirus
_____ 8. Carcinogen
_____ 9. ACV

a. A virus that is integrated into the host cell genome.
b. A form of virus that promotes tumor formation.
c. An environmental agent that causes cancer.
d. A gene that has the potential to incur a mutation and become an oncogene.
e. A gene that promotes cancer.
f. Cancer cells that migrate to other parts of the body.
g. These cells prevent cancerous growth.
h. Cells that have become cancerous.
i. Executioner cells that break down cellular components.

Match the following genes and proteins to their correct description.

_____ 10. *rb*

_____ 11. *ras*

_____ 12. *myc*

_____ 13. *p53*

_____ 14. Cyclin/Cdk

_____ 15. E2F

a. Activation of this gene leads to the activation of additional genes that promote cell division.
b. An oncogene that is associated with a number of human cancers.
c. The gene that confirmed the two-hit theory of tumor-suppressor gene activity.
d. The majority of human cancers are associated with defects in this tumor-suppressor gene.
e. Proteins that regulate the cell cycle.
f. A transcription factor that activates genes for a cell to progress through the cell cycle.

Chapter Quiz

1. Which of the following would be characterized by homozygotes that are more severely affected by a disorder?
    a. sex-linked inheritance
    b. autosomal dominant inheritance
    c. autosomal recessive inheritance
    d. all of the above

2. A self-replicating protein is called a(n) ________.
    a. tumor-suppressing gene
    b. caspase
    c. oncogene
    d. prion

3. p53 is an example of a __________.
    a. oncogene
    b. proto-oncogene
    c. tumor-suppressor gene
    d. prion
    e. caspase

4. A disease that may be the result of mutations in several different genes is called ______.
    a. heterogeneity
    b. concordance
    c. transformation
    d. homozygosity

5. Amniocentesis of a pregnant mother is an example of _______.
    a. genetic screening
    b. genetic testing

6. Which of the following are effectively clones?
   a. brother-sister
   b. dizygotic twins
   c. monozygotic twins
   d. father-son

7. Sex-linked inheritance has all of the following characteristics, except:
   a. males are much more likely to inherit the trait.
   b. the trait occurs with the same frequency in both sexes.
   c. mothers of affects males have brothers or fathers who are affected.
   d. the daughters of affected males will produce on average 50% affected sons.
   e. all of the above are associated with sex-linked inheritance.

8. Cyclin and Cdk are involved in which of the following?
   a. cell cycle regulation
   b. prion activity
   c. tumor suppressing
   d. oncogene activation
   e. transcription

9. Which of the following terms applies to a cell that has become cancerous and moved to another part of the body?
   a. benign
   b. malignant
   c. invasive
   d. metastatic

10. ACV is an example of a _____.
   a. virus
   b. prion
   c. oncogene
   d. transcription factor

Answer Key for Study Guide Questions

This answer key provides the answers to the exercises and chapter quiz for this chapter. Answers in parentheses ( ) represent possible alternate answers to a problem, while answers marked with an asterisk (*) indicate that the response to the question may vary.

22.1
1. c
2. d
3. e
4. b
5. a
6. f
7. g
8. F, genetic testing
9. T
10. F, PrP^{Sc}
11. b
12. b
13. c
14. a
15. d
16. a

22.2	1. d	6. g	11. b
	2. e	7. a	12. a
	3. f	8. c	13. d
	4. h	9. b	14. e
	5. i	10. c	15. f

Quiz

1. b	5. b	9. d
2. d	6. c	10. a
3. c	7. b	
4. a	8. a	

Answers to Conceptual and Experimental Questions

Conceptual Questions

C1. Recessive X-linked traits are distinguished from the other two by their prevalence in males. Dominant X-linked traits (which are exceedingly rare) are always passed from father to daughter and never from father to son. Autosomal recessive and dominant traits are distinguished primarily by the pattern of transmission from parents to offspring. A person with a dominant trait usually has an affected parent unless it is due to a new mutation or incomplete penetrance is observed. Also, two affected parents can have unaffected children, which should never occur with recessive traits. In the case of rare recessive traits, affected children usually have unaffected parents.

C2. When a disease-causing allele affects a trait, it is causing a deviation from normality, but the gene involved is not usually the only gene that governs the trait. For example, an allele causing hemophilia prevents the normal blood clotting pathway from operating correctly. It follows a simple Mendelian pattern because a single gene affects the phenotype. Even so, it is known that normal blood clotting is due to the actions of many genes.

C3. A heterogeneous disorder is one that can be caused by mutations in two or more different genes. Hemophilia and thalassemia are two examples. Heterogeneity can confound pedigree analysis because alleles in different genes that cause the same disorder may be inherited in different ways. For example, there are autosomal and X-linked forms of hemophilia. If a pedigree contained individuals with both types of alleles, the data would not be consistent with either an autosomal or X-linked pattern of inheritance.

C4. Changes in chromosome number, and unbalanced changes in chromosome structure, tend to affect phenotype because they create an imbalance of gene expression. For example, in Down syndrome, there are three copies of chromosome 21 and therefore three copies of all the genes on chromosome 21. This leads to a relative overexpression of genes that are located on chromosome 21 compared to the other chromosomes. Balanced translocations and inversions often are without phenotypic consequences because the total amount of genetic material is not altered, and the level of gene expression is not significantly changed.

C5. A. False. Dominant alleles require the inheritance of only one copy, and X-linked alleles require the inheritance of only one copy in males.

B. True.

C. True.

D. False. In many cases, it is difficult to assess the relative contributions of genetics and environment. In some cases, however, the environment seems most important. For example, with PKU, a low-phenylalanine diet will prevent harmful symptoms, even if an individual is homozygous for the mutant PKU alleles.

C6. There are lots of possible answers; here are a few. Dwarfism occurs in people and dogs. Breeds like the dachshund and basset hound are types of dwarfism in dogs. There are diabetic people and mice. There are forms of inherited obesity in people and mice. Hip dysplasia is found in people and dogs.

C7. Oftentimes, the age of onset coincides with a particular stage of development. Throughout the life span of an individual, from embryonic through adult life, the pattern of gene expression changes. Some genes are turned on only during embryonic stages, some are turned on during fetal stages, some are turned on at birth, some are turned on at adolescence, etc. An individual who is carrying a mutant gene may not manifest any disorder until the time in life when the gene is supposed to be expressed. For example, many genetic diseases manifest themselves after birth. Prior to birth, an individual may develop properly and be born a healthy baby. After birth, the baby begins a new developmental phase, and at this point, new genes may be turned on. If one or both copies of a "newborn" gene (i.e., a gene turned on at birth) are mutant, this may lead to a serious disorder, with an age of onset that occurs during infancy.

An infectious disease would probably not have a particular age of onset, because people could be exposed to the infectious agent at any time in their lives. Therefore, most infectious diseases do not have a predictable age of onset, although some diseases are more common in children or older people.

C8. A. Because a person must inherit two defective copies of this gene and it is known to be on chromosome 1, the mode of transmission is autosomal recessive. Both members of this couple must be heterozygous because they have one affected parent (who had to transmit the mutant allele to them) and their phenotypes are normal (so they must have received the normal allele from their other parent). Because both parents are heterozygotes, there is a 1/4 chance of producing a homozygote with Gaucher disease. If we let *G* represent the normal allele, and *g* the mutant allele:

♀ \ ♂	*G*	*g*
G	*GG* Normal	*Gg* Normal
g	*Gg* Normal	*gg* Gaucher's

B. From this Punnett square, we can also see that there is a 1/4 chance of producing a homozygote with both normal copies of the gene.

C. We need to apply the binomial expansion to solve this problem. See Chapter 2 for a description of this calculation. In this problem, $n = 5$, $x = 1$, $p = 0.25$, $q = 0.75$. The answer is 0.396, or 39.6%.

C9. In this pedigree, every affected member has an affected parent (except I-1, whose parentage is unknown). Also, the disease is found in both males and females. Therefore, this pattern suggests an autosomal dominant mode of transmission. Recessive transmission cannot be ruled out, but it is unlikely because it is a rare disorder and it would be extremely unlikely that unrelated individuals in this pedigree (i.e., I-2, II-1, II-7, III-4, and III-5) would all be heterozygous carriers.

C10. The mode of transmission is autosomal recessive. All of the affected individuals do not have affected parents. Also, the disorder is found in both males and females. If it were X-linked recessive, individual III-1 would have to have an affected father, which she does not.

C11. Since both members of this couple are phenotypically normal, and they have two sons with the disorder, the mother must be a heterozygous carrier, and the father has a normal allele (see the following Punnett square).

♀ \ ♂	X^F	Y
X^F	X^FX^F Normal	X^FY Normal
X^f	X^FX^f Carrier	X^fY Fabry

From this Punnett square, there is a 50% chance that their daughter is a carrier. If she is a carrier, there is a 50% chance that she will pass the mutant allele to her sons. So, if we multiply 0.5 times 0.5, this gives a value of 0.25, or 25%. This means that there is a 25% chance that she will have an affected son, if she gives birth to a son. Another way to view this calculation, however, is to consider that this female may have sons or daughters. When she gives birth, there is a 25% chance of having a male child with Fabry disease, if she is a carrier. If we do our calculation based on this perspective, the odds of this female having an affected son are 0.5 times 0.25, which equals 0.125, or 12.5%. In others words, there is a 12.5% chance that this female will have an affected son, each time she has a child.

C12. The 13 babies have acquired a new mutation. In other words, during spermatogenesis or oogenesis, or after the egg was fertilized, a new mutation occurred in the fibroblast growth factor gene. These 13 individuals have the same chances of passing the mutant allele to their offspring compared to the 18 individuals who inherited the mutant allele from a parent. The chance is 50%.

C13. Based on this pedigree, it appears to be an X-linked recessive trait (which it is known to be). All of the affected members are males, who had unaffected parents. From this pedigree alone, you could not rule out autosomal recessive, but it is less likely because there are no affected females.

C14. Because this is a dominant trait, the mother must have two normal copies of the gene, and the father (who is affected) is most likely to be a heterozygote. (Note: the father could be a homozygote, but this is extremely unlikely because the dominant allele is very rare.) If we let *M* represent the mutant Marfan allele, and *m* the normal allele, the following Punnett square can be constructed:

♀ \ ♂	*M*	*m*
m	*Mm* Marfan	*mm* Normal
m	*Mm* Marfan	*mm* Normal

A. There is a 50% chance that this couple will have an affected child.

B. We use the product rule. The odds of having an unaffected child are 50%. So if we multiply $0.5 \times 0.5 \times 0.5$, this equals 0.125, or a 12.5% chance of having three unaffected offspring.

C15. A. The mode of transmission is autosomal recessive. All of the affected individuals do not have affected parents. Also, the disorder is found in both males and females. If it were X-linked recessive, individual III-1 would have to have an affected father, which she does not.

B. If the disorder is autosomal recessive, individuals II-1, II-2, II-6, and II-7 must be heterozygous carriers because they have affected offspring. Since individuals II-2 and II-6 are heterozygotes, one of their parents (I-1 or I-2) must also be a carrier. Since this is rare disease, we would assume that only one of them (I-1or I-2) would be a carrier. Therefore, there is a 50% chance that II-3, II-4, and II-5 are also heterozygotes.

C16. A prion is a protein that behaves like an infectious agent. The infectious form of the prion protein has an abnormal conformation. This abnormal conformation is termed PrP^{Sc}, while the normal conformation of the protein is termed PrP^{C}. A prion protein in the PrP^{Sc} conformation can bind to a prion protein in the PrP^{C} conformation and convert it to the PrP^{Sc} form. An accumulation of the PrP^{Sc} form is what causes the disease symptoms. In the case of mad cow disease, an animal is initially exposed to a small amount of the prion protein in the PrP^{Sc} conformation. This PrP^{Sc} protein then converts the normal PrP^{C} proteins, which are normally found within the cells of the animal, into the PrP^{Sc} conformation. The disease symptoms rely on the conversion of the endogenous PrP^{C} conformation into the PrP^{Sc} form.

C17. The answer to this question is not actually known. However, we can guess about possible explanations based on the molecular mechanism of the disease. A prion protein can exist in the PrP^{Sc} conformation, which is an abnormal conformation that can convert the normal conformation (PrP^{C}) into the abnormal conformation (PrP^{Sc}). In people with Gerstmann-Straussler-Scheinker disease, the prion protein is predominantly in the PrP^{C} conformation, but at a very low rate, it may convert to the PrP^{Sc} conformation. One possibility is that the conversion to PrP^{Sc} does not occur very often in the cells of younger people.

This second explanation resembles the game of "tag" in which the person who is "it" can tag someone, and a person who has been tagged also is "it." Initially, there might be thousands of people playing, and the one person who is "it" might be slow at tagging anyone. However, once that person tags someone, then two people are "it" and then those two people can tag two more people, and then four people are "it." Initially, the number of people who are "it" will be low and will rise fairly slowly. However, once there are a substantial number of people who are "it," it becomes difficult for the people who are not "it" to evade the people who are "it." At this point, the number of people who are not "it" rapidly declines. This same analogy can be applied to the conformations of the prion protein. Initially, when the concentration of PrP^{Sc} is low, most of the PrP^{C} conformations can avoid an interaction with PrP^{Sc}. However, once the PrP^{Sc} concentration becomes high, this is more difficult, and all of the PrP^{C} conformations will be rapidly converted (in an exponential fashion).

C18. A. Keep in mind that a conformational change is a stepwise process. It begins in one region of a protein and involves a series of small changes in protein structure. The entire conformational change from PrP^{C} to PrP^{Sc} probably involves many small changes in protein structure that occur in a stepwise manner. Perhaps the conformational change (from PrP^{C} to PrP^{Sc}) begins in the vicinity of position 178 and then proceeds throughout the rest of the protein. If there is a methionine at position 129, the complete conformational change (from PrP^{C} to PrP^{Sc}) can take place. However, perhaps the valine at position 129 somehow blocks one of the steps that are needed to complete the conformational change. (Note: This answer is purely speculative. The actual biochemical mechanism is not known.)

B. Based on the answer to part A, once the PrP^{Sc} conformational change is completed, a PrP^{Sc} protein can bind to another prion protein in the PrP^{C} conformation. Perhaps it begins to convert it (to the PrP^{Sc} conformation) by initiating a small change in protein structure in the vicinity of position 178. The conversion would then proceed in a stepwise manner until the PrP^{Sc} conformation has been achieved. If a valine is at position 129, this could somehow inhibit one of the steps that are needed to complete the conformational change. If an individual had Val-129 in the polypeptide encoded by the second *PrP* gene, half of their prion proteins would be less sensitive to conversion by PrP^{Sc}, compared to individuals who had Met-129. This would explain why individuals with Val-129 in half of the prion proteins would have disease symptoms that would progress more slowly.

C19. An oncogene is abnormally activated to cause cancer while a tumor-suppressor gene is inactivated to cause cancer. *Ras* and *src* are examples of oncogenes and *Rb* and *p53* are tumor-suppressor genes.

C20. A proto-oncogene is a normal cellular gene that typically plays a role in cell division. It can be altered by mutation to become an oncogene and thereby cause cancer. At the level of protein function, a proto-oncogene can become an oncogene by synthesizing too much of a protein or synthesizing the same amount of a protein that is abnormally active.

C21. A retroviral oncogene is a cancer-causing gene found within the genome of a retrovirus. It is not necessary for viral infection and proliferation. Oncogene-defective viral strains are able to infect cells and multiply normally. It is thought that retroviruses have acquired oncogenes due to their life cycle. It may happen that a retrovirus will integrate next to a cellular proto-oncogene. Later in its life cycle, it may transcribe this proto-oncogene and thereby incorporate it into its viral genome. The high copy number of the virus or additional mutations may lead to the overexpression of the proto-oncogene and thereby cause cancer.

C22. The predisposition to develop cancer is inherited in a dominant fashion because the heterozygote has the higher predisposition. The mutant allele is actually recessive at the cellular level. But, because we have so many cells in our bodies, it becomes relatively likely that a defective mutation will occur in the single normal gene and lead to a cancerous cell. Some heterozygous individuals may not develop the disease as a matter of chance. They may be lucky and not get a defective mutation in the normal gene. Or, perhaps their immune system is better at destroying cancerous cells once they arise.

C23. Conversion of a proto-oncogene to an oncogene can occur by missense mutations, gene amplifications, translocations, and viral insertions. Examples are given in Table 22.7. The genetic changes are expected to increase the amount of the encoded protein or alter its function in a way that makes it more active.

C24. If an oncogene was inherited, it may cause uncontrollable cell growth at an early stage of development and thereby cause an embryo to develop improperly. This could lead to an early spontaneous abortion and thereby explain why we do not observe individuals with inherited oncogenes. Another possibility is that inherited oncogenes may adversely affect gamete survival, which would make it difficult for them to be passed from parent to offspring. A third possibility would be that oncogenes could affect the fertilized zygote in a way that would prevent the correct implantation in the uterus.

C25. A. No, because E2F is inhibited.

B. Yes, because E2F is not inhibited.

C. Yes, because E2F is not inhibited.

D. No, because there is no E2F.

C26. The role of p53 is to sense DNA damage and prevent damaged cells from proliferating. Perhaps, prior to birth, the fetus is in a protected environment so that DNA damage may be minimal. In other words, the fetus may not really need p53. After birth, agents such as UV light may cause DNA damage. At this point, p53 is important. A *p53* knockout is more sensitive to UV light because it cannot repair its DNA properly in response to this DNA-damaging agent, and it cannot kill cells that have become irreversibly damaged.

C27. A. True.

B. True.

C. False, most cancer cells are caused by mutations that result from environmental mutagens.

D. True.

C28. The p53 protein is a regulatory transcription factor; it binds to DNA and influences the transcription rate of nearby genes. This transcription factor (1) activates genes that promote DNA repair; (2) activates genes that arrest cell division and represses genes that are required for cell division; and (3) activates genes that promote apoptosis. If a cell is exposed to DNA damage, it has a greater potential to become malignant. Therefore, an organism wants to avoid the proliferation of such a cell. When exposed to an agent that causes DNA damage, a cell will try to repair the damage. However, if the damage is too extensive, the p53 protein will stop the cell from dividing and program it to die. This helps to prevent the proliferation of cancer cells in the body.

Experimental Questions

E1. A. False, this would suggest an infectious disease, because people living in the same area would be exposed to the same kinds of infectious agents. Relatives living together and apart would exhibit the same frequency for a genetic disease.

B. This could be true, because the individuals living in one area may be more genetically related. On the other hand, a particular infectious agent may be found only in southern Spain, and this might explain the high frequency in this region.

C. A specific age of onset is consistent with a genetic disease.

D. A higher likelihood of developing a disease among monozygotic twins compared to dizygotic twins is consistent with a genetic basis for a disease because monozygotic twins are more genetically similar (in fact, identical) compared to dizygotic twins.

E2. Perhaps the least convincing is the higher incidence of the disease in particular populations. Because populations living in specific geographic locations are exposed to their own unique environment, it is difficult to distinguish genetic versus environmental causes for a particular disease. The most convincing evidence might be the higher incidence of a disease in related individuals and/or the ability to correlate a disease with the presence of a mutant gene. Overall, however, the reliability that a disease has a genetic component should be based on as many observations as possible.

E3. The term *genetic testing* refers to the use of laboratory tests to determine if an individual is a carrier or affected by a genetic disease. Testing at the protein level means that the amount or activity of the protein is assayed. Testing at the DNA level means that the researcher tries to detect the mutant allele at the molecular or chromosomal level. Examples of approaches are described in Table 22.4.

E4. You would probably conclude that it is less likely to have a genetic component. If it were rooted primarily in genetics, it would be likely to be found in the Central American population. Of course, there is a chance that very few or none of the people who migrated to Central America were carriers of the mutant gene. This is somewhat unlikely for a large migrating population. By comparison, one might suspect that an environmental agent that is present in South America but not present in Central America may underlie the disease. Researchers could try to search for this environmental agent (e.g., pathogenic organism, etc.).

E5. A. As seen in lane 3, the α–galactosidase A polypeptide is shorter in cells obtained from Pete. This indicates that Pete's disease is caused by a mutation that either is a deletion in the gene or introduces an early stop codon. In Jerry's case (lane 6), there does not appear to be any of the α–galactosidase A polypeptide in his cells. This could be due to a deletion that removes the entire gene, a promoter mutation that prevents the expression of the gene, a mutation that prevents translation (e.g., a mutation in the start codon), or a mutation that results in a polypeptide that is very unstable and rapidly degraded.

B. Amy appears to have two normal copies of the α–galactosidase A gene. She will not pass a mutant allele to her offspring. Nan is a heterozygote. She has a 50% chance of passing the mutant allele. Half of her sons would be affected with the disease. Likewise, Aileen also appears to be a heterozygote because the amount of α–galactosidase A polypeptide seems to be about 50% of normal. She also would have a 50% chance of passing the mutant allele to her offspring; half of her sons would be affected.

E6. Males I-1, II-1, II-4, II-6, III-3, III-8, and IV-5 have a normal copy of the gene. Males II-3, III-2, and IV-4 are hemizygous for an inactive mutant allele. Females III-4, III-6, IV-1, IV-2, and IV-3 have two normal copies of the gene, whereas females I-2, II-2, II-5, III-1, III-5, and III-7 are heterozygous carriers of a mutant allele.

E7. You would not expect a high number of malignant foci. Mutations in tumor-suppressor genes that cause malignancy are due to an inactivation of the tumor-suppressor gene. The NIH3T3 cells must have normal (nondefective) tumor-suppressor genes, otherwise they would be malignant. If a defective tumor-suppressor gene was transformed into NIH3T3 cells, it would have no effect because the NIH3T3 cells already have normal tumor-suppressor genes that prevent malignant growth.

Note: In the experiment of Figure 22.9, the sources of DNA that led to a large number of malignant foci (e.g., MC5-5-0) must have contained oncogenes. Oncogenes have mutations that lead to the overexpression of genes that control cell division. When an oncogene is taken up by the NIH3T3 cells, it causes malignant growth due to gene overexpression. The NIH3T3 cells are not able to prevent this overexpression that leads to uncontrolled cell growth. In other words, the normal tumor-suppressor genes in NIH3T3 cells are not strong enough to overcome the effects of oncogenes.

E8. A transformed cell is one that has become malignant. In a laboratory, this can be done in three ways. First, the cells could be treated with a mutagen that would convert a proto-oncogene into an oncogene. Second, cells could be exposed to the DNA from a malignant cell line. Under the appropriate conditions, this DNA can be taken up by the cells and integrated into their genome so that they become malignant. A third way to transform cells is by exposure to an oncogenic virus.

E9. If the DNA sample had been treated with RNase or protease, the results would have been the same. If they had been treated with DNase, no transformation would have occurred.

E10. By comparing oncogenic viruses with strains that have lost their oncogenicity, researchers have been able to identify particular genes that cause cancer. This has led to the identification of many oncogenes. From this work, researchers have also learned that normal cells contain proto-oncogenes that usually play a role in cell division. This suggests that oncogenes exert their effects by upsetting the cell division process. In particular, it appears that oncogenes are abnormally active and keep the cell division cycle in a permanent "on" position.

E11. Most inherited forms of cancer are inherited in a dominant manner. This can oftentimes be revealed by a pedigree analysis, since affected individuals are much more likely to have affected offspring. Since cancer may be caused by the sequential accumulation of mutations, the correlation between affected parents and affected offspring may be relatively low because the offspring may not accumulate the other postzygotic mutations that are necessary for cancer to occur.

E12. One possible category of drugs would be GDP analogues (i.e., compounds that resemble the structure of GDP). Perhaps one could find a GDP analogue that binds to the Ras protein and locks it in the inactive conformation.

One way to test the efficacy of such a drug would be to incubate the drug with a type of cancer cell that is known to have an overactive Ras protein, and then plate the cells on solid media. If the drug locked the Ras protein in the inactive conformation, it should inhibit the formation of malignant growth or malignant foci.

There are possible side effects of such drugs. First, they might block the growth of normal cells, since Ras protein plays a role in normal cell proliferation. Second (if you have taken a cell biology course), there are many GTP/GDP-binding proteins in cells, and the drugs could somehow inhibit cell growth and function by interacting with these proteins.

E13. Mammalian cells grow as a monolayer on solid growth media, whereas malignant cells tend to pile on top of each other and form malignant foci. A malignant focus can be formed from a single cancer cell that has divided many times. (It is also possible that multiple independent cancer cells could form a malignant focus, but this would not be necessary.)

Questions for Student Discussion/Collaboration

1. Genetic testing may lead to better treatments of disease. There are many possible examples. If an individual discovered that he/she were carrying an allele that caused a higher predisposition to develop cancer, that person could be more careful about monitoring his/her health in this regard. Or if a person discovered that he/she were the carrier of a hypercholesterolemia allele, this person could try to eat a very low-fat diet. On the negative side, people may be discriminated against by their employers or by insurance companies because of the alleles that they carry.
2. There is no clearly correct answer to this question, but it should stimulate a large amount of discussion.

Chapter 23: Developmental Genetics

Student Learning Objectives

Upon completion of this chapter you should be able to:

1. Understand the key terms associated with the study of developmental genetics.
2. Understand patterns of development and the genetic factors that contribute to these processes.
3. Understand the genetic components of development in both plants and animals.

23.1 Invertebrate Development

Overview

The study of developmental genetics has focused primarily on two invertebrate animals, *Drosophila melanogaster* and the nematode *Caenorhabiditis elegans*. This section of the chapter uses these organisms to illustrate two important aspects of developmental genetics. The first, using *D. melanogaster* as the model organism, establishes how developmental patterns are established during embryogenesis. You should notice how patterns in gene expression are instrumental in forming the axis of orientation for the body plan. The second organisms, *C. elegans*, is used to demonstrate the concept of cell lineages, or the destiny of cells in an embryo. A knowledge of both of these systems is important in understanding developmental genetics.

Key Terms

Antero-posterior axis
Apoptosis
Cell adhesion
Cell adhesion molecules
Cell fate
Cell lineage
Cell migration
Developmental genetics
Dorso-ventral axis
Embryogenesis
Gain-of-function mutation
Gap genes
Heterochronic mutations
Homeobox
Homeodomain
Homeotic
Homeotic gene
Induction
Lineage diagram
Morphogens
Pair rule genes
Parasegments
Pattern
Positional information
Promixo-distal axis
Realizator genes
Right-left axis
Segmentation genes
Segment-polarity genes
Segments
Threshold concentration
Zygotic genes

Focal Points

- Invertebrate development in *Drosophila* (Figure 23.1)
- Segment and parasegment comparison (Figure 23.8)

Exercises and Problems

For questions 1 to 4, match the label from the diagram below with its correct description.

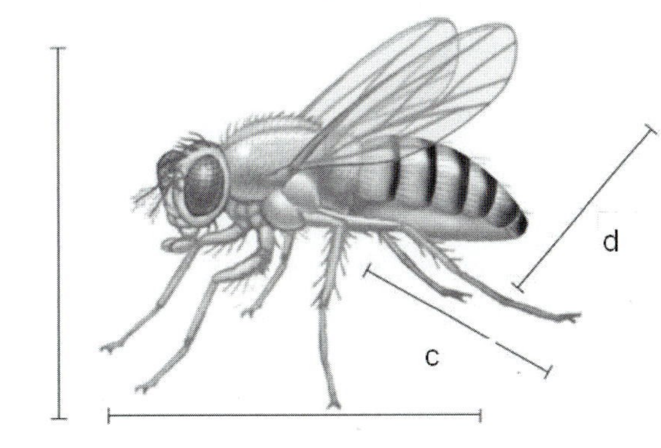

_____ 1. Dorso-ventral axis
_____ 2. Right-left axis
_____ 3. Antero-posterior axis
_____ 4. Promimo-distal axis

For questions 5 to 16, match the terms associated with *Drosophila* development with its correct definition.

_____ 5. Cell migration
_____ 6. Homeotic gene
_____ 7. Induction
_____ 8. Segmentation genes
_____ 9. Pattern
_____ 10. Positional information
_____ 11. Morphogens
_____ 12. Homeobox
_____ 13. Homeodomain
_____ 14. Cell fate
_____ 15. Realizator genes
_____ 16. Threshold level

a. The determination of cell type based upon its location.
b. The level at which a morphogen starts exerting its positional information.
c. A consensus sequence in developmental genes.
d. These genes produce the morphological characteristics of the segment.
e. The movement of cells to form the ectoderm, endoderm and mesoderm.
f. Gap genes, pair-rule genes, and segment-polarity genes are examples.
g. Spatial arrangements of different regions of the body.
h. Process by which cells govern the developmental fate of neighboring cells.
i. This gives developmental proteins the ability to bind to the major groove of the DNA.
j. A gene that plays a central role in deciding the final identity of a body region.
k. Molecules that convey positional information to promote developmental changes.
l. The morphological features that a group of cells will adopt.

The following questions relate to the formation of segment patterns in *Drosophila*. Place the letter of the first event in number 17, the second event in number 18, etc.

_____ 17.

_____ 18.

_____ 19.

_____ 20.

a. Maternal effect genes form a gradient in the oocyte.
b. Segment polarity genes are activated.
c. Gap genes and maternal effect genes activate the pair-rule genes.
d. Maternal effect genes activate zygotic genes.

For questions 21 to 23, match the term associated with cell lineages in *C. elegans* with its appropriate statement.

_____ 21. Cell lineage

_____ 22. Lineage diagram

_____ 23. Heterochronic mutations

a. A series of cells that are derived from each other by cell division.
b. Describes division patterns and cell fates for a given organism.
c. Due to irregular gene expression upsets the timing of cell fates within an organism.

23.2 Vertebrate Development

Overview

As expected, less is known about vertebrate development than is known concerning the invertebrates. However, there is a growing body of information on the developmental genetics of the mouse. This section examines the role of a specific group of genes, called the Hox genes, in the regulation of vertebrate development.

Key Terms

Basic domain
Determined
Differentiated
Gene knockout
Helix-loop-helix domain
Homologous genes
Hox complexes
Myogenic bHLH
Orthologs
Reverse genetics

Focal Points

- Hox genes (Figures 23.15 and 23.16)

Exercises and Problems

For each of the following, complete the sentence with the appropriate word or phrase.

1. Homeotic gene complexes in vertebrates are called ______ genes.
2. A cell that is destined to becomes a specific cell type is said to be ________.
3. Myogenic bHLH genes all contain a ______ and _____ domain.
4. Wild type genes are identified from clones using a procedure called _________.
5. The permanent specialization of a cell's morphology and function is called ________.
6. Two homologous genes that are found in different species are called _______.

23.3 Plant Development

Overview

The developmental genetics of plants is very different from that of animals. In this section you will examine some of the principles of development using the plant *Arabidopsis thaliana*. *A. thaliana* is to the plant kingdom as *Drosophila* is to animals, it is the model organism for the study of plant genetics. As you progress through this section, notice that the areas of development in plants, and the role of homeotic genes in controlling the differentiation of tissues in these areas.

Key Terms

ABC model
Apical region
Apical-basal-patterning genes
Basal region
Central region
Central zone
Organizing center
MADS box
Peripheral zone
Root meristem
Shoot meristem
Totipotent

Focal Points

- Overview of plant development (Figure 23.19)
- ABC model (Figure 23.22)

Exercises and Problems

Identify the following regions of the plant by matching the term to its appropriate function in *A. thaliana*.

_____ 1. Central zone

_____ 2. Apical region

_____ 3. Organizing center

_____ 4. Central region

_____ 5. Peripheral zone

_____ 6. Basal region

a. The area of the shoot meristem is where undifferentiated stem cells are maintained.
b. The area of the root meristem that produces the leaves and flowers of the plant.
c. The area of the root meristem that creates the stem of the plant.
d. The area of the shoot meristem that preserves the correct number of actively dividing stem cells.
e. The area of the shoot meristem that contains dividing cells that will become plant structures.
f. The area of the root meristem that produces the roots.

For questions 7 to 10, match the following genetic mechanisms of plant development with its function.

_____ 7. Totipotent

_____ 8. Apical-basal-patterning genes

_____ 9. ABC model

_____ 10. MADS box

a. The common feature of homeotic genes that encodes a DNA-binding domain.
b. Cells that have the ability to form almost any other cell type.
c. Genes that are important in the early stages of plant development to establish the orientation of the plant.
d. The genes that determine the formation of sepals, petals, stamens and carpels in the whorls of the flower.

23.4 Sex Determination in Animals and Plants

Overview

The final section of the chapter examines the genetic basis of sex determination in a variety of organisms. Since in many species there is a distinct difference in the morphology of the males and females, there must be elaborate genetic mechanisms to activate the formation of the sex. In this section, first examine the basic methods of sex determination (Table 23.3) and then focus on the major genetic mechanisms that are involved.

Key Terms

Dioecious

Focal Points

- Sex determination mechanisms (Table 23.3)

Exercises and Problems

Each of the following questions describe some aspect of sex determination. From the list below, select the type of organism to which the statement is associated with.

a. *C. elegans*
b. Plants
c. Mammals
d. *Drosophila*

_____ 1. The *DAX1* gene prohibits male development.

_____ 2. Sex is determined by the *Sry* gene.

_____ 3. In dioecious plants, genes such as Su^F promotes sexual development.

_____ 4. Expression of the *SXL* gene promotes female development.

_____ 5. Males and females are usually monomorphic.

_____ 6. Expression of the *MSL-2* gene promotes male development.

_____ 7. Males have one X chromosome, hermaphrodites have two X chromosomes.

Chapter Quiz

1. Which of the following terms describes cells that govern the developmental fate of neighboring cells?

a. threshold
b. adhesion
c. induction
d. homeotic
e. none of the above

2. Gap genes, pair-rule genes and segment polarity genes are example of this.

a. hox genes
b. segmentation genes
c. embryonic genes
d. morphogens

3. The consensus sequence of homeotic genes in invertebrates is called _______.
   a. MADS box
   b. homeobox
   c. Sry
   d. Hox

4. The fate of every cell in embryonic development was first accomplished through studies of what organism?
   a. *Drosophila*
   b. *Arabidopsis*
   c. mouse
   d. *C. elegans*

5. What area of the plant is where dividing cells are differentiating into plant structures?
   a. organizing center
   b. central region
   c. central zone
   d. peripheral zone
   e. basal region

6. Sry is associated with sex determination in which of the following?
   a. mammals
   b. plants
   c. *Drosophila*
   d. *C. elegans*
   e. none of the above

7. _____ refers to the ultimate morphological features that a cell or group of cells will adopt.
   a. Cell lineage
   b. Cell fate
   c. Cell destiny
   d. Cell programming
   e. Apoptosis

8. Morphogens must overcome a ______ before being effective.
   a. inhibitor
   b. protein barrier
   c. threshold concentration
   d. activator

9. Hox complexes are found in ______.
   a. vertebrates
   b. invertebrates
   c. plants
   d. fungi

10. The ABC model describes which of the following?
   a. segment formation
   b. activation of totipotent cells
   c. activity of morphogens
   d. sex determination in mammals
   e. flower development

Answer Key for Study Guide Questions

This answer key provides the answers to the exercises and chapter quiz for this chapter. Answers in parentheses ( ) represent possible alternate answers to a problem, while answers marked with an asterisk (*) indicate that the response to the question may vary.

23.1
1. a
2. d
3. b
4. c
5. e
6. j
7. h
8. f
9. g
10. a
11. k
12. c
13. i
14. l
15. d
16. b
17. a
18. d
19. c
20. b
21. a
22. b
23. c

23.2
1. Hox
2. determined
3. basic, helix-turn-helix
4. reverse genetics
5. differentiation
6. orthologs

23.3
1. a
2. b
3. d
4. c
5. e
6. f
7. b
8. c
9. d
10. a

23.4
1. c
2. c
3. b
4. d
5. b
6. d
7. a

Quiz

1. c
2. b
3. b
4. d
5. d
6. a
7. b
8. c
9. a
10. e

Answers to Conceptual and Experimental Questions

Conceptual Questions

C1. The four processes are cell division, cell differentiation, cell movement, and cell death. Cell division is needed to produce a multicellular organism. In other words, cell division is needed for growth. Cell differentiation is needed to create different cell types. Each cell type is differentiated to carry out its own specialized function. Cell movement is needed during embryonic development to create an embryo with the proper organization of cells and germ layers. And finally, cell death is needed to create certain bodily structures. For example, in the early embryonic development of mammals, the hand is initially a flattened, oval structure. The fingers are formed when cell death occurs in the regions between each finger.

C2. A. False, the head is anterior to the tail.

B. True.

C. False, the feet are posterior to the hips. Along the dorso-ventral axis, they are about the same.

D. True.

C3. A. This would likely be a mutation in a segmentation gene, because mutations in certain segmentation genes have fewer segments. In this case, it could be a loss-of-function mutation in a gap gene.

B. This would likely be a mutation in a homeotic gene because the characteristics of one segment have been converted to the characteristics of a different segment.

C. This would likely be a mutation in a homeotic gene because the characteristics of two segments have been converted to the characteristics of two different segments.

C4. A. True.

B. False, because gradients are also established after fertilization during embryonic development.

C. True.

C5. A parasegment is only a transient demarcation that divides the developing embryo. The segments become permanent regions that develop their own morphological characteristics. The expression of certain genes, such as *ftz* and *even-skipped,* occur in parasegments. *Ftz* is expressed in the odd-numbered parasegments and *even-skipped* is expressed in the even-numbered parasegments. This expression occurs prior to the formation of the segments.

C6. A. This is a mutation in *runt,* which is a pair-rule gene.

B. This is a mutation in *knirps,* which is a gap gene.

C. This is a mutation in *patched,* which is a segment-polarity gene.

C7. A morphogen is a molecule, such as a transcription factor, that influences the morphological fate of a cell or group of cells. A morphogen exerts its effects by affecting a genetic hierarchy that ultimately leads to the expression of genes that govern cell locations and morphologies. If a morphogen is expressed in the wrong place, an abnormal morphology results. An example is the phenotype called antennapedia in which a leg is found in place of an antenna. Examples of morphogens in *Drosophila* include Bicoid, Hunchback, Giant, Krüppel, and the homeotic gene products.

C8. Positional information refers to the phenomenon whereby the spatial locations of morphogens and CAMs provide a cell with information regarding its position relative to other cells. In *Drosophila,* the formation of a segmented body pattern relies initially on the spatial location of maternal gene products. These gene products lead to the sequential activation of the segmentation genes.

C9. The two processes are similar in that they set up concentration gradients that can lead to the spatial activation of genes in particular regions of the embryo. When the gradients are established in the oocyte, the morphogen becomes incorporated into cells during the cleavage and cell division stage of embryogenesis. In this case, the morphogen is already inside. Later in development, when a morphogen is secreted from a particular cell or group of cells, the morphogen must bind to cell surface receptors to elicit its effects.

C10. The anterior portion of the antero-posterior axis is established by the action of Bicoid. During oogenesis, the mRNA for Bicoid enters the anterior end of the oocyte and is sequestered there to establish an anterior (high) to posterior (low) gradient. Later, when the mRNA is translated, the Bicoid protein in the anterior region establishes a genetic hierarchy that leads to the formation of anterior structures. If Bicoid was not trapped in the anterior end, it is likely that anterior structures would not form.

C11. The Bicoid protein is a morphogen that causes the anterior portions of the embryo to form properly. It functions as a transcription factor. Other genes, such as *hunchback,* are stimulated by Bicoid in a concentration-dependent manner. In this case, *hunchback* is stimulated only in the anterior portion of the embryo. In a general sense, the concentration dependence of gene activation leads to the activation of genes in particular regions of the embryo, leading to the stimulation of a few broad bands and eventually to a segmented pattern.

C12. Maternal gene products influence the formation of the main body axes including the antero-posterior, dorso-ventral, and terminal regions. They are needed very early in development. Zygotic genes, particularly the three classes of the segmentation genes, are necessary after the axes have been established. The segmentation genes are expressed (after fertilization) during embryogenesis.

C13. A homeotic gene governs the final fate of particular segments in the adult animal. A gain-of-function mutation is due to an aberrant expression of a homeotic gene in the wrong place. This causes the region to develop inappropriate characteristics. A loss-of-function allele usually causes a segment to develop characteristics that are normally found in the anterior adjacent segment.

C14. The coding sequence of homeotic genes contains a 180 bp consensus sequence known as a homeobox. The protein domain encoded by the homeobox is called a homeodomain. The homeodomain contains three conserved sequences that are folded into α–helical conformations. The arrangement of these *a* helices promotes the binding of the protein to the major groove of the DNA. Helix III is called the recognition helix because it recognizes a particular nucleotide sequence within the major groove. In this way, homeotic proteins are able to bind to DNA in a sequence-specific manner and thereby activate particular genes.

C15. Since the *nanos* gene product plays a role in the development of posterior structures, such a larva would probably develop with two anterior ends.

C16. It would normally be expressed in the three thoracic segments that have legs (T1, T2, and T3).

C17. When a loss-of-function mutation occurs in a homeotic gene, the segment(s) where the homeotic gene is normally expressed will have the characteristics of the adjacent anterior segment(s). For a gain-of-function mutation, a homeotic gene is expressed in the wrong place. The segment where it is incorrectly expressed will have inappropriate characteristics (like legs where the antennae should be). Therefore, the phenotypic consequence of a gain-of-function mutation is that one or more segments will have characteristics of segments that could be anywhere else in the fly. (Note: It is possible that a gain-of-function mutation could resemble a loss-of-function mutation if the gain-of-function mutation resulted in the abnormal expression of a homeotic gene in the segment that is adjacent and posterior to the segment where the homeotic gene is normally expressed.)

A. Gain-of-function allele.

B. Either a loss-of-function or gain-of-function mutation could explain this phenotype; a loss-of-function mutation is more likely since it is easier to mutate a gene and eliminate its function.

C. Gain of function allele.

C18. A. When a mutation inactivates a gap gene, a contiguous section of the larva is missing.

B. When a mutation inactivates a pair-rule gene, some regions that are derived from alternating parasegments are missing.

C. When a mutation inactivates a segment-polarity gene, portions are missing at either the anterior or posterior end of the segments.

C19. A maternal effect gene is expressed in the nurse cells surrounding a developing oocyte, and the gene products (i.e., mRNA or protein) are transferred to the oocyte. Since the nurse cells are diploid, the phenotype of the resulting individual (after oocyte fertilization) actually depends on the diploid genotype of the nurse cells (which is the same as the diploid genotype of the mother). Zygotic genes are expressed after fertilization. (Note: Some genes are both maternal effect and zygotic; they are expressed in the nurse cell and after fertilization. For this question, however, let's assume that a gene is likely to be one or the other, but not both.)

A. *Nanos* is a maternal effect gene; the *nanos* mRNA accumulates in the oocyte.

B. *Antp* is a zygotic gene. It is a homeotic gene that is turned on in embryonic development after the segmentation genes have been expressed.

C. *Bicoid* is a maternal effect gene; the *bicoid* mRNA accumulates in the oocyte.

D. *Lab* is a zygotic gene. It is a homeotic gene that is turned on in embryonic development after the segmentation genes have been expressed.

C20. Proper development in mammals is likely to require the products of maternal effect genes that play a key role in initiating embryonic development. The adult body plan is merely an expansion of the embryonic body plan. Because the starting point for the development of an embryo is the oocyte, this explains why an enucleated oocyte is needed to clone mammals. The oocyte is needed to establish the embryonic body plan.

C21. A. In this case, the A-1 cell behaves like a B-1 cell. The A-1 cell produces the lineage of the B-1 cell.

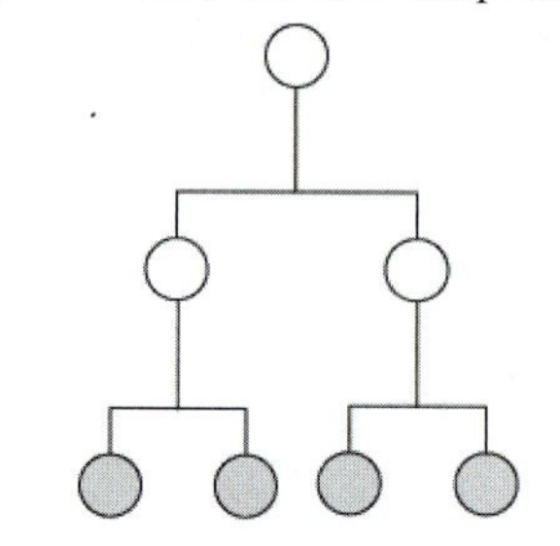

B. In this case, the B-1 cell behaves like an A-1 cell. The B-1 cell produces the lineage that the A-1 cell would normally produce.

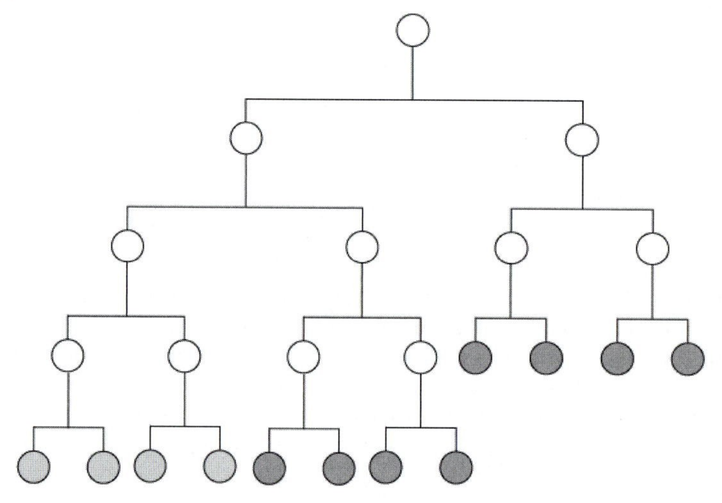

C22. A heterochronic mutation is one that alters the timing when a gene (involved in development) is normally expressed. The gene may be expressed too early or too late, which causes certain cell lineages to be out of sync with the rest of the animal. If a heterochronic mutation affected the intestine, the animal may end up with too many intestinal cells if it is a gain-of-function mutation or too few if it is a loss-of-function mutation. In either case, the effects might be detrimental because the growth of the intestine must be coordinated with the growth of the rest of the animal.

C23. *Drosophila* has eight homeotic genes located in two clusters (*antennapedia* and *bithorax*) on chromosome 3. The mouse has four *Hox* complexes designated *HoxA* (on chromosome 6), *HoxB* (on chromosome 11), *HoxC* (on chromosome 15), and *HoxD* (on chromosome 2). There are a total of 38 genes in the four complexes. Among the first six genes, five of them are homologous to genes found in the *antennapedia* complex of *Drosophila.* Among the last seven, three of them are homologous to the genes of the *bithorax* complex. Like the *bithorax* and *antennapedia* complexes in *Drosophila,* the arrangement of *Hox* genes along the mouse chromosomes reflects their pattern of expression from the anterior to the posterior end. With regard to differences, the mouse has a larger number of homeotic genes, and gene knockouts do not always lead to transformations that resemble the anterior adjacent segment.

C24. Cell differentiation is the specialization of a cell into a particular cell type. In the case of skeletal muscle cells, the bHLH proteins play a key role in the initiation of cell differentiation. When bHLH proteins are activated, they are able to bind to enhancers and activate the expression of many different muscle-specific genes. In this way, myogenic bHLH proteins turn on the expression of many muscle-specific proteins. When these proteins are synthesized, they change the characteristics of the cell into those of a muscle cell. Myogenic bHLH proteins are regulated by dimerization. When a heterodimer forms between a myogenic bHLH protein and an E protein, it activates gene expression. However, when a heterodimer forms between myogenic bHLH proteins and a protein called Id, the heterodimer is unable to bind to DNA. The Id protein is produced during early stages of development and prevents myogenic bHLH proteins from promoting muscle differentiation too soon. At later stages of development, the amount of Id protein falls, and myogenic bHLH proteins can combine with E proteins to induce muscle differentiation.

C25. Genes involved with cell differentiation and homeotic genes are similar in that they control genetic regulatory pathways. These types of genes typically encode transcription factors that regulate the expression of many genes. The main difference lies in the magnitude of the genetic control. *MyoD,* for example, controls the differentiation of skeletal muscle cells, and this cell type can be found in many different regions of the body. A homeotic gene controls the development of an entire segment of the body; a segment of a body has many different types of cells that are organized in a particular way.

C26. A totipotent cell is a cell that has the potential to create a complete organism.

A. In humans, a fertilized egg is totipotent, and the cells during the first few embryonic divisions are totipotent. However, after several divisions, embryonic cells lose their totipotency and, instead, are determined to become particular tissues within the body.

B. In plants, most living cells are totipotent.

C. Because yeast are unicellular, one cell is a complete individual. Therefore, yeast cells are totipotent; they can produce new individuals by cell division.

D. Because bacteria are also unicellular, one cell is a complete individual. Therefore, bacteria are totipotent; they can produce new individuals by cell division.

C27. A meristem is an organized group of actively dividing cells. A plant may have one or more shoot meristems that produce offshoots that give rise to structures such as leaves and flowers. The root meristem grows in the opposite direction and gives rise to the roots. The pattern of meristem growth is a primary determinant in the overall morphology of the plant. For example, if a shoot meristem produces offshoots at close intervals, leaves that are very close together result. In contrast, if a shoot meristem rarely produces offshoots giving rise to leaves, the plant will have very few leaves. This explains why some plants are bushy and others have sparsely located leaves.

C28. Animals begin their development from an egg and then form antero-posterior and dorso-ventral axes. The formation of an adult organism is an expansion of the embryonic body plan. Plants grow primarily from two meristems, a shoot and root meristem. At the cellular level, plant development is different in that it does not involve cell migration, and most plant cells are totipotent. Animals require the organization within an oocyte to begin development. At the genetic level, however, animal and plant development are similar in that they involve a genetic hierarchy of transcription factors that govern pattern formation and cell specialization.

C29. A. Carpel-stamen-stamen-carpel

B. Sepal-sepal-carpel-carpel

C. Carpel-carpel-carpel-carpel

Experimental Questions

E1. The expected result would be that the embryo would develop with two anterior ends. It is difficult to predict what would happen at later stages of development. At that point, the genetic hierarchy has already been established so its effects would be diminished. Also, at later stages of development, the embryo is divided into many cells so the injection would probably affect a smaller area.

E2. *Drosophila* is more advanced from the perspective that many more mutant alleles have been identified that alter development in specific ways. The hierarchy of gene regulation is particularly well understood in the fruit fly. *C. elegans* has the advantage of simplicity and a complete knowledge of cell fate. This enables researchers to explore how the timing of gene expression is critical to the developmental process.

E3. The term *cell fate* refers to the final cell type that a cell will become. For example, the fate of a cell may be a muscle cell. A lineage diagram depicts the cell lineages and final cell fates for a group of cells. In *C. elegans,* an entire lineage diagram has been established. A cell lineage is a description of the sequential division patterns that particular cells progress through during the developmental stages of an organism.

E4. To determine that a mutation is affecting the timing of developmental decisions, a researcher needs to know the normal time or stage of development when cells are supposed to divide, and what types of cells will be produced. With this information (i.e., a lineage diagram), one can then determine if particular mutations alter the timing when cell division occurs.

E5. A bag of worms phenotype can be due to an aberrancy in the development of several cell types. It is also an easy phenotype to observe and study. In some cases, the inability to lay eggs may be due to the abnormal timing of developmental steps in particular cell lineages (although this is not always the case). A researcher can watch the division patterns of cells that may be affected in the bag of worms phenotype to see if the timing of cell division is out of sync with the rest of the animal.

E6. Mutant 1 is a gain-of-function allele; it keeps reiterating the L1 pattern of division. Mutant 2 is a loss-of-function allele; it skips the L1 pattern and immediately follows an L2 pattern.

E7. These results indicate that the gene product is needed from 1 to 3 hours after fertilization for the embryo to develop properly and survive. The gene product is not needed at the other developmental stages that were examined in this experiment (0–1 hours, or 3–6 hours after fertilization).

E8. As discussed in Chapter 15, most eukaryotic genes have a core promoter that is adjacent to the coding sequence; regulatory elements that control the transcription rate at the promoter are typically upstream from the core promoter. Therefore, to get the *Antp* gene product expressed where the *abd-A* gene product is normally expressed, you would link the upstream genetic regulatory region of the *abd-A* gene to the coding sequence of the *Antp* gene. This construct would be inserted into the middle of a P element (see next). The construct shown here would then be introduced into an embryo by P element transformation.

	***abd-A* regulatory region**	/	***Antp* coding sequence**
P element			
P element			

The *Antp* gene product is normally expressed in the thoracic region and produces segments with legs, as illustrated in Figure 23.11. Therefore, because the *abd-A* gene product is normally expressed in the anterior abdominal segments, one might predict that the genetic construct shown above would produce a fly with legs attached to the segments that are supposed to be the anterior abdominal segments. In other words, the anterior abdominal segments would probably resemble thoracic segments with legs.

E9. A. Yes, because Krüppel protein acts as a transcriptional repressor, and its concentration is low in this region anyway.

B. Probably not, because Bicoid protein acts as a transcriptional activator.

C. Probably not, because Hunchback protein acts as a transcriptional activator.

D. Yes, because giant protein acts as a repressor, and its concentration is low in this region anyway.

E10. A. The female flies must have had mothers that were heterozygous for a (dominant) normal allele and the mutant allele. Their fathers were either homozygous for the mutant allele or heterozygous. The female flies inherited a mutant allele from both their father and mother. Nevertheless, because their mother was heterozygous for the normal (dominant) allele and mutant allele, and because this is a maternal effect gene, their phenotype is based on the genotype of their mother. The normal allele is dominant, so they have a normal phenotype.

B. *Bicoid-A* appears to have a deletion that removes part of the sequence of the gene and thereby results in a shorter mRNA. *Bicoid-B* could also be a deletion that removes all of the sequence of the *bicoid* gene or it could be a promoter mutation that prevents the expression of the *bicoid* gene. *Bicoid-C* seems to be a point mutation that does not affect the amount of the *bicoid* mRNA.

With regard to function, all three mutations are known to be loss-of-function mutations. *Bicoid-A* probably eliminates function by truncating the Bicoid protein. The Bicoid protein is a transcription factor. The *bicoid-A* mutation probably shortens this protein and thereby inhibits its function. The *bicoid-B* mutation prevents expression of the *bicoid* mRNA. Therefore, none of the Bicoid protein would be made, and this would explain the loss of function. The *bicoid-C* mutation seems to prevent the proper localization of the *bicoid* mRNA in the oocyte. There must be proteins within the oocyte that recognize specific sequences in the *bicoid* mRNA and trap it in the anterior end of the oocyte. This mutation must change these sequences and prevent these proteins from recognizing the *bicoid* mRNA.

C. If we only used the technique of Northern blotting, we would not have understood how *bicoid-C* was abnormal. Likewise, if we had only used the technique of *in situ* hybridization, we would not have understood how *bicoid-A* was abnormal.

E11. You could follow the strategy of reverse genetics. Basically, you would create a *HoxD-3* gene knockout. This inactivated *HoxD-3* gene would be introduced into a mouse by the technique of gene replacement described in Chapter 19. By making the appropriate crosses, homozygous mice would be obtained that carry the loss-of-function allele in place of the wild-type *HoxD-3* gene. The phenotypic characteristics of normal mice would then be compared to mice that were homozygous for a defective *HoxD-3* gene. This would involve an examination of the skeletal anatomies of mice at various stages of development. If the *HoxD-3* plays a role in development, you might see changes in morphology suggesting anterior transformations. In other words, a certain region of the mouse may have characteristics that are appropriate for more anterior segments.

E12. An egg-laying defect is somehow related to an abnormal anatomy. The *n540* strain has fewer neurons compared to a normal worm. Perhaps the *n540* strain is unable to lay eggs because it is missing neurons that are needed for egg laying. The *n536* and *n355* strains have an abnormal abundance of neurons. Perhaps this overabundance also interferes with the proper neural signals that are needed for egg laying.

E13. A. The larva would develop with two posterior ends. The larva would not survive to the adult stage.

B. The posterior end of the larva and adult fly would develop structures that were appropriate for thoracic segments. The adult fly may not survive.

C. The dorsal side of the larva and adult fly would develop structures that were appropriate for the ventral side. The adult fly may not survive.

E14. Geneticists who are interested in mammalian development have used reverse genetics because it has been difficult for them to identify mutations in developmental genes based on phenotypic effects in the embryo. This is because it is difficult to screen a large number of mammalian embryos in search of abnormal ones that carry mutant genes. It is easy to have thousands of flies in a laboratory, but it is not easy to have thousands of mice. Instead, it is easier to clone the normal gene based on its homology to invertebrate genes and then make mutations in vitro. These mutations can be introduced into a mouse to create a gene knockout. This strategy is opposite to that of Mendel, who characterized genes by first identifying phenotypic variants (e.g., tall vs. dwarf, green seeds vs. yellow seeds, etc.).

Questions for Student Discussion/Collaboration

1. *Drosophila:* Easy-to-identify developmental mutants, and easy to do *in situ* hybridization to follow their spatial expression.

   C. elegans: Also easy-to-identify mutants. A complete lineage diagram is its most important advantage over other invertebrates.

   Mammals: Difficult-to-identify developmental mutants except using a reverse genetics approach. An advantage might be interest, since many people like to know how things work in organisms that are closely related to humans.

   Arabidopsis: The organism of choice to study plant development because it is relatively small, has a short generation time, and has a relatively small genome. Many developmental mutations have also been identified.

2. In this problem, the students should try to make a flow diagram that begins with maternal effect genes, then gap genes, pair-rule genes, and segment-polarity genes. These genes then lead to homeotic genes and finally realizator genes. It is almost impossible to make an accurate flow diagram because there are so many gene interactions, but it is instructive to think about developmental genetics in this way. It is probably easier to identify mutant phenotypes that affect later stages of development because they are less likely to be lethal. However, modern methods can screen for conditional mutants as described in solved problem S3. To identify all of the genes necessary for chicken development, you may begin with early genes, but this assumes that you have some way to identify them. If they had been identified, you would then try to identify the genes that they stimulate or repress. This could be done using molecular methods described in Chapters 14, 15, and 18.

3. A gain-of-function mutation might be expressed in the wrong location because it may alter regulatory sequences that govern expression. For example, a mutation could occur that would make a regulatory sequence resemble a stripe-specific enhancer. This mutation could cause the adjacent gene to be expressed in a stripe where it normally is not expressed.

Chapter 24: Quantitative Genetics

Student Learning Objectives

Upon completion of this chapter you should be able to:

1. Understand the concept of a quantitative trait.
2. Understand the importance of statistical analyses to the study of quantitative traits.
3. Understand the concepts of polygenic inheritance.
4. Understand the principles of heritability and the effects of selective breeding and inbreeding.
5. Understand the principles of heterosis.

24.1 Quantitative Traits

Overview

The first section of this chapter introduces some important statistical methods that are used in the study of quantitative genetics. Although most genetics students have a background in statistics, it is important that you review the descriptions of standard deviation, variance, mean and correlations before proceeding into the next sections. In addition, be sure that you understand the variables in the formulas and can apply them as needed.

Key Terms

Additive
Biometric field
Correlation coefficient
Covariance
Discontinuous traits
Frequency distribution
Mean
Normal distribution curve
Quantitative genetics
Quantitative trait
Regression analysis
Standard deviation
Variance

Focal Points

- Mathematical explanations of standard deviation, variance and correlation coefficient

Exercises and Problems

For each of the following, match the term with its appropriate definition.

_____ 1. Frequency distribution

_____ 2. Quantitative traits

_____ 3. Biometric field

_____ 4. Normal distribution curve

_____ 5. Discontinuous traits

a. Traits that can be placed into specific phenotypic categories.
b. A distribution is divided into convenient, discrete units for analysis.
c. A distribution that varies in a symmetric manner around an average value.
d. Traits that demonstrate a continuum of phenotypic variation.
e. The study of the variation of quantitative traits within groups of individuals.

For questions 6 to 11, choose the correct term that matches the statistical statement provided.

_____ 6. How two factors vary in relation to one another.

_____ 7. The sum of all of the values in the group divided by the number of individuals.

_____ 8. A measure of how much one variable will change in response to a second variable.

_____ 9. The sum of the squared deviations from the mean divided by the degrees of freedom.

_____ 10. The degree of variation between two variables in a group.

_____ 11. The square root of the variance.

a. variance
b. standard deviation
c. covariance
d. correlation coefficient
e. regression analysis
f. mean

24.2 Polygenic Inheritance

Overview

Many quantitative traits can be explained by the fact that more than one gene is contributing to the phenotype. This is called polygenic inheritance and is the focus of this section of the chapter. As you proceed through this section, notice how phenotypes based on polygenic mechanisms vary from monogenic mechanisms (Figure 24.3). In addition, examine how geneticists map quantitative traits loci (QTLs).

Key Terms

Backcross
Genetic markers
Polygenic inheritance
QTL mapping
Quantitative trait loci

Focal Points

- Distribution of polygenic traits (Figure 24.4)

Exercises and Problems

Complete the statements below with the correct word or phrase.

1. The first QTL to be identified was associated with ____ resistance in *Drosophila*.
2. An association between genetically determined phenotypes and molecular markers is called _______.
3. When assessing a QTL, a _____ is often performed to produce F2 generation individuals that differ with regard to their combinations of the parental chromosomes.
4. A trait that is governed by two or more genes is called __________.
5. The pigmentations of the husks in wheat is an example of __________.

24.3 Heritability

Overview

The role of the environment in determining the phenotype of the organism has been introduced several times in the text. This last section of the chapter examines the methods of measuring the amount that genetic and environmental conditions are contributing to the observed phenotypes. As with the previous sections, familiarize yourself with the terms and formulas associated with the study of heritability.

This section also makes the distinction between artificial selection and natural selection and how it relates to heritability. Finally, the concept of hybrid vigor, or heterosis, is discussed with relation to heritability.

Key Terms

Artificial selection
Broad sense heritability
Heritability
Heterosis
Hybrid vigor
Inbreeding
Monomorphic
Narrow sense heritability
Pseudo-overdominance
Realized heritability
Selective breeding

Focal Points

- Formulas associated with heritability studies (pgs. 679-679)

Exercises and Problems

For each of the following, match the term to its correct definition.

_____ 1. Inbreeding

_____ 2. Broad sense heritability

_____ 3. Narrow sense heritability

_____ 4. Natural selection

_____ 5. Artificial selection

_____ 6. Heterosis

a. Heritability due to the additive effect of alleles.
b. All genetic variation that may influence the phenotype.
c. The practice of mating between genetically related individuals.
d. This is based upon naturally occurring variation in reproductive success.
e. A hybrid strain that is more vigorous than the true-breeding parental strains.
f. Selection in which specific traits are favored due to economic importance.

Chapter Quiz

1. Which of the following provides a measure of the amount of variation within the group?
   a. mean
   b. variance
   c. standard deviation
   d. regression
   e. correlation coefficient.

2. The term that describes the variation of a trait in a symmetric manner around an average value is called _______.
   a. biometric
   b. frequency distribution
   c. mean
   d. normal distribution

3. The square root of the variance is called the ________.
   a. standard deviation
   b. mean
   c. covariance
   d. correlation coefficient

4. This measures the strength of the association between two variables.
   a. mean
   b. variance
   c. standard deviation
   d. regression
   e. correlation coefficient

5. _____ determine(s) the amount of phenotypic variation within a group of individuals that is due to genetic variation.

a. Backcrosses
b. Regression analyses
c. Heritability
d. QTLs

6. Mendel's work on pea plants described which of the following?

a. QTLs
b. inbreeding
c. heritability
d. discontinuous traits

7. The selection of traits that provide economic importance is an example of _____.

a. natural selection
b. inbreeding
c. artificial selection
d. heritability

8. A hybrid that has a more vigorous phenotype than the parental strains is an example of _____.

a. artificial selection
b. natural selection
c. heritability
d. quantitative traits
e. heterosis

9. A _____ trait varies measurably in a species.

a. discontinuous
b. quantitative
c. additive
d. subtractive

10. A trait that is the same in 99% of the members of a species is said to be_____

a. quantitative
b. discontinuous
c. polymorphic
d. monomorphic

Answer Key for Study Guide Questions

This answer key provides the answers to the exercises and chapter quiz for this chapter. Answers in parentheses ( ) represent possible alternate answers to a problem, while answers marked with an asterisk (*) indicate that the response to the question may vary.

24.1
1. b
2. d
3. e
4. c
5. a
6. d
7. f
8. e
9. a
10. c
11. b

24.2	1. DDT 2. QTL mapping	3. backcross 4. polygenic inheritance	5. polygenic inheritance
24.3	1. c 2. b	3. a 4. d	5. f 6. e

Quiz

1. c 2. d 3. a 4. e	5. c 6. d 7. c 8. e	9. b 10. d

Answers to Conceptual and Experimental Questions

Conceptual Questions

C1. Quantitative traits are described numerically. Examples include height, weight, speed, and metabolic rate. In a population, a trait may be given a mean value, and the degree of variation may be described by the variance and standard deviation.

C2. At the molecular level, quantitative traits often exhibit a continuum of phenotypic variation because they are usually influenced by multiple genes that exist as multiple alleles. A large amount of environmental variation will also increase the phenotypic overlaps among different genotypic categories.

C3. A normal distribution varies in a symmetrical way around a mean value. When graphed, it exhibits a bell-shaped curve. It is common for quantitative traits in a population to have a normal distribution around a mean value. If a distribution is normal, the standard deviation can predict the percentage of individuals that will fall within certain limits. For example, approximately 68% of all individuals in a population will be within one (plus or minus) standard deviation unit from the mean.

C4. A discontinuous trait is one that falls into discrete categories. Examples include brown eyes versus blue eyes in humans or purple versus white flowers in pea plants. A continuous trait is one that does not fall into discrete categories. Examples include height in humans and fruit weight in tomatoes. Most quantitative traits are continuous; the trait falls within a range of values. The reason why quantitative traits are continuous is because they are usually polygenic and greatly influenced by the environment. As shown in Figure 24.4*b*, this tends to create ambiguities between genotypes and a continuum of phenotypes.

C5. With regard to genetics, a frequency distribution is a graph that compares the number of individuals and their phenotypes for a given trait. This is a common way to depict the distribution of a quantitative trait in a population. To make a frequency distribution for a continuous quantitative trait, it is necessary to divide individuals into arbitrary groups. For example, if the trait is weight, individuals could be divided into groups that are in multiples of 10 lb. Such a grouping would begin with individuals in the 0–10 lb range, 10–20 lb, 20–30 lb, etc. To make a frequency distribution, a population of individuals would be chosen, and then the number of individuals in each grouping would be determined. The groupings (weight categories) would be plotted on the x-axis, and the number of individuals in each group would be plotted on the y-axis.

C6. To be in the top 2.5% is about two standard deviation units. If we take the square root of the variance, the standard deviation would be 20.6 lb. To be in the top 2.5% an animal would have to weigh 41.2 lb heavier than the mean, which equals 465.2 lb. To be in the bottom 0.13%, an animal would have to be three standard deviations lighter, which would be 61.8 lb lighter than the mean or 362.2 lb.

C7. A. The differences in variance could be explained by variation in the way the potatoes are affected by environmental variation, or they could be explained by genetic differences. The strain with a higher variance may have more genetic variation.

B. On the one hand, the strain with a higher variance may be better because the farmer may want to select for individuals who produce larger potatoes. On the other hand, if the farmer wants the size of his/her potatoes to be uniform, the population with a smaller variance would be better.

C. You would use the strain with a higher variance, hoping that a significant proportion of the variance was due to genetic variation.

C8. There is a positive correlation, but it could have occurred as a matter of chance alone. You would need to conduct more experimentation to determine if there is a significant correlation such as examining a greater number of pairs of individuals. If $N = 500$, the correlation would be statistically significant and you would conclude that the correlation did not occur as a matter of random chance. However, you could not conclude cause and effect.

C9. A negative correlation means that two variables are related to each other in opposite ways. For example, a high amount of snowfall in the winter might be negatively correlated with the survival of wild species.

C10. When a correlation coefficient is statistically significant, it means that the association is likely to have occurred for reasons other than random sampling error. It may indicate cause and effect but not necessarily. For example, large parents may have large offspring due to genetics (cause and effect). However, the correlation may be related to the sharing of similar environments rather than cause and effect.

C11. Polygenic inheritance occurs when two or more different genes influence the outcome of a single trait. Such traits are difficult to study because there may be multiple genes with multiple alleles and environmental effects may be significant. Therefore, the discrete genotypes do not fall within discrete phenotypic categories. Instead, there tends to be overlap and this overlap confounds a genetic analysis. Molecular approaches, such as QTL mapping, can greatly help the situation.

C12. Quantitative trait loci are sites within chromosomes that contain genes that affect a quantitative trait. It is possible for a QTL to contain one gene, or it may contain two or more closely linked genes. QTL mapping, which involves linkage to known molecular markers, is commonly used to determine the locations of QTLs.

C13. A. The probability of inheriting two light alleles from a parent would be $0.5 \times 0.5 = 0.25$. The probability of inheriting two light alleles from both parents would be $0.25 \times 0.25 = 0.0625$, or 6.25%.

B. The probability of inheriting three light alleles from a parent would be $0.5 \times 0.5 \times 0.5 = 0.125$. The probability of inheriting three light alleles from both parents would be $0.125 \times 0.125 = 0.0156$, or 1.56%.

C. The probability of inherited four light alleles from a parent would be $0.5 \times 0.5 \times 0.5 \times 0.5 = 0.0625$. The probability of inheriting four light alleles from both parents would be $0.0625 \times 0.0625 = 0.0039$, or 0.39%.

C14. If the broad sense heritability equals 1.0, it means that all of the variation in the population is due to genetic variation rather than environmental variation. It does not mean that the environment is unimportant in the outcome of the trait. Under another set of environmental conditions the trait may have turned out quite differently.

C15. The dominance hypothesis proposes that the inbred strains are homozygous for harmful recessive alleles whereas the hybrid is not. The overdominance hypothesis suggests that certain combinations of alleles are particularly beneficial in the heterozygote. Overall, it is difficult to judge which hypothesis is correct, although geneticists rarely observe overdominance in single genes. An exception is the sickle-cell allele, which we have discussed throughout this textbook. On the other hand, the formation of recessive alleles that are harmful is a widespread phenomenon. These observations tend to support the dominance hypothesis, but the situation is by no means resolved.

C16. Hybrid vigor is the phenomenon in which an offspring produced from two inbred strains is more vigorous than the corresponding parents. Tomatoes and corn are often the products of hybrids.

C17. The primary advantage of selective breeding in agriculture is that it improves the magnitude of quantitative traits. For example, selective breeding can increase fruit size, fruit number, etc. A disadvantage is that selective breeding will diminish the overall genetic diversity in the population, and this may result accidentally in the prevalence of traits that are not desirable. For example, as a tomato breeder selects for bigger and juicier tomatoes, he/she may unwittingly obtain a variety that has bigger and juicier tomatoes, but the plants may be more susceptible to pathogens, compared to the original (nonselected) parental strain.

Undesirable traits, like pathogen sensitivity, are often inherited in a recessive manner. These are due to rare recessive alleles that happen to become monomorphic, as a random result of selective breeding. The creation of hybrids may overcome the adverse effects of such rare recessive alleles. Two different varieties that have been independently subjected to selective breeding are not likely to be monomorphic for the same recessive alleles. Therefore, when two varieties are crossed, the hybrids become heterozygous and carry one copy of the normal allele at each gene and one copy of a detrimental recessive allele. If the allele is recessive, it will not have an adverse effect on the phenotype of the organism.

C18. When a species is subjected to selective breeding, the breeder is focusing his/her attention on improving one particular trait. In this case, the rose breeder is focused on the size and quality of the flowers. Since the breeder usually selects a small number of individuals (e.g., the ones with best flowers) as the breeding stock for the next generation, this may lead to a decrease in the allelic diversity at other genes. For example, several genes affect flower fragrance. In an unselected population, these genes may exist as "fragrant alleles" and "nonfragrant alleles." After many generations of breeding for large flowers, the fragrant alleles may be lost from the population, just as a matter of random chance. This is a common problem of selective breeding. As you select for an improvement in one trait, you may inadvertently diminish the quality of an unselected trait.

Other people have suggested that the lack of fragrance may be related to flower structure and function. Perhaps the amount of energy that a flower uses to make beautiful petals somehow diminishes its capacity to make fragrance.

C19. When you examine a trait in a population, it often varies among different members. Sometimes the variation is large and sometimes it is small. The variation can be caused by genetic differences among the members of the population, and it can be caused by the fact that each member of the population is exposed to a slightly or greatly different environment. The proportion of the variation that is caused by genetic differences among the members of the population is the heritability of the trait. It is applicable only to a particular population raised in a particular environment because each population has its own unique amount of genetic variation (based on the genetic heritage of its members), and the members of each population are exposed to their own unique environmental conditions.

C20. Broad sense heritability takes into account all genetic factors that affect the phenotypic variation in a trait. Narrow sense heritability considers only alleles that behave in an additive fashion. In many cases, the alleles affecting quantitative traits appear to behave additively. More importantly, if a breeder assumes that the heritability of a trait is due to the additive effects of alleles, it is possible to predict the outcome of selective breeding. This is also termed the *realized heritability*.

C21. A. False. The environment has very little impact on the amount of variation in the trait. However, environment always contributes greatly to the outcome of every trait. You could not have a living organism without an environment.

B. True.

C. Probably true, but we cannot say for sure. Quantitative traits are usually polygenic.

D. False. You cannot say anything about the heritability of egg weight in Montana chickens. It depends on the amount of genetic variation in the population and on the type of environment in which they are raised.

C22. A. Because of their good nutrition, you may speculate that they would grow to be taller.

B. If the environment is rather homogeneous, then heritability values tend to be higher because the environment contributes less to the amount of variation in the trait. Therefore, in the commune, the heritability might be higher, because they uniformly practice good nutrition. On the other hand, since the commune is a smaller population, the amount of genetic variation might be less, so this would make the heritability lower. However, since the problem states that the commune population is large, we would probably assume that the amount of genetic variation is similar to the general population. Overall, the best guess would be that the heritability in the commune population is higher because of the uniform nutrition standards.

C. As stated in part B, the amount of variation would probably be similar, since the commune population is large. As a general answer, larger populations tend to have more genetic variation. Therefore, the general population probably has a bit more variation, but it may not be much more than the commune population.

C23. Eventually, the alleles that are being selected will become monomorphic in the population, and further selection will have no effect.

C24. A natural population of animals is more likely to have a higher genetic diversity compared to a domesticated population. This is because domesticated populations have been subjected to many generations of selective breeding, which decreases the genetic diversity. Therefore, V_G is likely to be higher for the natural population. The other issue is environment. It is difficult to say which group would have a more homogeneous environment. In general, natural populations tend to have a more heterogeneous environment, but not always. If the environment is more heterogeneous, this tends to cause more phenotypic variation, which makes V_E higher.

$$\text{Heritability} = V_G/V_T$$
$$= V_G/(V_G + V_E)$$

When V_G is high, this increases heritability. When V_E is high, this decreases heritability. In the natural wolf population, we would expect that V_G would be high. In addition, we would guess that V_E might be high as well (but that is less certain). Nevertheless, if this were the case, the heritability of the wolf population might be similar to the domestic population. This is because the high V_G in the wolf population is balanced by its high V_E. On the other hand, if V_E is not that high in the wolf population, or if it is fairly high in the domestic population, then the wolf population would have a higher heritability for this trait.

C25. A. Dominance hypothesis

B. Both hypotheses

C. Overdominance hypothesis

Experimental Questions

E1. A. We first need to calculate the standard deviations for height and weight, and then the covariance for both traits:

Height: variance = 140.02; standard deviation = 11.8

Weight: variance = 121.43; standard deviation = 11.02

Covariance = 123.3

$$r_{(X,Y)} = \frac{CoV_{(X,Y)}}{SD_X SD_Y}$$

$$r_{(X,Y)} = (123.3)/(11.8)(11.02)$$

$$r_{(X,Y)} = 0.948$$

B. With 8 degrees of freedom, this value is statistically significant. This means the association between these two variables occurs more frequently than would be expected by random sampling error. It does not necessarily imply cause and effect.

E2. To calculate the mean, we add the values together and divide by the total number.

$$\text{Mean} = \frac{1.9 + 2(2.4) + 2.(2.1) + 3(2.0) + 2(2.2) + 1.7 + 1.8 + 2(2.3) + 1.6}{15}$$

$$\text{Mean} = 2.1$$

The variance is the sum of the squared deviations from the mean divided by $N - 1$. The mean value of 2.1 must be subtracted from each value, and then the square is taken. These 15 values are added together and then divided by 14 (which is $N - 1$).

$$\text{Variance} = \frac{0.85}{14} = 0.061$$

The standard deviation is the square root of the variance.

Standard deviation = 0.25

E3.

$$n = \frac{D^2}{8V_G}$$

$$n = \frac{(514 - 621)^2}{8(382)}$$

$$n = 3.7$$

So there would be a minimum of about 4 genes that cause weight to vary among these two varieties of cattle.

E4. The results are consistent with the idea that there are QTLs for this trait on chromosomes 2 and 3 but not on the X chromosome.

E5. The heritability reflects the amount of genetic variation that influences a trait. In the strain with seven QTLs, there are at least seven different genes that exist in two or more alleles that are influencing the outcome of the trait. In the other strain, there are the same types of genes, but three of them are monomorphic and therefore do not contribute to the variation in the outcome of the trait.

E6. When we say that an RFLP is associated with a trait, we mean that a gene that influences a trait is closely linked to an RFLP. At the chromosomal level, the gene of interest is so closely linked to the RFLP that a crossover almost never occurs between them.

Note: Each plant inherits four RFLPs, but it may be homozygous for one or two of them.

Small: 2,700 and 4,000 (homozygous for both)
Small-medium: 2,700 (homozygous), 3,000, and 4,000; or 2,000, 2,700, and 4,000 (homozygous)
Medium: 2,000 and 4,000 (homozygous for both); or 2,700 and 3,000 (homozygous for both); or 2,000, 2,700, 3,000, and 4,000
Medium-large: 2,000 (homozygous), 3,000, and 4,000; or 2,000, 2,700, and 3,000 (homozygous)
Large: 2,000 and 3,000 (homozygous for both)

E7. These results suggest that there are (at least) three different genes that influence the size of pigs. This is a minimum estimate because a QTL may have two or more closely linked genes. Also, it is possible that the large and small strains have the same RFLP band that is associated with one or more of the genes that affect size.

E8. Let's assume that there is an extensive molecular marker map for the rice genome. We would begin with two strains of rice, one with a high yield and one with a low yield, that greatly differ with regard to the molecular markers that they carry. We would make a cross between these two strains to get F_1 hybrids. We would then backcross the F_1 hybrids to either of the parental strains and then examine hundreds of offspring with regard to their rice yields and molecular markers. In this case, our expected results would be that six different markers in the high-producing strain would be correlated with offspring that produce higher yields. We might get fewer than six bands if some of these genes are closely linked and associate with the same marker. We also might get fewer than six if the two parental strains have the same marker that is associated with one or more of the genes that affect yield.

E9. This answer assumes that a mouse must be homozygous for all the sensitive alleles in order to be susceptible to the leukemia virus. Let's call the viral-resistance genes *V1, V2, V3,* etc. *V1* is a resistant allele and *v1* is an allele of the same gene that confers sensitivity. *V2* is an allele of a different gene compared to *V1,* and *V2* confers resistance while *v2* confers sensitivity. With these ideas in mind, we can calculate the probability that an F_2 mouse would be homozygous for all of the sensitivity alleles.

If there are two genes involved, the parental cross is

$$V1V1V2V2 \times v1v1v2v2$$

This cross produces F_1 offspring that are *V1v1V2v2.*

If you construct a Punnett square, there is a 1/16 chance that an F_2 offspring will be *v1v1v2v2.* Or, we can use the product rule. The chances of an F_1 parent passing a recessive allele at each gene to their offspring is 1/2. The chances of an offspring inheriting all four recessive alleles is: $1/2 \times 1/2 \times 1/2 \times 1/2 = 1/16$

Expected number of viral-sensitive F_2 offspring $= 1/16 \times 120 = 7.5$

We can follow this same approach to determine the likelihood that offspring will be homozygous for all the recessive alleles if three or more genes are involved.

For three genes: $1/2 \times 1/2 \times 1/2 \times 1/2 \times 1/2 \times 1/2 = 1/64$

Expected number of sensitive F_2 offspring $= 1/64 \times 120 = 1.9$

For four genes: $1/2 \times 1/2 \times 1/2 \times 1/2 \times 1/2 \times 1/2 \times 1/2 \times 1/2 = 1/256$

Expected number of sensitive F_2 offspring $= 1/256 \times 120 = 0.5$

For 120 offspring, we expect about two mice that would be sensitive to the virus if three genes are involved. Therefore, three genes would appear to be the most likely number of genes. If we had not obtained any sensitive mice, we would probably conclude that four or more genes are involved. It would be unlikely to have no sensitive mice if three genes or less were involved.

E10. A. If we assume that the highly inbred strain has no genetic variance:

$$V_G \text{ (for the wild strain)} = 3.2\text{ g}^2 - 2.2\text{ g}^2 = 1.0\text{ g}^2$$

B. $H_B{}^2 = 1.0\text{ g}^2/3.2\text{ g}^2 = 0.31$

C. It is the same as $H_B{}^2$ so it also equals 0.31.

E11.

$$h_N^2 = \frac{R}{S}$$
$$R = \overline{X}_O - \overline{X}$$
$$S = \overline{X}_P - \overline{X}$$

In this problem:

$\overline{X}$ equals 1.01 (as given in the problem)

$\overline{X}_O$ equals 1.11 (by calculating the mean for the parents)

$\overline{X}_P$ equals 1.09 (by calculating the mean for the offspring)

$R = 1.09 - 1.01 = 0.08$

$S = 1.11 - 1.01 = 0.10$

$h_N{}^2 = 0.08/0.10 = 0.8$ (which is a pretty high heritability value)

E12. A

$$h_N{}^2 = \frac{\overline{X}_o - \overline{X}}{\overline{X}_P - \overline{X}}$$
$$0.21 = (26.5\text{ g} - 25\text{ g})/(27\text{ g} - 25\text{ g})$$
$$\overline{X}_O - 25\text{ g} = 2\text{ g }(0.21)$$
$$\overline{X}_O = 25.42g$$

B.

$$0.21 = (26.5\text{ g} - 25\text{ g})/(\overline{X}_p - 25\text{ g})$$
$$(\overline{X}_p - 25\text{ g})(0.21) = 1.5\text{ g}$$
$$\overline{X}_p = 32.14\text{ g parents}$$

However, since this value is so far from the mean, there may not be 32.14 g parents in the population of mice that you have available.

E13. In this problem, you need to set up Punnett squares based on genotypes.

A. Let's call the alleles *A1* and *A2*. The F_1 would be genotypically *A1A2.* The F_2 results would be a 1 *A1A1* : 2 *A1A2* : 1 *A1A1.* Since *A1A1* and *A2A2* plants produce 1 lb fruit and *A1A2* produce 2 lb fruit, this would yield a phenotypic ratio of 50% 2 lb : 50% 1 lbs.

B. Let's call the alleles *A* and *a* and *B* and *b* and assume they contribute additively. The F_1 offspring will be *AaBb.* The F_2 ratio will be

1 *AABB* (2 lb) : 2 *AABb* (2 lb) : 1 *AAbb* (1.5 lb) :
2 *AaBB* (2 lb) : 2 *Aabb* (1.5 lb) : 4 *AaBb* (2 lb) :
1 *aaBB* (1.5 lb) : 2 *aaBb* (1.5 lb) : 1 *aabb* (1 lb)

The phenotypic ratio would be 9 (2 lb) : 6 (1.5 lb) : 1 (1 lb).

C. Let's call the alleles *A1* and *A2* and *B1* and *B2* and assume they contribute additively. The F_1 offspring will be *A1A2B1B2.* The F_2 ratio will be

1 *A1A1B1B1* (1 lb) : 2 *A1A1B1B2* (1.5 lb) : 1 *A1A1B2B2* (1 lb) : 2 *A1A2B1B1* (1.5 lb) : 2 *A1A2B2B2* (1.5 lb) : 4 *A1A2B1B2* (2 lb) : 1 *A2A2B1B1* (1 lb) : 2 *A2A2B1B2* (1.5 lb) : 1 *A2A2B2B2* (1 lb)

The phenotypic ratio would be 4 (1 lb) : 8 (1.5 lb) : 4 (2 lb), which can be reduced to a 1:2:1 ratio.

D. You cannot calculate a precise ratio. As the number of genes increases, it becomes more unlikely to be heterozygous at all of the loci so it becomes less likely to produce 2 lb fruit. With a very large number of genes, most of the F_2 offspring would be in the intermediate (1.5 lb) range.

E14. We first need to calculate *a* and *b*. In this calculation, *X* represents the height of fathers and *Y* represents the height of sons.

$$b = \frac{144}{112} = 1.29$$

$$a = 69 - (1.29)(68) = -18.7$$

For a father who is 70 in tall,

$$Y = (1.29)(70) + (-18.7) = 71.6$$

The most likely height of the son would be 71.6 in.

E15. If we assume that this correlation coefficient is statistically significant, and if we assume that all the genetic variance is due to the additive effects of alleles, then

$$h_N{}^2 = r_{obs}/r_{exp}$$

In this problem, the r_{obs} is 0.15 and the expected correlation for siblings is 0.5. Therefore

$$h_N{}^2 = 0.15/0.5 = 0.3$$

E16. The identical and fraternal twins, which probably share very similar environments, but who differ in the amount of genetic material they share, are a strong argument against an environmental bias. The differences in the observed correlations (0.49 vs 0.99) are consistent with the differences in the expected correlations (0.5 vs 1.0).

E17. A. After six or seven generations, the selective breeding seems to have reached a plateau. This suggests that the tomato plants have become monomorphic for the alleles that affect tomato weight.

B. There does seem to be heterosis since the first generation has a weight of 1.7 lb, which is heavier than Mary's and Hector's tomatoes. This partially explains why Martin has obtained tomatoes that are heavier than 1.5 lb. However, heterosis is not the whole story; it does not explain why Martin obtained tomatoes that weigh 2 lb. Even though Mary's and Hector's tomatoes were selected for heavier weight, they may not have all of the "heavy alleles" for each gene that controls weight. For example, let's suppose that there are 20 genes that affect weight, with each gene existing in a light and heavy allele. During the early stages of selective breeding, when Mary and Hector picked their 10 plants as seed producers for the next generation, as a matter of random chance, some of these plants may have been homozygous for the light alleles at a few of the 20 genes that control weight. Therefore, just as a matter of chance, they probably "lost" a few of the heavy alleles that affect weight. So, after 12 generations of breeding, they have predominantly heavy alleles but also have light alleles for some of the genes. If we represent heavy alleles with a capital letter, and light alleles with a lowercase letter, Mary's and Hector's strains could be the following:

Mary's strain: *AA BB cc DD EE FF gg hh II JJ KK LL mm NN OO PP QQ RR ss TT*
Hector's strain: *AA bb CC DD EE ff GG HH II jj kk LL MM NN oo PP QQ RR SS TT*

As we see here, Mary's strain is homozygous for the heavy allele at 15 of the genes but carries the light allele at the other 5. Similarly, Hector's strain is homozygous for the heavy allele at 15 genes and carries the light allele at the other 5. It is important to note, however, that the light alleles in Mary's and Hector's strains are not in the same genes. Therefore, when Martin crosses them together, he will initially get

Martin's F_1 offspring: *AA Bb Cc DD EE Ff Gg Hh II Jj Kk LL Mm NN Oo PP QQ RR Ss TT*

If the alleles are additive and contribute equally to the trait, we would expect about the same weight (1.5 lb), because this hybrid has a total of 10 light alleles. However, if heterosis is occurring, genes (which were homozygous recessive in Mary's and Hector's strains) will become heterozygous in the F_1 offspring, and this may make the plants healthier and contribute to a higher weight. If Martin's F_1 strain is subjected to selective breeding, the 10 genes that are heterozygous in the F_1 offspring may eventually become homozygous for the heavy allele. This would explain why Martin's tomatoes achieved a weight of 2.0 pounds after five generations of selective breeding.

E18. $h_N^2 = r_{obs}/r_{exp}$

The value for r_{exp} comes from the known genetic relationships.

Mother/daughter	$r_{obs} = 0.36$	$r_{exp} = 0.5$	$h_N^2 = 0.72$
Mother/granddaughter	$r_{obs} = 0.17$	$r_{exp} = 0.25$	$h_N^2 = 0.68$
Sister/sister	$r_{obs} = 0.39$	$r_{exp} = 0.5$	$h_N^2 = 0.78$
Twin sisters (fraternal)	$r_{obs} = 0.40$	$r_{exp} = 0.5$	$h_N^2 = 0.80$
Twin sisters (identical)	$r_{obs} = 0.77$	$r_{exp} = 1.0$	$h_N^2 = 0.77$
			0.75

The average heritability is 0.75.

E19. A.

$$h_N^{\,2} = \frac{\overline{X}_O - \overline{X}}{\overline{X}_P - \overline{X}}$$

$$h_N^{\,2} = \frac{269 - 254}{281 - 254} = 0.56$$

B.

$$0.56 = \frac{275 - 254}{\overline{X}_P - 254}$$

$$\overline{X}_P = 291.5 \text{ lb}$$

E20. These data suggest that there might be a genetic component to blood pressure, because the relatives of people with high blood pressure also seem to have high blood pressure themselves. Of course, more extensive studies would need to be conducted to determine the role of environment. To calculate heritability, the first thing to do is to calculate the correlation coefficient between relatives to see if it is statistically significant. If it is, then you could follow the approach described in the experiment of figure 24.9. You would determine the correlation coefficients between genetically related individuals as a way to determine the heritability for the trait. In this approach, heritability equals r_{obs}/r_{exp}. It would be important to include genetically related pairs that were raised apart (e.g., uncles and nieces) to see if they had a similar heritability value compared to genetically related pairs raised in the same environment (e.g., brothers and sisters). If their values were similar, this would give you some confidence that the heritability value is due to genetics and not due to fact that relatives often share similar environments.

Questions for Student Discussion/Collaboration

1. Heritability is an important phenomenon in agriculture because a population with high heritability can be subjected to selective breeding as a way to improve quantitative traits. People misunderstand heritability because they think that it is the proportion of a trait that is governed by genetics. It is the proportion of the *variation in a particular population* that is governed by genetics. We can never tell the relative impact of genetics and environment on the final outcome of a trait.

2. Most traits depend on the influence of many genes. Also, genetic variation is a common phenomenon in most populations. Therefore, most individuals have a variety of alleles that contribute to a given trait. For quantitative traits, some alleles may make the trait bigger and other alleles may make the trait turn out smaller. If a population contains many different genes and alleles that govern a quantitative trait, most individuals will have an intermediate phenotype because they will have inherited some large and some small alleles. Fewer individuals will inherit a predominance of large alleles or a predominance of small alleles. An example of a quantitative trait that does not fit a normal distribution is snail pigmentation. The dark snail and light snails are favored rather than the intermediate colors because they are less susceptible to predation.

3. Heterosis (or hybrid vigor) refers to the observation that hybrid offspring are more vigorous than the corresponding inbred parental strains. Theoretically, it could be caused by effects of single genes or many genes. This would depend on the strains involved. In highly inbred strains, it is probably related to many genes that have become homozygous recessive and mildly harmful (the dominance hypothesis) or several alleles that are mildly beneficial in the heterozygous condition (the overdominance hypothesis).

Chapter 25: Population Genetics

Student Learning Objectives

Upon completion of this chapter you should be able to:

1. Understand the concept of a population and polymorphism in populations.
2. Use the Hardy-Weinberg equation to examine the frequency of alleles in a population.
3. Understand the factors that change allele frequencies in a population and how they influence Hardy-Weinberg equilibrium.

25.1 Genes in Populations

Overview

The majority of the text has examined the relationship between genes and the individual. This chapter explores the bigger picture of populations genetics. In the first section of this chapter, we explore the concept of genetic variation in a population and the mathematical calculations of allele and genotype frequencies.

Key Terms

Allele frequencies	Genotype frequencies	Polymorphism
Allozymes	Local populations	Population
Demes	Monomorphic	Population genetics
Gene pool	Polymorphic	Subpopulations

Focal Points

- Calculations of allele and genotype frequencies (pg. 698)

Exercises and Problems

For each of the following, match the term to its correct definition.

_____ 1. Population

_____ 2. Gene pool

_____ 3. Allozymes

_____ 4. Allele frequencies

_____ 5. Genotype frequencies

a. The sum of all of the genes within a population.
b. A group of members of the same species that can interbreed with one another.
c. The percent that an allele is represented in a population.
d. The percent of individuals that are homozygous recessive, homozygous dominant or heterozygous in a population.
e. Enzymes that vary slightly in their amino acid sequences.

25.2 The Hardy-Weinberg Equilibrium

Overview

In order to study population genetics we need to first develop a mathematical model that examines the allele frequencies of a population that is stable from one generation to the next. Once that is established it is possible to assess the influences of factors that may alter the allele frequency of a population. This is the basis of the Hardy-Weinberg equilibrium.

While the Hardy-Weinberg formula (pg. 699) appears to mathematical, in reality it is a theoretical model of a stable population. As you will observe in the next sections, stable populations are rare in the natural world since there are many factors that influence the frequencies of alleles in a population. However, an understanding of the equation is needed in order to assess the influence of these factors.

The last portion of this chapter examines the influence of inbreeding on the allele frequency in a population (pgs. 700-701). You should familiarize yourself with these calculations before proceeding.

Key Terms

Assortative mating
Coefficient of inbreeding
Disassortative mating
Disequilibrium
Equilibrium
Fixation coefficient
Hardy-Weinberg equation
Inbred
Inbreeding
Outbreeding

Focal Points

- Conditions of the Hardy-Weinberg equation (pg. 699)
- Equations associated with the Hardy-Weinberg equilibrium and the coefficient of inbreeding (pgs. 699-701)

Exercises and Problems

Each of the following pertain to the Hardy-Weinberg equilibrium. For each, provide the correct definition of the term.

_____ 1. p^2

_____ 2. q^2

_____ 3. 2pq

a. The genotype frequency of homozygous dominant individuals.
b. The genotype frequency of heterozygous individuals.
c. The genotype frequency of homozygous recessive individuals.

For each of the following, match the term to its correct definition.

_____ 4. Disassortative mating

_____ 5. Diequilibrium

_____ 6. Coefficient of inbreeding

_____ 7. Inbreeding

_____ 8. Outbreeding

_____ 9. Assortative mating

a. the mating of two genetically related individuals
b. has the ability to create hybrids that are heterozygous for many genes
c. individuals who mate due to similar phenotypes
d. individuals who mate based on dissimilar phenotypes
e. allele and genotype frequencies are not in Hardy-Weinberg equilibrium
f. the quantification of the degree of inbreeding

25.3 Factors That Change Allele Frequencies in Populations

Overview

The previous section introduced the fact that a stable population possesses certain characteristics under the Hardy-Weinberg equilibrium. Yet in the natural world, few populations are in equilibrium. The factors that contribute to disequilibrium is the topic of this section of the chapter. As you proceed through this section you should familiarize yourself with the influence of each of these factors on allele frequency.

Key Terms

Adaptive forces
Balanced polymorphism
Bottleneck effect
Conglomerate
Darwinian fitness
Founder effect
Gene flow
Genetic load
Heterozygote advantage
Mean fitness of the population
Mutation rate
Natural selection
Neutral processes
Overdominance
Random genetic drift
Selection coefficient

Focal Points

- The effects of genetic drift (Figure 25.7)
- Patterns of selection (Figure 25.9)

Exercises and Problems

For questions 1 to 10, choose the most appropriate term for the definition.

_____ 1. A population is not evolving towards fixation of an allele.

_____ 2. The degree to which a genotype is selected against.

_____ 3. A fitness calculation in which the terms do not add up to 1.0.

_____ 4. The drastic reduction in size of a population.

_____ 5. Random changes in allele frequencies due to chance events.

_____ 6. Genetic variation that decreases the average fitness of a population.

_____ 7. Occurs due to the movement of individuals between populations.

_____ 8. The probability that a gene will be changed by mutation.

_____ 9. The relative likelihood that a phenotype will survive and contribute to the next generation's gene pool.

_____ 10. Occurs when a small group of individuals establish a new population.

a. mutation rate
b. founder effect
c. genetic drift
d. bottleneck effect
e. gene flow
f. mean fitness of the population
g. balanced polymorphism
h. Darwinian fitness
i. genetic load
j. selection coefficient

For each of the following, choose the model of selection from the diagram below:

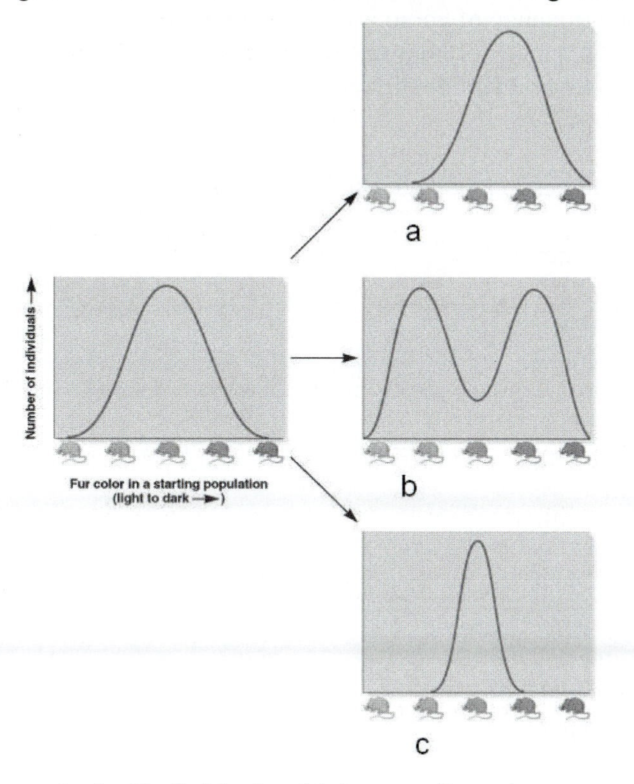

_____ 11. Favors the survival of individuals with intermediate phenotypes.

_____ 12. Industrial melanism is an example of this.

_____ 13. Separates a population into distinct phenotypic classes.

_____ 14. Favors one extreme of the phenotype.

_____ 15. Occurs in species that occupy diverse environments.

Chapter Quiz

1. Which of the following forms of selection would best be described as favoring an intermediate phenotype?

   a. directional selection
   b. disruptive selection
   c. stabilizing selection
   d. none of the above

2. Which of the following is an example of genetic drift in which catastrophic events influence the allele frequencies?

   a. bottleneck effect
   b. inbreeding
   c. founder effect
   d. migration

3. In the Hardy-Weinberg equation, what does the term 2pq represent?
   a. the genotype frequency of homozygous dominant individuals
   b. the genotype frequency of homozygous recessive individuals
   c. the genotype frequency of heterozygous individuals
   d. none of the above

4. The sum of all of the alleles in a population is called the_____.
   a. deme
   b. genotype frequency
   c. gene pool
   d. fitness coefficient

5. Individuals who prefer dissimilar phenotypes for mating are said to be exhibiting _____.
   a. inbreeding
   b. assortative breeding
   c. diequilibrium
   d. disassortative mating

6. Which of the following is the measure of inbreeding?
   a. mutation frequency
   b. Darwinian fitness
   c. fixation coefficient
   d. selection coefficient

7. A population in which two alleles are not evolving towards fixation is said to be an example of _________.
   a. diequilibrium
   b. balanced polymorphism
   c. monomorphism
   d. fixation

8. Industrial melanism is an example of ________.
   a. Hardy-Weinberg equilibrium
   b. balanced polymorphism
   c. stabilizing selection
   d. directional selection
   e. neutralizing selection

9. The degree to which a geneotype is being selected against is called the ________.
   a. selection coefficient
   b. mean fitness of the population
   c. Darwinian fitness
   d. fixation coefficient

10. The variation that decreases the average fitness of a population is called the ______.
   a. genetic load
   b. mean fitness of the population
   c. Darwinian fitness
   d. fixation coefficient

Answer Key for Study Guide Questions

This answer key provides the answers to the exercises and chapter quiz for this chapter. Answers in parentheses ( ) represent possible alternate answers to a problem, while answers marked with an asterisk (*) indicate that the response to the question may vary.

25.1	1. b 2. a	3. e 4. c	5. d
25.2	1. a 2. c 3. b	4. d 5. e 6. f	7. a 8. b 9. c
25.3	1. g 2. j 3. f 4. d 5. c	6. i 7. e 8. a 9. h 10. b	11. c 12. a 13. b 14. a 15. b

Quiz

1. c 2. a 3. c 4. c	5. d 6. c 7. b 8. d	9. a 10. a

Answers to Conceptual and Experimental Questions

Conceptual Questions

C1. A gene pool is all of the genes present in a particular population. Each type of gene within a gene pool may exist in one or more alleles. The prevalence of an allele within the gene pool is described by its allele frequency. If a gene is monomorphic, the allele frequency is close to 100%. If it is polymorphic, each allele has a frequency that is between 1 and 99%. The sum of all the allele frequencies for a particular gene will add up to 100%.

C2. A population is a group of interbreeding individuals. Let's consider a squirrel population in a forested area. Over the course of many generations, several things could happen to this population that may change its gene pool. A forest fire, for example, could dramatically decrease the number of individuals and thereby cause a bottleneck. This would decrease the genetic diversity of the population. A new predator may enter the region and natural selection may select for the survival of squirrels that are best able to evade the predator. Another possibility is that a group of squirrels within the population may migrate to a new region and found a new squirrel population.

C3. The term *genetic polymorphism* refers to the phenomenon in which a gene is found in two or more alleles within a population. Mutation is the ultimate source of genetic variation.

C4. A. Phenotype frequency and genotype frequency

B. Genotype frequency

C. Allele frequency

C5. When a trait is polymorphic, this means that different individuals show phenotypic variation with regard to the trait. For example, petunias can have red or white flowers. Flower color is polymorphic in petunias. When a gene exists in two or more alleles, it is polymorphic. At the molecular level, alleles of a given gene have different DNA sequences. These differences could be very slight (e.g., a single-base change) or they could involve significant additions or deletions. Different alleles may cause differences in phenotype. For example, in Mendel's pea plants, the *T* allele resulted in tall plants while the *t* allele caused dwarf plants. However, alleles do not always cause differences in phenotype. A single-base substitution in a gene may not affect the amino acid sequence of the encoded polypeptide (e.g., the base change may affect the wobble base) or a single-base change might alter the amino acid sequence, but not in a way that alters the protein's function. Therefore, gene polymorphism does not always result in phenotypic polymorphism.

C6. A. The genotype frequency for the *CF* homozygote is 1/2,500, or 0.004. This would equal q^2. The allele frequency is the square root of this value, which equals 0.02. The frequency of the corresponding normal allele equals 1 – 0.02 = 0.98.

B. *CF* homozygote is 0.004. Normal homozygote is $(098)^2 = 0.96$. Heterozygote is 2(0.98)(0.02), which equals 0.039.

C. If a person is known to be a heterozygous carrier, the chances that this particular person will happen to choose another as a mate is equal to the frequency of heterozygous carriers in the population, which equals 0.039, or 3.9%. The chances that two randomly chosen individuals will choose each other as mates equals 0.039 × 0.039 = 0.0015, or 0.15%.

C7. In the absence of other forces, inbreeding does not alter allele frequencies since it does not favor the transmission of one allele over another. It simply increases the likelihood of homozygosity (which is a genotype frequency). In human populations, inbreeding increases the frequency of individuals who are homozygous for rare recessive alleles that cause human disease.

C8. For two alleles, the heterozygote is at a maximum when they are 0.5 each. For three alleles, the two heterozygotes are at a maximum when each allele is 0.33.

C9. If we apply the Hardy-Weinberg equation:

$$BB = (0.67)^2 = 0.45, \text{ or } 45\%$$

$$Bb = 2(0.67)(0.33) = 0.44, \text{ or } 44\%$$

$$bb = (0.33)^2 = 0.11, \text{ or } 11\%$$

The actual data show a higher percentage of homozygotes (compare 45% with 50% and 11% with 13%) and a lower percentage of heterozygotes (compare 44% with 37%). Therefore, these data would be consistent with inbreeding, which increases the percentage of homozygotes and decreases the percentage of heterozygotes.

C10. Since this is a recessive trait, only the homozygotes for the rolling allele will be able to roll their tongues. If *p* equals the rolling allele and *q* equals the nonrolling allele, the Hardy-Weinberg equation predicts that the frequency of homozygotes who can roll their tongues would be p^2. In this case, $p^2 = (0.6)^2 = 0.36$, or 36%.

C11. A. Yes.

B. The common ancestors are I-1 and I-2.

C.

$$F = \Sigma(1/2)^n (1 + F_A)$$

$$F = (1/2)^9 + (1/2)^9$$

$$F = 1/512 + 1/512 = 2/512 = 0.0039$$

D. Not that we can tell.

C12. A. The inbreeding coefficient is calculated using the formula

$$F = \Sigma (1/2)^n (1 + F_A)$$

In this case there is one common ancestor, I-2. Because we have no prior history on I-2, we assume she is not inbred, which makes $F_A = 0$. The inbreeding loop for IV-3 contains five people, III-4, II-2, I-2, II-5, and III-5. Therefore, $n = 5$.

$$F = (1/2)^5(1 + 0) = 1/32 = 0.031$$

B. Based on the data shown in this pedigree, individual IV-4 is not inbred.

C13. A. The inbreeding coefficient is calculated using the formula

$$F = \Sigma\,(1/2)^n\,(1 + F_A)$$

In this case there are two common ancestors, I-1 and I-2. Since we have no prior history on I-1 or I-2, we assume they are not inbred, which makes $F_A = 0$. The two inbreeding loops for IV-2 contain five people, III-4, II-2, I-1, II-6, and III-5, and III-4, II-2, I-2, II-6, and III-5. Therefore, $n = 5$ for both loops.

$$F = (1/2)^5\,(1 + 0) + (1/2)^5\,(1 + 0)$$

$$F = 0.031 + 0.031 = 0.062$$

B. Based on the data shown in this pedigree, individual III-4 is not inbred.

C14. Migration, genetic drift, and natural selection are the driving forces that alter allele frequencies within a population. Natural selection acts to eliminate harmful alleles and promote beneficial alleles. Genetic drift involves random changes in allele frequencies that may eventually lead to elimination or fixation of alleles. It is thought to be important in the establishment of neutral alleles in a population. Migration is important because it introduces new alleles into neighboring populations. According to the neutral theory, genetic drift is largely responsible for the variation seen in natural populations.

C15. In genetic drift, allele frequencies are drifting. It is an appropriate term because the word *drift* implies a random process. Nevertheless, drift can be directional. A boat may drift from one side of a lake to another. It would not drift in a straight path, but the drifting process will alter its location. Similarly, allele frequencies can drift up and down and eventually lead to the elimination or fixation of particular alleles within a population.

C16. A neutral force is one that alters allele frequencies without any regard to whether the changes are beneficial or not. Genetic drift and migration are the two main ways this can occur. An example is the founder effect, when a group of individuals migrate to a new location such as an island. Adaptive forces increase the reproductive success of a species. Natural selection is the adaptive force that tends to eliminate harmful alleles from a population and increase the frequency of beneficial alleles. An example would be the long neck of the giraffe, which enables it to feed in tall trees. At the molecular level, beneficial mutations may alter the coding sequence of a gene and change the structure and function of the protein in a way that is beneficial. For example, the sickle-cell anemia allele alters the structure of hemoglobin, and in the heterozygous condition, this inhibits the sensitivity of red blood cells to the malaria pathogen.

C17.

$$(1-u)^t = \frac{p_t}{p_o}$$

$$(1-10^{-4})^t = 0.5/0.6 = 0.833$$

$$(0.9999)^t = 0.833$$

$$t = 1{,}827 \text{ generations}$$

C18. Genetic drift is due to sampling error, and the degree of sampling error depends on the population size. In small populations, the relative proportion of sampling error is much larger. If genetic drift is moving an allele toward fixation, it will take longer in a large population because the degree of sampling error is much smaller.

C19. A. Probability of fixation = $1/2N = 1/2(4) = 1/8$, or 0.125

B. $\bar{t} \simeq 4N = 4(4) = 16$ generations

C. The preceding calculations assume a constant population size. If the population grows after it has been founded by these four individuals, the probability of fixation will be lower and the time it takes to reach fixation will be longer.

C20. During the bottleneck effect, allele frequencies are dramatically altered due to genetic drift. In extreme cases, some alleles are lost while others may become fixed at 100%. The overall effect is to decrease the genetic diversity within the population. This may make it more difficult for the species to respond in a positive way to changes in the environment. Species that are approaching extinction are also facing a bottleneck as their numbers decrease. The loss of genetic diversity may make it even more difficult for the species to rebound.

C21. When two populations intermix, both populations tend to have more genetic variation because each population introduces new alleles into the other population. In addition, the two populations tend to have similar allele frequencies, particularly when a large proportion migrates.

C22. Directional selection favors the phenotype at one phenotypic extreme. Over time, natural selection is expected to favor the fixation of alleles that cause these phenotypic characteristics. Disruptive selection favors two or more phenotypic categories. It will lead to a population with a balanced polymorphism for the trait. Stabilizing selection favors individuals with intermediate phenotypes. It will also tend to promote polymorphism, since the favored individuals may be heterozygous for particular alleles.

C23. Darwinian fitness is the relative reproductive potential of an individual in a population. The most successful individuals are given a value of 1.0. Characteristics that promote survival, ability to attract a mate, or an enhanced fertility would be expected to promote Darwinian fitness. Examples are the thick fur of a polar bear, which helps it to survive in a cold climate; the bright plumage of male birds, which helps them to attract a mate; and the high number of gametes released by certain species of fish, which helps their fertility.

C24. The intuitive meaning of the mean fitness of a population is the relative likelihood that members of a population will reproduce. If the mean fitness is high, it is likely that an average member will survive and produce offspring. Natural selection increases the mean fitness of a population.

C25. The genetic load refers to all the genetic variation within a population that has the potential to be deleterious. It is a common consequence of genetic variation since genetic balances can arise. For example, heterozygote advantage can lead to the prevalence of alleles that are deleterious in the homozygous condition. In general, it happens that alleles, or combinations of alleles, that are advantageous in one circumstance are not advantageous in another. Since populations shuffle their genes from one generation to the next, and may move to new environments, genetic load may occur.

C26. A. Random mutation is the source of genetic variation that may lead to antibiotic resistance. A random mutation may create an antibiotic-resistance allele. This could occur in different ways. Two possibilities are:

1. Many antibiotics exert their effects by binding to an essential cellular protein within the microorganism and inhibiting its function. A random mutation could occur in the gene that encodes such an essential cellular protein; this could alter the structure of the protein in a way that would prevent the antibiotic from binding to the protein or inhibiting its function.
2. As another possibility, microorganisms, which are killed by antibiotics, possess many enzymes, which degrade related compounds. A random mutation could occur in a gene that encodes a degradative enzyme so that the enzyme now recognizes the antibiotic and degrades it.

B. When random mutations occur, they may be lost due to genetic drift. This is particularly likely when the frequency of the mutation is very low in a large population. Alternatively (and much less likely), a random mutation that confers antibiotic resistance could become fixed in a population.

C. If a random mutation occurs that confers antibiotic resistance, and if the mutation is not lost by genetic drift, natural selection will favor the growth of microorganisms that carry the antibiotic-resistance allele if the organisms are exposed to the antibiotic. Therefore, if antibiotics are widely used, this will kill microorganisms that are sensitive and favor the proliferation of ones that happen to carry antibiotic-resistance alleles.

C27. A. True.

B. True.

C. False; it causes allele loss or fixation, which results in less diversity.

D. True.

C28. A. Migration will increase the genetic diversity in both populations. A random mutation could occur in one population to create a new allele. This new allele could be introduced into the other population via migration.

B. The allele frequencies between the two populations will tend to be similar to each other, due to the intermixing of their alleles.

C. Genetic drift depends on population size. When two populations intermix, this has the effect of increasing the overall population size. In a sense, the two smaller populations behave somewhat like one big population. Therefore, the effects of genetic drift are lessened when the individuals in two populations can migrate. The net effect is that allele loss and allele fixation are less likely to occur due to genetic drift.

C29. A. Disruptive. There are multiple environments that favor different phenotypes.

B. Directional. The thicker the fur, the more likely that survival will occur.

C. Stabilizing. Low birth weight is selected against because it results in low survival. Also, very high birth weight is selected against because it could cause problems in delivery, which also could decrease the survival rate.

D. Directional. Sturdy stems and leaves will promote survival in windy climates.

Experimental Questions

E1. Allozymes can be detected by gel electrophoresis. However, this technique underestimates the total amount of genetic variation because some sequence differences will not alter the mobility of an enzyme within a gel. By comparison, DNA sequencing will reveal all the sequence differences between different individuals.

E2. 1. One hypothesis is that the population having only two allozymes was founded from a small group that left the other population. When the small founding group left, it had less genetic diversity than the original population.

2. The population with more genetic diversity may be in a more diverse environment so it may select for a greater variety of phenotypes.

3. It may just be a matter of chance that one population had accumulated more neutral alleles than the other.

E3. A. At the DNA level, a clear-cut way to determine genetic variation is to clone and sequence genes. If the same gene is cloned from two different individuals and the sequences are different, this shows that there is genetic variation. In addition, several other methods can be used to detect genetic variation. For example, a comparison of Southern blots using samples from different individuals might reveal that a gene exists in different sizes or that it contains different restriction sites.

B. At the RNA level, a Northern blot may reveal genetic variation. If the RNA encoded by two different alleles has a different size, this can be detected in a Northern blot.

C. At the protein level, gel electrophoresis may reveal genetic variation. This method was described in your textbook. Different allozymes may migrate at different rates during gel electrophoresis. Another approach is to study the function of an enzyme using a biochemical assay of its activity. Allozymes may have different levels of enzymatic activity, and this may be detected using an enzyme assay.

E4. Glutamic acid is a negatively charged amino acid and valine is neutral. The Hb^A polypeptide has a glutamic acid at the sixth position while Hb^S has a valine. Therefore, the Hb^A polypeptide will move a little more quickly toward the positive end of the gel.

Lane 1—Hb^SHb^S
Lane 2—Hb^AHb^A
Lane 3—Hb^AHb^S

E5. Note: You need to look at solved problem S5 and realize that the Hardy-Weinberg equation can be extended to a gene existing in four alleles. In this case:

$$(p + q + r + s)^2 = 1$$

$$p^2 + q^2 + r^2 + s^2 + 2pq + 2qr + 2qs + 2rp + 2rs + 2sp = 1$$

Let $p = C$, $q = c^{ch}$, $r = c^h$, and $s = c$.

A. The frequency of albino rabbits is s^2.

$$s^2 = (0.05) = 0.0025 = 0.25\%$$

B. Himalayan is dominant to albino but recessive to full and chinchilla. Therefore, Himalayan rabbits would be represented by r^2 and by $2rs$.

$$r^2 + 2rs = (0.44)^2 + 2(0.44)(0.05) = 0.24 = 24\%$$

Among 1,000 rabbits, about 240 would have a Himalayan coat color.

C. Chinchilla is dominant to Himalayan and albino but recessive to full coat color. Therefore, heterozygotes with chinchilla coat color would be represented by $2qr$ and by $2qs$.

$$2qr + 2qs = 2(0.17)(0.44) + 2(0.17)(0.05) = 0.17, \text{ or } 17\%$$

Among 1,000 rabbits, about 170 would be heterozygotes with chinchilla fur.

E6. A. Let W represent the white fat allele and w represent the yellow fat allele. Assuming a Hardy-Weinberg equilibrium, we can let p^2 represent the genotype frequency of WW animals, and then Ww would be $2pq$ and ww would be q^2. The only genotype frequency we know is that of the ww animals.

$$ww = q^2 = \frac{76}{5,468}$$

$$q^2 = 0.014$$

$q = 0.12$, which is the allele frequency of w

$$p = 1 - q$$

$p = 0.88$, which is the allele frequency of W

B. The heterozygous carriers are represented by $2pq$. If we use the values of p and q, which were calculated in part A:

$$2pq = 2(0.88)(012) = 0.21$$

Approximately 21% of the animals would be heterozygotes with white fat.

If we multiply 0.21 times the total number of animals in the herd:

$$0.21 \times 5,468 = 1,148 \text{ animals}$$

E7. A.

Eskimo	$M = 0.913$ $N = 0.087$
Navajo	$M = 0.917$ $N = 0.083$
Finns	$M = 0.673$ $N = 0.327$
Russians	$M = 0.619$ $N = 0.381$
Aborigines	$M = 0.176$ $N = 0.824$

B. To determine if these populations are in equilibrium, we can use the Hardy-Weinberg formula and calculate the expected number of individuals with each genotype.

Eskimo $MM = (0.913)^2 = 83.3$

$MN = 2\,(0.913)(0.087) = 15.9$

$NN = (0.087)^2 = 0.76$

In general, the values agree pretty well with an equilibrium. The same is true for the other four populations.

C. Based on similar allele frequencies, the Eskimo and Navajo Indians seem to have interbred as well as the Finns and Russians.

E8. The first thing we need to do is to determine the allele frequencies. Let's let p represent i, q represent I^A, and r represent I^B.

p^2 is the genotype frequency of ii
q^2 is the genotype frequency of I^AI^A
r^2 is the genotype frequency of I^BI^B
$2pq$ is the genotype frequency of I^Ai
$2pr$ is the genotype frequency of I^Bi
$2qr$ is the genotype frequency of I^AI^B

$$p^2 = \frac{721}{721 + 932 + 235 + 112}$$

$$p^2 = 0.36$$

$$p = 0.6$$

Next, we can calculate the allele frequency of I^A. Keep in mind that there are two genotypes (I^AI^A and I^Ai) that result in type A blood.

$$q^2 + 2pq = \frac{932}{721 + 932 + 235 + 12}$$

$$q^2 + 2(0.6)q = 0.47$$

$$q = 0.31$$

Now it is easy to solve for r,

$$p + q + r = 1$$

$$0.6 + 0.31 + r = 1$$

$$r = 0.09$$

Based on these allele frequencies, we can compare the observed and expected values. To determine the expected values, we multiply the genotype frequencies times 2,000, which was the total number of individuals in this population.

p^2 is the genotype frequency of $ii = (0.6)^2(2{,}000) = 720$
q^2 is the genotype frequency of $I^AI^A = (0.31)^2(2{,}000) = 192$
r^2 is the genotype frequency of $I^BI^B = (0.09)^2(2{,}000) = 16$
$2pq$ is the genotype frequency of $I^Ai = 2(0.31)(0.6)(2{,}000) = 744$
$2pr$ is the genotype frequency of $I^Bi = 2(0.09)(0.6)(2{,}000) = 216$
$2qr$ is the genotype frequency of $I^AI^B = 2(0.31)(0.09)(2{,}000) = 111$

	Expected Numbers	**Observed Numbers**
Type O	720	721
Type A	192 + 744 = 936	932
Type B	16 + 216 = 232	235
Type AB	111	112

The observed and expected values agree quite well. Therefore, it does appear that this population is in Hardy-Weinberg equilibrium.

E9. A. $\Delta p_C = m(p_D - p_R)$

With regard to the sickle-cell allele:

$$\Delta p_C = (550/10{,}550)(0.1 - 0.01) = 0.0047$$

$$p_C = p_R + \Delta p_C = 0.01 + 0.0047 = 0.0147$$

B. We need to calculate the genotypes separately:

For the 550 migrating individuals,

$Hb^AHb^A = (0.9)^2 = 0.81$, or 81% We expect (0.81)550 = 445.5 individuals to have this genotype
$Hb^AHb^S = 2(0.9)(0.1) = 0.18$ We expect (0.18)550 = 99 heterozygotes
$Hb^SHb^S = (0.1)^2 = 0.01$ We expect (0.01)550 = 5.5 Hb^SHb^S

For the original recipient population,

$Hb^A\text{Hb}^A = (0.99)^2 = 0.98$ We expect 9,801 individuals to have this genotype
$Hb^AHb^S = 2(0.99)(0.01) = 0.0198$ We expect 198 with this genotype
$Hb^SHb^S = (0.01)^2 = 0.0001$ We expect 1 with this genotype

To calculate the overall population:

(445.5 + 9801)/10,550 = 0.971 Hb^AHb^A homozygotes
(99 + 198)/10,550 = 0.028 heterozygotes
(5.5 + 1)/10,550 = 0.00062 Hb^SHb^S homozygotes

C. After one round of mating, the allele frequencies in the conglomerate (calculated in part A), should yield the expected genotype frequencies according to the Hardy-Weinberg equilibrium.

$$\text{Allele frequency of } Hb^S = 0.0147, \text{ so } Hb^A = 0.985$$

$$Hb^A Hb^A = (0.985)^2 = 0.97$$

$$Hb^A Hb^S = 2(0.985)(0.0147) = 0.029$$

$$Hb^S Hb^S = (0.0147)^2 = 0.0002$$

E10. Let's assume that the relative fitness values are 1.0 for the dominant homozygote and the heterozygote and 0 for the recessive homozygote. The first thing we need to do is to calculate the mean fitness for the population.

$$p^2 W_{AA} + 2pqW_{Aa} + q^2 W_{aa} = \overline{W}$$
$$(0.78)^2 + 2(0.78)(0.22) = \overline{W}$$
$$\overline{W} = 0.95$$

The genotype frequency in the next generation for AA equals

$$\frac{p^2 W_{AA}}{\overline{W}}$$
$$\frac{(0.78)^2}{0.95} = 0.64$$
$$p = \sqrt{0.64} = 0.80 \text{ and } q = 0.20$$

For the second generation, we first need to calculate the mean fitness of the population, which now equals 0.96. Using the preceding equation, the genotype frequency of *AA* in the second generation equals 0.67 and the allele frequency equals 0.816. The frequency of the recessive allele in the second generation would equal 0.184 and the mean fitness would now equal 0.967. The genotype frequency of *AA* in the third generation would be 0.688 and the allele frequency would be 0.83. The frequency of the recessive allele would be 0.17.

E11. A. Probability of fixation = 1/2*N* (Assuming equal numbers of males and females contributing to the next generation)

Probability of fixation = 1/2(2,000,000)

= 1 in 4,000,000 chance

B. where t = the average number of generations to achieve fixation

N = the number of individuals in population, assuming that males and females contribute equally to each succeeding generation

C. If the blue allele had a selective advantage, the value calculated in part A would be slightly larger; there would be a higher chance of allele fixation. The value calculated in part B would be smaller; it would take a shorter period of time to reach fixation.

E12. If we let *C* represent the *carbonaria* allele and *c* represent the *typical* allele:

$$W_{CC} = 1.0$$
$$W_{Cc} = 1.0$$
$$W_{cc} = 0.47$$

In the next generation, we expect that the Hardy-Weinberg equilibrium will be modified by the following amount:

$$p^2 W_{CC} + 2pqW_{Cc} + q^2 W_{cc}$$

In a population that is changing due to natural selection, these three genotypes will not add up to 1.0 as in the Hardy-Weinberg equilibrium. Instead, the three genotypes will add up to the mean fitness of the population.

$$p^2 W_{CC} + 2pqW_{Aa} + q^2 W_{aa} = \overline{W}$$
$$(0.7)^2(1.0) + 2(0.7)(0.3)(1.0) + (0.3)^2(0.47) = \overline{W}$$
$$\overline{W} = 0.95$$

After one generation of selection:

Allele frequency of *C:*

$$p = \frac{p^2 W_{CC}}{\overline{W}} + \frac{pqW_{Cc}}{\overline{W}}$$

$$p = \frac{(0.7)^2(1.0)}{0.95} + \frac{(0.7)(0.3)(1.0)}{0.95}$$

$$p = 0.74$$

Allele frequency of *c:*

$$q = \frac{q^2 Wcc}{\overline{W}} + \frac{pqW_{Cc}}{\overline{W}}$$

$$q = \frac{(0.3)^2(0.47)}{0.95} + \frac{(0.7)(0.3)(1.0)}{0.95}$$

$$q = 0.27$$

After one generation, the allele frequency of *C* has increased from 0.7 to about 0.74 while the frequency of *c* has decreased from 0.3 to about 0.27. This is because the homozygous, *cc,* genotype has a lower fitness compared to the heterozygous, *Cc,* and homozygous, *CC*, genotypes.

E13. The selection coefficients are

$$s_{ww} = 1 - 0.19 = 0.81$$

$$s_{WW} = 1 - 0.37 = 0.63$$

If the rats are not exposed to warfarin, the equilibrium will no longer exist and natural selection will tend to eliminate the warfarin-resistance allele because the homozygotes are vitamin K deficient.

E14. Each area that he tested had its own endogenous population of moths. For example, the polluted areas had many more darkly colored moths, so we would expect to capture many more of these simply because there are more of them in the first place. Kettlewell wanted to release an equal number of moths of both types and then recapture them as a way to examine how well each type of moth could survive in polluted and unpolluted environments.

E15. Let's use the data for bird predation, but we could also carry out a chi square analysis for the percentage of recapture.

Hypothesis: Color has nothing to do with predation by birds. Note: We need to propose this hypothesis to obtain expected values. According to this hypothesis, we would expect an equal number of dark and light moths to be eaten by birds.

$$\chi^2 = \frac{\Sigma(O - E)^2}{E}$$

In the Dorset woods, there were (43 + 15) = 58 moths that were eaten. We would expect 29 to be *carbonaria* and 29 to be *typical* according to our hypothesis. In the Birmingham woods, there were (26 + 164) = 190 moths eaten, so we would expect 95 to be *carbonaria* and 95 to be *typical.*

$$\chi^2 = \frac{(43-29)^2}{29} + \frac{(15-29)^2}{29} + \frac{(26-95)^2}{95} + \frac{(164-95)^2}{95}$$

$$\chi^2 = 113.8$$

If we look in the chi square table in Chapter 2 with 3 degrees of freedom, this rather high chi square value is very unlikely to occur as a matter of chance (less than 1% of the time). Therefore, we reject our hypothesis that color does not affect predation. As an alternative, we would propose that the color of the moths does have a significant effect on their likelihood of predation.

E16. Fitness based on the number eaten by birds:

The number of moths eaten by birds is really a measure of the selection coefficient (s), not a measure of fitness. $s = 1 - W$

We first need to compare the *carbonaria* and the *typicals*.

$$carbonaria = \frac{43}{43+15}$$
$$carbonaria = 0.74$$
$$typical = \frac{15}{43+15}$$
$$typcial = 0.26$$

If we wish to give the *typical* moths a fitness value of 1.0, this means the selection coefficient for *typical* moths must be zero. Therefore, to calculate the selection coefficient for the *carbonaria* moths:

$$s_{carbonaria} = 0.74 - 0.26 = 0.48$$
$$W_{carbonaria} = 1 - 0.48 = 0.52$$

Fitness based on the number of moths recaptured:

$$W_{carbonaria} = \frac{7.0}{12.5}$$
$$W_{carbonaria} = 0.56$$

The two values (0.52 and 0.56) agree reasonably well. The fitness value based on recapture data is probably more reliable since it seems to be an unbiased measure of the survival rate. The fitness based on the number eaten by birds is somewhat biased because it assumes that this is the only factor that affects the survival of the two types of moths. However, there could be other factors. For example, animals other than birds may eat moths.

E17. If we assume that the different phenotypes of snails have the same level of fertility, we can estimate their fitness values by determining their survival percentages in the different habitats. This calculation assumes that the snails freely migrate through these different environments during their lifetimes, and that the percentages of snails in any given environment reflect their likelihood of survival. This assumption may not be valid if the snails are fairly sedentary and do not migrate very far from where they were born.

A. In beechwoods, the pink snails have the greatest survival rate. If we assign a fitness value of 1 for the pink snails, then the brown snails have a fitness of 0.23/0.61 = 0.38, and the yellow snails have a fitness value of 0.16/0.61 = 0.26.

B. In deciduous woods, the pink snails also have the greatest survival rate. If we assign a fitness value of 1 for the pink snails, then the brown snails have a fitness of 0.05/0.68 = 0.07, and the yellow snails have a fitness value of 0.27/0.68 = 0.40.

C. In hedgerows, the yellow snails have the greatest survival rate. If we assign a fitness value of 1 for the yellow snails, then the brown snails have a fitness of 0.05/0.64 = 0.08, and the pink snails have a fitness value of 0.31/0.64 = 0.48.

E18. One could follow an analogous protocol as conducted by Kettlewell. You could mark snails with a dye and release equal numbers of dark and light snails into dimly lit forested regions and sunny fields. At a later time, recapture the snails and count them. It would be important to have a method of unbiased recapture because the experimenter would have an easier time locating the light snails in a forest and the dark snails in a field. Perhaps one could bait the region with something that the snails like to eat and only collect snails that are at the bait. In addition to this type of experiment, one could also sit in a blind and observe predation as it occurs.

Questions for Student Discussion/Collaboration

1. Assortative mating: In natural populations, individuals with similar markings may tend to recognize each other as being members of the same species. In human populations, people tend to choose mates who are similar with regard to size, race, etc. Inbreeding in agriculture is used to produce larger tomatoes, greater beef yield, etc.

   Disassortative mating: This is less common, although it does occur. In certain plant species, there are alleles that prevent self-hybridization. Therefore, outcrossing is favored, which is somewhat like disassortative mating. Some people tend to prefer mates who are very dissimilar in their characteristics.

2. Mutation is responsible for creating new alleles, but the rate of new mutations is so low that it cannot explain allele frequencies in this range. Let's call the two alleles *B* and *b* and assume that *B* was the original allele and *b* is more recent allele that arose as a result of mutation. Three scenarios explain the allele frequencies:

    1. The *b* allele is neutral and reached its present frequency by genetic drift. It has not reached elimination or fixation yet.
    2. The *b* allele is beneficial and its frequency is increasing due to natural selection. However, there has not been enough time to reach fixation.
    3. The *Bb* heterozygote is at a selective advantage, leading to a balanced polymorphism.

3. It is a difficult question to answer. Over the short term, the elimination of harmful alleles is probably more significant at keeping the population healthy because the frequency of new mutations that are deleterious is much higher than the frequency of new mutations that are beneficial. Over the long term, however, every species needs variation that promotes its adaptation to its environment. In this regard, the selection of beneficial alleles is much more important. In human populations, the selection of beneficial alleles may be less important because humans are becoming better at controlling their environment and enhancing their reproductive capabilities. On the other hand, humans are also getting better at curing genetic diseases, so perhaps the elimination of harmful diseases may not be so critical either. Overall, it may be true that natural selection does not play as important of a role in human populations as it once did.

Chapter 26: Evolutionary Genetics

Student Learning Objectives

Upon completion of this chapter you should be able to:

1. Understand what the term species means to biology.
2. Recognize the various models of speciation and the factors that contribute to each model.
3. Understand the models of the rates of evolutionary change.
4. Understand the concept of a phylogenetic tree, the basis principles of their formation.
5. Analyze a phylogenetic tree.
6. Understand the different evolutionary theories.

Introduction

26.1 Origin of Species

Overview

The final chapter of the text examines some of the big picture perspectives of genetics. Specifically this involves the role of genetics in the study of speciation and evolutionary processes.

Key Terms

Adaptive peaks
Allopatric speciation
Anagenesis
Biological evolution
Biological species concept
Cladogenesis
Founder effect
Gradualism
Hybrid zones
Isolating mechanisms
Macroevolution
Microevolution
Molecular evolution
Neo-Darwinian theory
Parapatric speciation
Postzygotic mechanisms
Prezygotic mechanisms
Punctuated equilibrium
Retrospective testing
Sexual selection
Species
Species recognition complex
Sympatric speciation

Focal Points

- Isolation mechanisms (Table 26.1)
- Comparison of anagenesis and cladogenesis (Figure 26.2)
- Comparisons of graduate versus punctuated equilibrium (Figure 26.7)

Exercises and Problems

For each of the following statements on species and speciation, choose the appropriate term that best describes the statement.

_____ 1. Reproductive communities that are isolated from other species.

_____ 2. Mechanisms that establish a species.

_____ 3. Based upon characteristics that lead to selection by one sex of the species.

_____ 4. A single species is transformed into another species over time.

_____ 5. A group of individuals who have the potential to interbreed and produce viable, fertile offspring.

_____ 6. The division of an original species into two or more separate species.

a. biological species concept
b. isolating mechanisms
c. cladogenesis
d. species recognition concept
e. sexual selection
f. anagenesis

For each of the following, choose the pattern of divergent evolution that is best described by the statement.

a. parapatric speciation
b. sympatric speciation
c. allopatric speciation

_____ 7. May occur as a result of founder effect.

_____ 8. Occurs as a result of partial separation of two species.

_____ 9. Polyploidy in plants is an example.

_____ 10. Occurs with two species that occupy the same habitat or range.

_____ 11. Occurs as a result of geographic separation.

_____ 12. Hybrid zones are a characteristic of this model.

26.2 Evolution at the Molecular Level

Overview

The final section of the text examines a few of the principles regarding molecular evolution. Molecular evolution is the foundation for all other discussions of evolutionary processes. The information provided in this section is simply given to provide an overview of evolutionary genetics.

Key Terms

Ancient DNA analysis
Archaebacteria
Eubacteria
Eukaryotes
Gene family
Homologous
Horizontal gene transfer
Interspecies homology
Intraspecies homology
Molecular clock
Molecular paleontology
Neutral mutations
Neutral theory of evolution
Non-Darwinian evolution
Nonneutral mutation
Paralogous
Paralogs
Selectionists
Synteny groups
UPGMA method
Vertical evolution

Focal Points

- Evolution of paralogs and orthologs (Figure 26.9)

Exercises and Problems

For each of the following, complete the statement using a word or phrase associated with molecular evolution.

1. The constant rate of evolutionary change at the molecular level can be measured by the _______.
2. The three domains of life are the _______, ________, and ________.
3. The movement of genes between species is called _______________.
4. _____ mutations do not affect the phenotype of an organism.
5. Phylogenetic trees using molecular data are frequently done using the _____ method.

Chapter Quiz

1. The most prevalent model of speciation is believed to be ____ speciation.
   a. sympatric
   b. parapatric
   c. horizontal
   d. allopatric

2. A mutation that does not affect the phenotype of the individual is called a ____ mutation.
   a. stabilizing
   b. selective
   c. neutral
   d. selective
   e. none of the above

3. Which of the following is not a domain of life?
   a. eubacteria
   b. animals
   c. archaebacteria
   d. eukaryotes
   e. all of the above are domains

4. Hybrid zones are associated with which of the following?
   a. sympatric speciation
   b. parapatric speciation
   c. sexual selection
   d. allopatric speciation

5. The theory that evolution occurs in bursts, followed by periods of inactivity is called the ________ model of evolution.
   a. microevolution
   b. neutral theory
   c. gradualism
   d. punctuated equilibrium

Answer Key for Study Guide Questions

This answer key provides the answers to the exercises and chapter quiz for this chapter. Answers in parentheses ( ) represent possible alternate answers to a problem, while answers marked with an asterisk (*) indicate that the response to the question may vary.

26.1

1. d
2. b
3. e
4. f
5. a
6. c
7. c
8. a
9. b
10. b
11. c
12. a

26.2

1. molecular clock
2. archaebacteria, eubacteria, eukaryotes
3. horizontal gene transfer
4. neutral
5. UPGMA

Quiz

1. d
2. c
3. b
4. b
5. d

Answers to Conceptual and Experimental Questions

Conceptual Questions

C1. The first principle is that there is genetic variation within natural populations. Therefore, offspring can inherit different alleles, which may affect their phenotype. The second principle is natural selection. This process selects for individuals that have phenotypes that make them reproductively superior. Reproductive superiority may be related to survival because certain alleles may be favored under particular environmental conditions. In addition, natural selection may be a sexual selection process whereby phenotypes that are more likely to mate and produce offspring are at a reproductive advantage.

C2. Evolution is unifying because all living organisms on this planet evolved from the same primordial organism. At the molecular level, all organisms have a great deal in common. With the exception of a few viruses, they all use DNA as their genetic material. This DNA is found within chromosomes, and the sequence of the DNA is organized into units called genes. Most genes are structural genes that encode the amino acid sequence of polypeptides. Polypeptides fold to form functional units called proteins. At the cellular level, all living organisms also share many similarities. For example, living cells share many of the same basic features including a plasma membrane, ribosomes, enzymatic pathways, etc. In addition, as discussed in Chapter 7, the mitochondria and chloroplasts of eukaryotic cells are evolutionarily derived from bacterial cells.

C3. Reproductive isolation occurs when two species are unable to mate and produce viable offspring. As discussed in Table 26.1, several prezygotic and postzygotic mechanisms can prevent interspecies matings. According to the biological species concept, reproductive isolation is the underlying cause of speciation. Speciation occurs when a group becomes reproductively isolated from another group. The recognition species concept is similar except that it places a greater emphasis on the forces of natural selection in promoting reproductive isolation. According to the species recognition concept, natural selection promotes reproductive isolation by selecting for groups of individuals who are reproductively compatible.

C4. Sexual selection is a form of natural selection that favors traits that make it more likely for an organism to reproduce. In many species of animals, natural selection selects for male traits that make it more likely to find a mate and successfully produce offspring. For example, the bright plumage of male birds makes it easier for the female to identify the males of their own species. Many species of deer have antlers so that males may spar with each other as a way to achieve reproductive superiority.

C5. A. Postzygotic

B. Prezygotic

C. Prezygotic

D. Postzygotic

C6. Anagenesis is the evolution of one species into another, while cladogenesis is the divergence of one species into two or more species. Cladogenesis is more prevalent. There may be many reasons why. It is common for an abrupt genetic change such as alloploidy to produce a new species from a preexisting one. Also, migrations of a few members of species into a new region may lead to the formation of a new species in the new region (i.e., allopatric speciation).

C7. Sometimes a single mutation or a few mutations will dramatically alter the phenotypic characteristics of an individual. If the phenotypic changes are beneficial, natural selection may lead to a rapid change in the gene pool. Similarly, when a small group of individuals migrate to a new area that has a different environment, natural selection may quickly lead to a change in the gene pool by favoring adaptations that are beneficial. A third mechanism that does not rely on natural selection is allotetraploidy. When this occurs, the allotetraploid is reproductively isolated from the corresponding diploid species because a cross between the allotetraploid and the diploid will produce infertile offspring.

C8. A. Allopatric

B. Sympatric

C. At first, it may involve parapatric speciation with a low level of intermixing. Eventually, when smaller lakes are formed, allopatric speciation will occur.

C9. Allotetraploids are usually reproductively isolated from the two original species due to hybrid sterility. The hybrid may survive, but it will not have an even number of sets of chromosomes. Therefore, when it undergoes meiosis, each chromosome will not have a homologue to pair with. Therefore, the cells that are produced will be highly aneuploid and lead to inviable offspring. Allotetraploids tend to be fertile because they have an even number of chromosomes that can pair during meiosis and do not form aneuploid cells.

C10. The main evidence in favor of punctuated equilibrium is the fossil record. Paleontologists rarely find a gradual transition of fossil forms. The transition period in which environment pressure and genetic changes cause a previous species to evolve into a new species is thought to be so short that few, if any, of the transitional members would be preserved as fossils. Therefore, the fossil record primarily contains representatives from the long equilibrium periods. Also, rapid evolutionary change is consistent with known genetic phenomena, including single-gene mutations that have dramatic effects on phenotypic characteristics, the founder effect, and genetic events such as changes in chromosome structure (e.g., inversions and translocations) or chromosome number, which may abruptly create individuals with new phenotypic traits. In some cases, however, gradual changes are observed in certain species over long periods of time. In addition, the gradual accumulation of mutations is known to occur from the molecular analyses of DNA.

C11. Reproductive isolation does not really apply to bacteria. Two different bacteria of the same species do not produce gametes that have to fuse to produce an offspring, although bacteria can exchange genetic material (as described in Chapter 6). For this reason, it becomes more difficult to distinguish different species of bacteria. A geneticist would probably divide bacteria into different species based on the sequences of their DNAs. When the sequence differences had reached some arbitrary level, two populations of bacteria would be considered separate species. Historically, bacteria were first categorized as different species based on morphological and physiological differences. Later, when genetic tools such as DNA sequencing became available, the previously identified species could be categorized based on genetic sequences. One issue that makes categorization rather difficult is that a species of bacteria can exist as closely related strains that may have a small number of genetic differences.

C12. A. Yes, it may help the females identify the males of this species.

B. Yes, it helps the males to reproduce.

C. No, it may help the females survive, but it probably does not help the males to identify the females.

D. Yes, it may help the females identify the males of this species.

C13. Allopatric speciation involves a physical separation of a species into two or more separate populations. Over time, each population accumulates mutations that alter the characteristics of each population. Since the populations are separated, each will evolve different characteristics and eventually become distinct species. Sympatric speciation is just the opposite. Members of a population are not physically separated, but something happens (e.g., polyploidy) that abruptly results in reproductive isolation between members of the population. For example, a species could be diploid and a member of the population could become tetraploid. The tetraploid member would be reproductively isolated from the diploid members because hybrid offspring would be triploid and sterile. Therefore, the tetraploid individual has become a separate species. Parapatric speciation is a mixture of allopatric and sympatric. There is some physical separation of two or more populations, but the separation is not absolute. On occasion, members of different populations can interbreed. Even so, the (somewhat) separated populations will tend to accumulate different genetic changes (e.g., inversions) that will ultimately lead to reproductive isolation among the different populations.

C14. One of the major goals in the field of molecular evolution is to understand, at the molecular level, how changes in the genetic material have led to the formation of present-day species. Along these same lines, molecular biologists would like to understand why genetic variation is prevalent within a single species, and also to examine the degree of differences in genetic variation among different species. Comparisons of the genetic material at the molecular level can help to elucidate evolutionary relationships. The field of molecular evolution also is aimed at understanding how genes change at the molecular level. Geneticists would like to know the rate of genetic change and whether changes are neutral or adaptive.

C15. Line up the sequences where the two Gs are underlined.

TTGCATAGGCATACCGTATGATATCGAAAACTAGAAAAATAGGGCGATAGCTA

GTATGTTATCGAAAAGTAGCAAAATAGGGCGATAGCTACCCAGACTACCGGAT

C16. The rate at which a gene evolves depends on whether mutations in the gene will affect the function of the encoded RNA or protein. If a gene can tolerate many mutations without inhibiting the function of the encoded RNA or protein, it will evolve rapidly. Some genes, however, are unable to tolerate many mutations because most mutations inhibit function. When examining evolutionary relationships between species, it is helpful to have a moderate number of differences between the species, but not too many and not too few. Rapidly evolving genes are useful to examine relationships between closely related species because they will probably have a significant number of differences. More distantly related species can be compared by analyzing slowly evolving genes.

C17. A gene sequence can evolve more rapidly. The purpose of structural genes is to encode a polypeptide with a defined amino acid sequence. Many nucleotide changes will have no effect on the amino acid sequence of the polypeptide. For example, mutations in introns and mutations at the wobble base may not affect the amino acid sequence of the encoded polypeptide. These neutral mutations will happen rather rapidly on an evolutionary timescale because natural selection will not remove them from the population. In contrast, changes in the amino acid sequence may alter the structure and function of the polypeptide. Most random mutations that affect the polypeptide sequence are more likely to be detrimental than beneficial, and detrimental mutations will be eliminated by natural selection. This makes it more difficult for the amino acid sequence of the polypeptide to evolve. Only neutral changes and beneficial changes will happen rapidly, and these are less likely to occur in the amino acid sequence compared to the gene sequence.

C18. Some regions of a polypeptide are particularly important for the structure or function of a protein. For example, a region of a polypeptide may form the active site of an enzyme. The amino acids that are found within the active site are likely to be exquisitely located for the binding of the enzyme's substrate and/or for catalysis. Changes in the amino acid sequence of the active site usually have a detrimental effect on the enzyme's functions. Therefore, these types of polypeptide sequences (like those found in active sites) are not likely to change. If they did change, natural selection would prevent the change from being transmitted to future generations. In contrast, other regions of a polypeptide are less important. These other regions would be more tolerant of changes in amino acid sequence and therefore would evolve more rapidly. When comparing related protein sequences, regions that are important for function can often be identified based on less sequence variation.

C19. You would expect the sequences of plant storage proteins to evolve rapidly. The polypeptide sequence is not particularly important for the structure or function of the protein. The purpose of the protein is to provide nutrients to the developing embryo. Changing the sequence would be tolerated. However, major changes in the amino acid composition (not the sequence) may be selected against. For example, the storage protein would have to contain some cysteines in its amino acid sequence because the embryo would need some cysteine to grow. However, the location of cysteines within the amino acid sequence would not be important; it would only be important that the gene sequence have some cysteine codons.

C20. There are lots of possibilities. A few are listed here.

1. DNA is used as the genetic material.
2. The semiconservative mechanism of DNA replication is the same.
3. The genetic code is fairly universal.
4. Certain genes are found in all forms of life (such as 16S rRNA genes).
5. Gene structure and organization is pretty similar among all forms of life.
6. RNA is transcribed from genes.
7. mRNA is used as a messenger to synthesize polypeptides.

C21. The α–globin sequences in humans and horses are more similar to each other, compared to the α–globin in humans and the β-globin in humans. This suggests that the gene duplication that produced the α– and β-globin gene occurred first. After this gene duplication occurred, each gene accumulated several different mutations that caused the sequences of the two genes to diverge. At a much later time, during the evolution of mammals, a split occurred that produced different branches in the evolutionary tree of mammals. One branch led to the formation of horses and a different branch led to the formation of humans. During the formation of these mammalian branches (which has been more recent), some additional mutations could occur in the α– and β-globin genes. This would explain why the α–globin gene in humans and horses is not exactly the same. However, it is more similar than the α– and β-globin genes within humans because the divergence of humans and horses occurred much more recently that than the gene duplication that produced the α– and β-globin genes. In other words, there has been much less time for the α–globin gene in humans to diverge from the α–globin gene in horses.

C22. Both theories propose that random mutations occur in populations. The primary difference between the neutral theory and the selectionist theory is the relative contributions of neutral mutations and nonneutral mutations for explaining present-day variation. Are most forms of genetic variation neutral or brought about by natural selection? Both sides agree that natural selection has been an important force in shaping the phenotypes of all species. It is natural selection that favors the traits that allow organisms to survive in their environments. Within each species, however, there is still a great deal of variation (i.e., not all members of the same species are genetically identical). The neutral theory would argue that most of this variation is neutral. Random mutations occur, which have no effect on the phenotype of the individual, and genetic drift causes the allele frequency to rise to significant levels. The neutral theory suggests that most (but certainly not all) variation can be explained in this manner. In contrast, the selectionist theory suggests that most of the variation seen in natural populations is due to natural selection. The neutral theory of evolution is sometimes called non-Darwinian evolution because it isn't based on natural selection, which was a central tenet in Darwin's theory.

C23. A. This is an example of neutral mutation. Mutations in the wobble base are neutral when they do not affect the amino acid sequence.

B. This is an example of natural selection. *Carbonaria* moths can avoid predation in polluted woods while *typical* moths can avoid predation in unpolluted woods.

C. This is an example of natural selection. Random mutations that occur in vital regions of a polypeptide sequence are likely to inhibit function. Therefore, these types of mutations are eliminated by natural selection. That is why they are relatively rare.

D. This is a combination of neutral mutation and natural selection. The prevalence of mutations in introns is due to the accumulation of neutral mutations. Most mutations within introns do not have any effect on the expression of the exons, which contain the polypeptide sequence. In contrast, mutations within the exons are more likely to be affected by natural selection. As mentioned in the answer to part C, mutations in vital regions are likely to inhibit function. Natural selection tends to eliminate these mutations. Therefore, mutations within exons are less likely than mutations within introns.

C24. Natural selection plays a dominant role in the elimination of deleterious mutations and the fixation of beneficial mutations. Genetic drift also would affect the frequencies of deleterious and beneficial alleles, particularly in small populations. Neutral alleles are probably much more frequent than beneficial alleles. Neutral alleles are not acted upon by natural selection. Nevertheless, they can become prevalent within a population due to genetic drift. The neutral theory suggests that most of the genetic variation in natural populations is due to the accumulation of neutral mutations. A large amount of data supports this theory, although some geneticists still oppose it.

C25. Generally, one would expect a similar number of chromosomes with very similar banding patterns. However, there may be a few notable differences. An occasional translocation could change the size or chromosomal number between two different species. Also, an occasional inversion may alter the banding pattern between two species.

C26. The rate of deleterious and beneficial mutations would probably not be a good molecular clock. Their rate of formation might be relatively constant, but their rate of elimination or fixation would probably be quite variable. These alleles are acted upon by natural selection. As environmental conditions change, the degree to which natural selection would favor beneficial alleles and eliminate deleterious alleles would also change. For example, natural selection favors the sickle-cell allele in regions where malaria is prevalent but not in other regions. Therefore, the prevalence of this allele does not depend solely on its rate of formation and random genetic drift.

Experimental Questions

E1. The use of colchicine would make a one-step process more feasible. Müntzing could have crossed *G. pubescens* and *G. speciosa* to produce an alloploid. At high doses, colchicine can cause complete nondisjunction. To produce an allotetraploid, Müntzing could have treated the alloploid with colchicine to obtain a segment of the plant that would be allotetraploid. This allotetraploid segment could be removed to regenerate an entire plant that is an allotetraploid.

E2. Since the artificial and natural *G. tetrahit* strains can produce fertile offspring, it means that the chromosomes in these strains form homologous sets. When the offspring make gametes, each chromosome has a homologous partner to pair with. Therefore, the offspring are able to make gametes that have a complete set of chromosomes rather than making highly aneuploid gametes that would be inviable. If the two strains had not produced fertile offspring, it would mean that reproductive isolating mechanisms exist between them. Such mechanisms could have been due to mutations that occurred after the formation of the natural species. Alternatively, it could suggest that the natural species was not really an allotetraploid of *G. speciosa* and *G. pubescens.*

E3. Not necessarily. The *G. tetrahit* species may have been created a very long time ago. If this were the case, the *G. tetrahit* species and the *G. pubescens* and *G. speciosa* species would have had lots of time to accumulate different mutations. Therefore, using modern-day *G. pubescens* and *G. speciosa* species to create a *G. tetrahit* species may not have yielded the same results. This is because the modern-day *G. pubescens* and *G. speciosa* species could have accumulated mutations that would not be found in the naturally made *G. tetrahit* species, and/or the natural *G. tetrahit* species may have accumulated mutations that would not be found in the modern-day *G. pubescens* and *G. speciosa* species. These mutations could have affected the phenotype so that the natural and artificial *G. tetrahit* may not be entirely similar. Actually, this did not happen, so it appears that the three species have not accumulated mutations that dramatically affect the phenotype of the artificial *G. tetrahit* species. Perhaps the natural *G. tetrahit* arose fairly recently on an evolutionary timescale.

E4. Some possible experimental approaches are listed here.

1. You could karyotype some members of each population. Different species can often be distinguished by changes in chromosome structure and number.
2. You could see if eastern and western snakes could mate with each other in captivity. If they cannot, this would suggest reproductive isolation. If they can, they still might be different species, if mating never occurs in nature.
3. You could sit in a blind (a box where someone cannot be seen by animals) and watch both populations to see if eastern and western snakes ever mate with each other in nature.

E5. Perhaps the easiest way to determine allotetraploidy is by the chromosomal examination of closely related species. A researcher could karyotype the chromosomes from many different species and look for homeologous chromosomes that have similar banding patterns. This may enable them to identify allotetraploids that contain a diploid set of chromosomes from two different species.

E6. If we compare the percentages of sequence identity among the five genes, we see that *A1* is very similar to *B2,* and *A2* is very similar to *B1* and *B3.* This suggests that the most recent ancestor had two copies of the gene prior to the formation of these two species. After the formation of species A and B, one of the *B* genes must have duplicated again in species B to form *B1* and *B3.*

E7. We begin by identifying the pair that is the most closely related. In this case, it is species B and D, because they have the fewest nucleotide differences. We divide 423 by 2 to yield the average number of substitutions, which equals 211.5. The percentage of nucleotide differences between species B and D is

$$\frac{211.5}{20,000} \times 100 = 1.06$$

Species A is the next most closely related species to B and D. The percentage of nucleotide difference between species A compared to B and D is computed as follows:

There are 465 differences between A and D, and 443 differences between A and B. So the average number of differences is:

$$\frac{443 + 465}{2} = 454$$

We divide 454 by 2, which equals 227 to yield the number of nucleotide differences that occur in species A and B or D. The average percentage of nucleotide differences between species A compared to B and D is

$$\frac{227}{20,000} \times 100 = 1.14$$

Finally, we consider species C. The percentage of nucleotide difference between species C compared to A, B, and D is computed as follows. There are 719 differences between C and A, 744 differences between C and B, and 723 differences between C and D. So the average number of differences is

$$\frac{719 + 744 + 723}{3} = 729$$

We divide 729 by 2, which equals 364.5, to yield the number of nucleotide differences that occurred in C or A, B, or D. The average percentage of nucleotide differences between species C compared to A, B, and D is

$$\frac{364.5}{20,000} \times 100 = 1.82$$

With these data, we can construct the phylogenetic tree shown here.

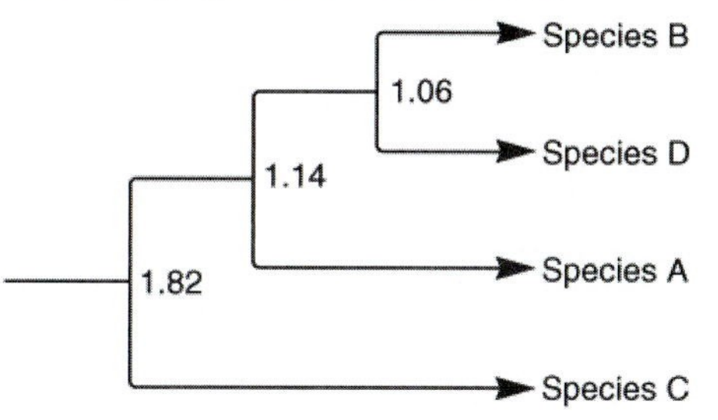

E8. The mutations that have occurred in this sequence are neutral mutations. In all cases, the wobble base has changed, and this change would not affect the amino acid sequence of the encoded polypeptide. Therefore, a reasonable explanation is that the gene has accumulated random neutral mutations over the course of many generations. This observation would be consistent with the neutral theory of evolution. A second explanation would be that one of these two researchers made a few experimental mistakes when determining the sequence of this region.

E9. Two closely related species oftentimes carry the same kinds of genes and a similar total genetic complement of DNA. Therefore, interspecies F_1 hybrids are frequently viable. However, due to changes in the arrangement of the genetic material within the chromosomes, the F_1 offspring may not be fertile. For example, the two species may have different numbers and types of chromosomes. If this is the case, the chromosomes of the F_1 offspring will not be able to pair properly during meiosis, and this will create highly aneuploid gametes. When these gametes participate in fertilization, the embryos will either not develop, or they will result in offspring that are not very healthy. Similarly, the chromosomes of closely related species may differ with regard to inversions and translocations. As discussed in Chapter 8, these changes in chromosome structure are likely to produce gametes that have imbalances in their genetic material. This will also result in embryos that do not survive or have detrimental consequences.

E10. Inversions do not affect the total amount of genetic material. Usually, inversions do not affect the phenotype of the organism. Therefore, if members of the two populations were to interbreed, the offspring would probably be viable because they would have inherited a normal amount of genetic material from each parent. However, such offspring would be inversion heterozygotes. As described in Chapter 8 (see Figure 8.12), crossing over during meiosis may create chromosomes that have too much or too little genetic material. If these unbalanced chromosomes are passed to the next generation of offspring, the offspring may not survive. For this reason, inversion heterozygotes (that are phenotypically normal) may not be very fertile because many of their offspring will die. Since inversion heterozygotes are less fertile, this would tend to keep the eastern and western populations reproductively isolated. Over time, this would aid in the independent evolution of the two populations and would ultimately promote the evolution of the two populations into separate species.

E11. If a mutation is neutral, it will have no effect on the phenotype of the organism. Therefore, you cannot look at the traits of an organism and conclude that it carries neutral mutations. However, molecular techniques can determine if a group of individuals have genetic variation due to the accumulation of neutral mutations. For example, allozymes can be detected by gel electrophoresis, as discussed in chapter 25. The best method to detect genetic variation (both neutral and nonneutral) is DNA sequencing. By analyzing the DNA sequences from different individuals, it is possible to identify changes that are likely to be phenotypically neutral. For example, changes in the wobble base that do not affect the polypeptide sequence would be expected to be neutral.

E12. The technique of PCR is used to amplify the amount of DNA in a sample. To accomplish this, one must use oligonucleotide primers that are complementary to the region that is to be amplified. For example, as described in the experiment of Figure 26.14, PCR primers that were complementary to and flank the 12S rRNA gene can be used to amplify the 12S rRNA gene. The technique of PCR is described in Chapter 18.

E13. As shown here, regions are boxed where the kiwis and cassowary are more similar to each other compared to the moas.

E14. We would expect the probe to hybridize to the natural *G. tetrahit* and also the artificial *G. tetrahit* since both of these strains contain two sets of chromosomes from *G. pubescens.* We would expect two bright spots in the *in situ* experiment. Depending on how closely related *G. pubescens* and *G. speciosa* are, the probe may also hybridize (to two sites) in the *G. speciosa* genome, but this is difficult to predict *a priori.* If so, the *G. tetrahit* species would show four spots.

E15. One way to establish credibility would be to analyze *Tyrannosaurus rex* DNA from samples that have been obtained from different locations. Samples from different locations would be less likely to be contaminated with the same kind of DNA.

Questions for Student Discussion/Collaboration

1. In some ways it is and in some ways it is not. The accumulation of neutral mutations within a population seems to be a random process that is primarily driven by genetic drift. However, phenotypic changes that are adaptive do not occur randomly. Natural selection occurs because certain phenotypic changes favor the reproduction of individuals that carry particular traits. Over the course of many generations, these favorable changes may become the predominant trait within a population.

2. The founder effect and allotetraploidy are examples of rapid forms of evolution. In addition, some single-gene mutations may have a great impact on phenotype and lead to the rapid evolution of new species by cladogenesis. Geological processes may promote the slower accumulation of alleles and alter a species' characteristics more gradually. In this case, it is the accumulation of many phenotypically minor genetic changes that ultimately leads to reproductive isolation. Slow and fast mechanisms of evolution have the common theme that they result in reproductive isolation. This is a prerequisite for the evolution of new species. Fast mechanisms tend to involve small populations and a small number of genetic changes. Slower mechanisms may involve larger populations and involve the accumulation of a large number of genetic changes that each contributes in a small way to reproductive isolation.

3. It is difficult to say if Darwin would object. During Darwin's time, the random nature of mutations was not understood. Therefore, neutral change was not something that he would have considered. The neutral theory does not refute Darwin's major idea, which was that natural selection leads to adaptation. The neutral theory simply suggests that the percentage of adaptive changes in DNA sequences is relatively small compared to neutral ones.